AF358717

Signal Processing for Brain–Computer Interfaces

Editors

Noman Naseer
Imran Khan Niazi
Hendrik Santosa

Basel • Beijing • Wuhan • Barcelona • Belgrade • Novi Sad • Cluj • Manchester

Editors

Noman Naseer
Department of Mechatronics
and Biomedical Engineering
Air University, Islamabad
Islamabad
Pakistan

Imran Khan Niazi
Centre for Chiropractic
Research
New Zealand College of
Chiropractic
Auckland
New Zealand

Hendrik Santosa
Department of Radiology
University of Pittsburgh
Pittsburgh, PA
United States

Editorial Office
MDPI
St. Alban-Anlage 66
4052 Basel, Switzerland

This is a reprint of articles from the Special Issue published online in the open access journal *Sensors* (ISSN 1424-8220) (available at: www.mdpi.com/journal/sensors/special_issues/SignalProcessing_BCI).

For citation purposes, cite each article independently as indicated on the article page online and as indicated below:

Lastname, A.A.; Lastname, B.B. Article Title. *Journal Name* **Year**, *Volume Number*, Page Range.

ISBN 978-3-7258-0520-4 (Hbk)
ISBN 978-3-7258-0519-8 (PDF)
doi.org/10.3390/books978-3-7258-0519-8

Contents

About the Editors

Noman Naseer

Dr. Noman Naseer is a Senior Member of IEEE and Tenured Associate Professor and Chairman of the Department of Mechatronics and Biomedical Engineering at Air University, Islamabad. He is also the founding Head of the Neurorobotics Research Group at the Department of Mechatronics Engineering, Air University, Islamabad, and has been serving as the Vice President of the IEEE Robotics and Automation Society, Pakistan, since 2018. Dr. Noman Naseer was originally a Mechatronics Engineer, who completed his Bachelors, Masters and PhD in Mechatronics Engineering. Dr. Naseer has published more than 120 peer-reviewed books, book chapters, journals, and conference papers and has received more than 4000 citations from all around the globe. Dr. Naseer serves as a reviewer of more than 100 SCI indexed journals and is serving as an Associate Editor of 6 SCI indexed journals. He has won research and travel grants worth more than USD 500,000. Dr. Naseer is a recipient of the prestigious Bill and Malinda Gates foundation award. His name has been included in the directory of productive scientists of Pakistan for three consecutive years. He is also ranked among the top 1% of reviewers in his research field by Publons, owned by Clarivate Analytics.

Imran Khan Niazi

Dr Imran Khan Niazi is currently working as the Dean of Innovation and Technology and the Director of the Centre for Chiropractic Research at New Zealand College of Chiropractic, New Zealand. He leads a multidisciplinary team. He has been working as an active researcher in physical and neural rehabilitation, focusing on non-pharmacological/conservative modes of treatment to optimize patients' rehabilitation journey and improve the quality of life. He has extensive experience working in multidisciplinary teams, including chiropractors, physiotherapists, medical doctors, engineers, and neuroscientists, with a solid patient-centred, evidence-informed approach. He holds an adjunct position at Auckland University of Technology, New Zealand, and Aalborg University, Denmark. He has authored 91 peer-reviewed journal papers and more than 100 conference papers (proceedings and extended abstracts). According to Google Scholar, his work has been cited more than 2800 times and has an h-index of 27.

Hendrik Santosa

Dr Hendrik Santosa is a Faculty Member in the Department of Radiology, University of Pittsburgh. He received his Ph.D. in Cogno-Mechatronics Engineering from Pusan National University, Korea, in 2016. Currently, he is working at the University of Pittsburgh Medical Center on the development of experimental paradigms, data collection, and analysis methods in brain imaging including NIRS, EEG/MEG, MRI, and multimodal techniques. His research interests include a statistical method focused particularly on these three signal-processing techniques: brain–computer interfaces, hyper scanning, and advanced brain signal processing (e.g., ICA/PCA, PLS).

Editorial

Editorial: Signal Processing for Brain–Computer Interfaces—Special Issue

Noman Naseer [1,*], **Imran Khan Niazi** [2] **and Hendrik Santosa** [3]

[1] Department of Mechatronics Engineering, Air University, Islamabad 44000, Pakistan
[2] Centre for Chiropractic Research, New Zealand College of Chiropractic, Auckland 1060, New Zealand; imran.niazi@nzchiro.co.nz
[3] Department of Radiology, University of Pittsburgh, Pittsburgh, PA 15213, USA; hendrik.santosa@pitt.edu
[*] Correspondence: noman.naseer@mail.au.edu.pk

Citation: Naseer, N.; Niazi, I.K.; Santosa, H. Editorial: Signal Processing for Brain–Computer Interfaces—Special Issue. *Sensors* **2024**, *24*, 1201. https://doi.org/10.3390/s24041201

Received: 24 January 2024
Accepted: 8 February 2024
Published: 12 February 2024

1. Introduction

With the astounding ability to capture a wealth of brain signals, Brain–Computer Interfaces (BCIs) have the potential to revolutionize humans' quality of life [1] by processing these brain signals for controlling external devices [2]. Being an emerging and innovative field, BCI offers numerous applications in various fields of life, including robotics, education, prosthetics, security and communication technologies [3]. Processing the neurophysiological signals, a major component of BCI, involves further procedures of (1) noise removal, (2) feature extraction and (3) classification [4]. Pre-processed signals are subject to various noises, including power line noises, physiological noises, motion artifacts and interference noises. These noises can affect the efficiency of the entire BCI procedure. For this reason, noise removal algorithms are utilized for noise removal or reduction [5]. Next, the process of feature extraction starts where algorithms are used to acquire relevant task-based features. This phase acquires data based on spectral and spatial and temporal domains [6]. The last step for signal processing is classification, whereby the acquired and processed features are converted into viable commands, which ultimately control external devices [7]. This Special Issue of *Sensors* focused particularly on these three signal-processing techniques. We invited scientists to share their work conducted on improving performance, information transfer rate, reliability and accuracy of the BCI systems' signal processing. These works could be based on either non-invasive or invasive techniques, for instance, functional near-infrared spectroscopy (fNIRS), electroencephalography (EEG) and other hybrid brain-imaging techniques.

2. Overview of the Contributions

This issue's first article uses mental tasks and binary classification tests by employing features of power spectral density (PSD). The authors were able to enhance the classifier's performance through their peculiar feature set, which can be evidently used in the field of neurophysiology. The second article of this issue applies the P300-BCI Paradigm for controlling home appliances through the brain. The proposed system achieves a high accuracy and shows potential for future application on smart homes. The third article makes use of EEG modality for proposing the multi-kernel temporal and spatial convolution network (MultiT-S ConvNet) based on deep learning and the end-to-end convolutional neural network (ConvNet). The model uses temporal filtering and, thus, can be used to enhance the learning capacity. The fourth article proposes a classification approach using fNIRS-based biomarkers. The data were acquired from both neurotypical and neurodivergent individuals. The four-class classification performances were promising and offered potential for applicability in the field of neuropsychiatry. The fifth article of this issue proposes the design of an unmanned vehicle, using a biomedical sensor, which would be able to monitor health and aid physicians in medical emergencies.

Furthermore, the sixth article of this issue uses fNIRS-based BCI to enhance classification accuracy through Least Absolute Shrinkage and Selection Operator (LASSO) homotopy-based sparse representation classification. This methodology is helpful for rehabilitation purposes in particular. The seventh article conducts brain signal classification acquired through fNIRS signals based on motor execution tasks. The results show enhanced improvement in classification accuracy through deep learning. This classification can be helpful in the field of gait rehabilitation. The eighth article detects ErrPs through EEG in participants with cerebral palsy, amputation or stroke. It also deciphers the discriminative information which is held by different brain regions. This study can be beneficial in formulating adaptive BCIs. The ninth article of this issue utilizes the fNIRS-based approach for upper limb motion detection. The results showed promising accuracy and can be utilized for real-time control. The tenth article works on complicated dexterous task of finger-tapping and acquires data using the fNIRS approach. The classification accuracies appear promising and can be further enriched to generate control commands for BCI application. The eleventh article analyzes the methodologies of 92 studies which utilize fNIRS-EEG-based data and analyses. The review highlights gaps, future directions and potential challenges.

All of the aforementioned studies accelerate foundational and practical knowledge and application in the field of BCI signal processing. We, the editorial team, appreciate all of these innovative research endeavors and would like to thank the authors for their diligent incorporation of feedback, critical assessment of their work and attentiveness to following the timeline, because of which we have been able to successfully publish this Special Issue. We hope the readers feel inspired by and are able to learn more from the research articles.

List of Contributions

1. Rosanne, O.; de Oliveira, A.A.; Falk, T.H. EEG Amplitude Modulation Analysis across Mental Tasks: Towards Improved Active BCIs. *Sensors* **2023**, *23*, 9352.
2. Akram, F.; Alwakeel, A.; Alwakeel, M.; Hijji, M.; Masud, U. A Symbols Based BCI Paradigm for Intelligent Home Control Using P300 Event-Related Potentials. *Sensors* **2022**, *22*, 10000.
3. Emasawas, T.; Morita, T.; Kimura, T.; Fukui, K.; Numao, M. Multi-Kernal Temporal and Spatial Convolution for EEG-Based Emotion Classification. *Sensors* **2022**, *22*, 8250.
4. Erdoğan, S.B.; Yükselen, G. Four-Class Classification of Neuropsychiatric Disorders by Use of Functional Near-Infrared Spectroscopy Derived Biomarkers. *Sensors* **2022**, *22*, 5407.
5. Masud, U.; Saeed, T.; Akram, F.; Malaikah, H.; Akbar, A. Unmanned Aerial Vehicle for Laser Based Biomedical Sensor Development and Examination of Device Trajectory. *Sensors* **2022**, *22*, 3413.
6. Gulraiz, A.; Naseer, N.; Nazeer, H.; Khan, M.J.; Khan, R.A.; Khan, U.S. LASSO Homotopy-Based Sparse Representation Classification for fNIRS-BCI. *Sensors* **2022**, *22*, 2575.
7. Hamid, H.; Naseer, N.; Nazeer, H.; Khan, M.J.; Khan, R.A.; Khan, U.S. Analyzing Classification Performance of fNIRS-BCI for Gait Rehabilitation Using Deep Neural Networks. *Sensors* **2022**, *22*, 1932.
8. Usama, N.; Niazi, I.K.; Dremstrup, K.; Jochumsen, M. Single-Trial Classification of Error-Related Potentials in People with Motor Disabilities: A Study in Cerebral Palsy, Stroke and Amputees. *Sensors* **2022**, *22*, 1676.
9. Sattar, N.Y.; Kausar, Z.; Usama, S.A.; Farooq, U.; Shah, M.F.; Muhammad, S.; Khan, R.; Badran, M. fNIRS-Based Upper Limb Motion Intention Recognition Using an Artificial Neural Network for Transhumeral Amputees. *Sensors* **2022**, *22*, 726.
10. Khan, H.; Noori, F.M.; Yazidi, A.; Zia Uddin, M.; Khan, M.N.A.; Mirtaheri, P. Classification of Individual Finger Movements from Right Hand Using fNIRS Signals. *Sensors* **2021**, *21*, 7943.
11. Li, R.; Yang, D.; Fang, F.; Hong, K.-S.; Reiss, A.L.; Zhang, Y. Concurrent fNIRS and EEG for Brain Function Investigation: A Systematic, Methodology-Focused Review. *Sensors* **2022**, *22*, 5865.

Funding: This research received no external funding.

Conflicts of Interest: The authors declare no conflict of interest.

References

1. Värbu, K.; Muhammad, N.; Muhammad, Y. Past, Present, and Future of EEG-Based BCI Application. *Sensors* **2022**, *22*, 3331. [CrossRef] [PubMed]
2. Abdulkader, S.N.; Atia, A.; Mostafa, M.-S.M. Brain computer interface advancement in neurosciences: Applications and issues. *Eqyptian Inform. J.* **2015**, *16*, 213–230. [CrossRef]
3. Bi, L.; Fan, X.-A.; Liu, Y. EEG-Based Brain-Controlled Mobile Robots: A Survey. *IEEE Trans. Hum. Mach. Syst.* **2013**, *43*, 161. [CrossRef]
4. Naseer, N.; Hong, K.-S. fNIRS-based brain-computer interfaces: A review. *Front. Hum. Neurosci.* **2015**, *9*, 3. [CrossRef] [PubMed]
5. Kamran, M.A.; Hong, K.-S. Reduction of physiological effects in fNIRS waveforms for efficient brain-state decoding. *Neurosci. Lett.* **2014**, *580*, 130–136. [CrossRef] [PubMed]
6. Hwang, H.-J.; Kim, S.; Choi, S.; Im, C.-H. EEG-based brain-computer interfaces: A thorough literature survey. *Int. J. Hum.-Comput. Interface* **2013**, *29*, 814–826. [CrossRef]
7. Coyle, S.M.; Ward, T.E.; Markham, C.M. Brain-computer interface using a simplified functional near-infrared spectroscopy system. *J. Neural Eng.* **2007**, *4*, 219–226. [CrossRef] [PubMed]

Article

EEG Amplitude Modulation Analysis across Mental Tasks: Towards Improved Active BCIs

Olivier Rosanne [1], Alcyr Alves de Oliveira [2] and Tiago H. Falk [1,*]

[1]   Institut National de la Recherche Scientifique, University of Quebec, Montreal, QC H5A 1K6, Canada; olivier.rosanne@inrs.ca

[2]   Graduate Program in Psychology and Health, Federal University of Health Sciences of Porto Alegre, Porto Alegre 90050-170, Brazil; alcyr@ufcspa.edu.br

*   Correspondence: tiago.falk@inrs.ca

Abstract: Brain–computer interface (BCI) technology has emerged as an influential communication tool with extensive applications across numerous fields, including entertainment, marketing, mental state monitoring, and particularly medical neurorehabilitation. Despite its immense potential, the reliability of BCI systems is challenged by the intricacies of data collection, environmental factors, and noisy interferences, making the interpretation of high-dimensional electroencephalogram (EEG) data a pressing issue. While the current trends in research have leant towards improving classification using deep learning-based models, our study proposes the use of new features based on EEG amplitude modulation (AM) dynamics. Experiments on an active BCI dataset comprised seven mental tasks to show the importance of the proposed features, as well as their complementarity to conventional power spectral features. Through combining the seven mental tasks, 21 binary classification tests were explored. In 17 of these 21 tests, the addition of the proposed features significantly improved classifier performance relative to using power spectral density (PSD) features only. Specifically, the average kappa score for these classifications increased from 0.57 to 0.62 using the combined feature set. An examination of the top-selected features showed the predominance of the AM-based measures, comprising over 77% of the top-ranked features. We conclude this paper with an in-depth analysis of these top-ranked features and discuss their potential for use in neurophysiology.

Keywords: active BCI; mental state; modulation features

Citation: Rosanne, O.; Alves de Oliveira, A.; Falk, T.H. EEG Amplitude Modulation Analysis across Mental Tasks: Towards Improved Active BCIs. *Sensors* **2023**, *23*, 9352. https://doi.org/10.3390/s23239352

Academic Editors: Imran Khan Niazi, Noman Naseer and Hendrik Santosa

Received: 30 September 2023
Revised: 15 November 2023
Accepted: 20 November 2023
Published: 23 November 2023

1. Introduction

Active brain–computer interfaces (BCIs) have emerged as powerful communication tools for users with severe and multiple disabilities [1]. In recent years, BCIs have dropped in price and become portable, thus allowing for so-called passive applications to also emerge, e.g., in entertainment, marketing, and mental state monitoring for safety, to name a few [2–4]. Medical applications, particularly in neurorehabilitation [5,6], still remain a predominant use of active brain–computer interfaces (BCIs) as they can improve the quality of life for patients suffering from amputations or paralysis due to neuronal damage, such as stroke [7,8]. Typical interventions range from upper limb rehabilitation [9] to gait enhancement [10], communication support [11], and interactive engagement [12] via the modulation of sensorimotor rhythms to aid motor function restoration and drive neuroplasticity [13].

Despite the great potential for both active and passive BCIs, there are still several major challenges that need to be overcome. From the user perspective, psychological state and familiarity with BCI technology influence the efficacy of rehabilitation. Studies have shown that mental state, such as fatigue, frustration, and attention level, can significantly affect BCI performance [14]. Since learning to use a BCI system demands considerable mental effort, user fatigue is a significant psychological factor, underscoring the importance of user motivation in the successful adoption of BCI systems [15]. Furthermore, other factors, such as individual attention span and spatial ability, also contribute to the variable reliability of BCIs in practical scenarios [16,17]. Overall, several major challenges relating to the robustness of BCIs still exist, such as inter- and intra-subject variability [18], as well as varying signal quality (e.g., signals obtained from gel-based versus dry electrodes) and artifacts, which can bury specific task-related brain activity within noise [19,20]. In fact, BCI performance is highly dependent on the settings that the system has been trained for; thus, any out-of-domain test settings can drastically reduce accuracy.

Over the years, several approaches have been proposed to improve the robustness of BCIs. Multimodal systems, or so-called hybrid BCIs [21,22], take advantage of different neurophysiological modalities (e.g., eye and facial movements or hemodynamics via functional near-infrared spectroscopy (fNIRS)) to improve BCI accuracy. Other approaches have included the development of new signal processing and feature extraction tools to help sift out important brain patterns from noise. For example, the last decades have seen developments in features such as fractal dimension and entropy measures [23,24], as well as the use of amplitude envelope-based features [25,26] to complement traditional spectral power [27–29], spectral coherence [30,31], and time domain statistics [32] features. More recently, deep learning models have emerged as data-driven methods that can help to advance BCI technologies [33–36]. While such data-driven methods have shown improved accuracy on specific datasets, they are known to poorly generalize across datasets [35,37] and may introduce new vulnerabilities (e.g., susceptibility to adversarial attacks [38,39]).

In this paper, we propose to improve active BCI robustness by incorporating a new signal representation that allows for the measurement of amplitude modulation (AM) dynamics and cross-frequency coupling using electroencephalography (EEG) signals. While such a representation has been used in the past for Alzheimer's disease biomarker development [40,41] and passive BCI monitoring [42], it is explored here as a new feature for active BCIs. The proposed method has several advantages over conventional power spectral density (PSD) features [43–45], which motivated this exploration. First, it addresses the inherent non-stationary nature of EEG signals, which traditionally complicates the detection of neural activity patterns. Second, AM dynamics features have been shown to be more robust to artifacts in passive BCI applications (e.g., [46]); thus, they may be able to assist with artifact robustness for active BCIs. Third, cortical hemodynamics measured with fNIRS have been found to be correlated with amplitude modulations measured from EEG signals [47], suggesting that amplitude modulation is a good indication of local neural processing. Since multimodal EEG–fNIRS systems have been shown to outperform EEG systems alone [48], the proposed features may be able to capture multiple signal modalities using a single electrode, thus also improving user experience while making BCIs more robust. Lastly, cross-frequency amplitude and phase coherence features have been linked to different cognitive processes; thus, they may further assist with inter- and intra-subject variability [49–52]. To validate the proposed method, we used the multitask BCI database described in [48] and provide an in-depth discussion of the potential neural processes captured by these new features, providing insights into their importance in the active BCI field.

2. Materials and Methods

In this section, we describe the dataset, extracted features, and feature ranking and classification algorithms used.

2.1. Experimental Protocol

The present study used the open-source BCI database described in [48]. This dataset investigates the discrimination of distinct neural response patterns associated with seven different mental tasks using features extracted from both EEG and fNIRS modalities. The seven mental tasks include mental rotation (ROT), word generation (WORD), mental subtraction (SUB), mental singing (SING), mental navigation (NAV), motor imagery (MI), and face imagery (FACE). Mental rotation, word generation, and mental subtraction are classified as brainteasers, while mental singing, mental navigation, and motor imagery are classified as dynamic imagery tasks and face imagery is classified as a static imagery task. Table 1 describes each mental task.

Multimodal data were collected from 12 participants who were fluent in English and/or French, had no history of neurological disorders, and had no previous experience with BCIs. The participants consented to participating in the study and monetary compensation was provided after each completed session. All participants agreed with the terms and conditions of the study, which was approved by the INRS Research Ethics Committee.

The data were collected over three recording sessions of 2 to 3 h each, during a period of 3 to 5 weeks. Each session consisted of four sub-sessions, in which each mental task was randomly repeated four times. Each sub-session began and ended with a 30 s baseline period, during which the participants were asked to remain in a neutral mental state and fixate on the cross at the center of their screen. Before each trial, a 3 s countdown screen identified the task to be performed. Once the countdown was over, the participants had to execute the required mental task for a period of 15 s and were instructed to carry out the task as many times as possible during that period. Each trial was followed by a rest period of random duration, sampled from a uniform distribution of between 10 and 15 s, in which participants were asked to continue minimizing movements but were allowed to blink, swallow, etc. The participants were also required to complete a subjective evaluation questionnaire between the second and third sub-session of each session. The stimuli and questionnaire were implemented using Presentation software (Neural Behavioral Systems, USA). More details about the experimental protocol can be found in [48].

EEG data were recorded using a BioSemi ActiveTwo system with 62 electrodes and 4 electrooculography (EOG) electrodes. The experiment used a standard 10-10 system for electrode placement, but without AF7 and AF8, where holders were used for fNIRS probes instead. Only data from nine participants were analyzed as data from Participants 2 and 8 were rejected due to excessive artifacts and higher overall drowsiness levels observed during the recordings. Data from Participant 12 were rejected because the three required sessions were not completed. These participant exclusions matched those suggested in [48] and were replicated here to facilitate comparisons. Moreover, while 60 fNIRS optodes were also included in the dataset, they were not analyzed in this study.

Table 1. Short description of each mental task.

Mental Task	Task Description
Mental Rotation (ROT)	Participants had to imagine the 3D rotation of two objects and determine whether the objects were identical
Word Generation (WORD)	A letter was presented randomly and the participants needed to find as many words as possible beginning with this letter
Subtraction (SUB)	Participants had to execute the mental subtraction of 1 to 2 digit numbers from a 3 digit number
Singing (SING)	Participants had to choose a song and then mentally sing it while paying attention to the emotions that they felt
Navigation (NAV)	Participants had to imagine walking from one room to another in their past or current residence
Motor Imagery (MI)	Participants had to imagine moving their fingers
Face Imagery (FACE)	Participants had to remember the face of a friend

2.2. Dataset Pre-Processing

Utilizing the publicly available EEGLAB MATLAB toolbox [53], we first pre-processed the raw EEG signals. Initially, the EEG dataset was re-referenced to the Cz channel, which was later removed. To eliminate electrical grid noise, we applied a notch filter of between 59 and 61 Hz. To tackle high-frequency noise and signal drift arising from electrode impedance variation, a band-pass finite impulse response (FIR) filter, ranging from 0.1 Hz to 50 Hz, was employed.

To facilitate comparisons to the results in [48], channels P8 and O1 were also discarded. Moreover, the first 2 and 14 markers from subject 3 (session 1) and subject 5 (session 1) showed discrepancies with the experimental protocol and were also discarded. We then used the fastICA algorithm [54] to extract independent components. These components were further evaluated using the ADJUST algorithm [55] that is available in the EEGLAB toolbox, allowing us to identify and automatically remove components associated with artifacts. To mitigate the potential impact of lost data, all channels removed in the earlier stages of pre-processing were restored using spherical interpolation. Following this, the EOG channels were removed and the dataset was re-referenced to the average. For detailed examination and replication purposes, the pre-processing script is available at the following GitHub repository: https://github.com/OlivierRS/EEG-Preprocessing-with-ADJUST, accessed on 19 November 2023.

2.3. Feature Extraction

2.3.1. Power Spectral Density (Baseline) Features

As a means of comparison to the proposed amplitude modulation features, we utilized power spectral density (PSD) features as the baseline. Initially, the pre-processed EEG signals were segmented into epochs, starting 1 s before the beginning of each mental task and continuing for 15 s thereafter, generating a total epoch duration of 16 s. Subsequently, these epochs were subdivided into non-overlapping 1 s time windows. The PSD was calculated from each of these windows using Welch's method, employing the MNE toolbox [56], where the power of each frequency was normalized by the sum of the entire power spectrum. PSD features were extracted from several conventional frequency bands, including theta (4–8 Hz), alpha1 (8–10 Hz), alpha2 (10–12 Hz), beta1 (12–21 Hz), beta2 (21–30 Hz), theta to beta (8–30 Hz), and delta to gamma (0–50 Hz), for each EEG electrode. In addition to the PSD features, we also extracted power ratios, notably alpha (8–12 Hz) to beta (12–30 Hz) and theta to beta ratios, for each electrode. Subsequent to the extraction process, all features underwent log-scaling.

2.3.2. Proposed Amplitude Modulation Power Features

The extraction of the amplitude modulation power (AMP) features is presented in Figure 1. The process involved the decomposition of EEG signals into five spectral bands utilizing a filter bank (FB). The bands selected for this study included the conventional delta (1–4 Hz), theta (4–8 Hz), alpha (8–12 Hz), beta (12–30 Hz), and gamma (30–50 Hz), given their established roles in representing neuronal dynamics [57]. We used a zero-phase FIR filter to execute the signal filtering, which consisted of two successive filtering steps in opposing directions on a mirror-padded version of the input signal. The outcome was a series of signals that represented the power dynamics across each frequency band over time. We subsequently obtained the envelope using the absolute value of the Hilbert transform. This envelope was then filtered using the same filter bank, culminating in a set of 5-by-5 modulation signals (shown in the bottom part of the figure) that represented the spectral decomposition of the power dynamics within each specific EEG frequency band, i.e., the measure of the amplitude–amplitude coupling of two EEG frequency bands.

For notation, we refer to these 25 signals as '<modulated band>-m<modulant>', where 'modulated band' refers to one of the five bands from the first filtering step and 'modulant' pertains to the band applied during the second filtering step. A visual representation of the entire set of the AM time series according to this convention is exhibited in the bottom matrix of Figure 1. However, as per Bedrosian's theorem [58], not all of these 25 signals could be possible; in fact, only 14 of them could, as low-frequency bands modulated by high-frequency bands were invalid. A more in-depth description of the amplitude modulation-based features can be found in [46,59].

As mentioned previously, while AMP features have been explored for use as biomarkers for neurodegenerative diseases or passive BCIs, they were explored here for active BCI control. One important aspect of active BCIs is being as close to real-time as possible in order to maximize the information transfer rate. However, AMP features have an inherent latency in that envelopes greater than 1 s (i.e., the window sizes typically used for PSD features) are needed. In the past, 8 s windows have been shown to be optimal for biomarker development [59]. Here, we proposed to reduce this to 4 s but with a sliding window of 3 s, suggesting that decisions can be made every second after the initial 'buffer' period. This approach yielded 11 distinct feature time 'frames' per trial for each frequency band, covering the total trial duration from -1 to 15 s. The features were normalized by the total spectral power at the corresponding time frame.

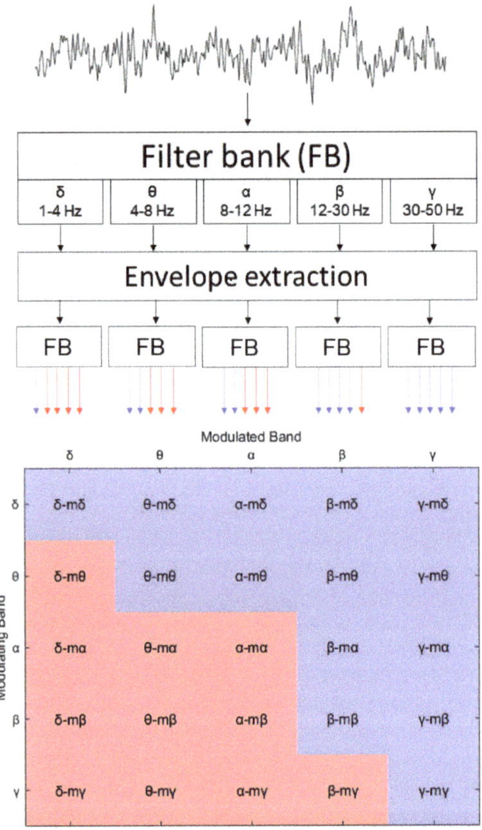

Figure 1. Flow chart illustrating the procedure for amplitude modulation time series extraction from EEG signals. In the top left corner, the straight line denotes the original and band-filtered EEG signal. The dashed lines and adjoining blocks represent the envelope extraction and processing stages. Based on Bedrosian's theorem, the bottom left matrix differentiates between valid (blue) and invalid (red) amplitude modulation time series.

2.3.3. Phase Circular Correlation of Amplitude Modulated Signals

In addition to analyzing power time dynamics using AMP features, we also extracted features to quantify the connectivity between cortical sites. Although metrics such as phase locked value and magnitude squared coherence have traditionally been used for this purpose, they have been criticized for their vulnerability to coincidental phase synchrony, which can result in the misinterpretation of the results [60]. To address this, we used the phase circular correlation of AM signals (CCORAM) method [60] to examine the connections between different electrode pairs, a choice motivated by the increased robustness of the method, as well as its ability to account for phase co-variance within electrode pairs. To reduce computational demands and avoid excessive dimensionality, we only used a subset of electrodes, specifically F7, F8, T7, T8, C3, C4, P7, P8, O1, and O2. This selection was based on a strategy to maximize the spatial span using a 19-channel montage and minimize the effects of source propagation by excluding neighboring channels [61,87]. With this approach, we calculated the CCORAM for each possible pair within the selected 14 AM bands.

2.4. Feature Selection, Classification, and Figures of Merit

Feature selection methods are essential in removing redundant features and avoiding the curse of dimensionality. Methods such as minimum redundancy maximum relevance (mRMR) [62] and recursive feature elimination (RFE) [63] account for interactions among features in their selection process, aiming to minimize redundancy within the selected feature set. Despite their effectiveness, these techniques pose computational challenges, especially in scenarios involving datasets with extensive numbers of features. In contrast, the Fisher linear discriminant (FLD) [64], a filter-based selection method, offers a more practical solution as it evaluates individual features iteratively with significantly lower computational complexity.

For classification, we employed a stratified shuffle split cross-validation approach. In this approach, the aggregated data from all participants were divided, maintaining a 90-10 split for training and testing, respectively, ensuring the proportional representation of both classes in each subset. This procedure was reiterated 100 times, using a different random split each time, to foster a robust estimation of the model's performance. The hyperparameters of the SVM, with radial basis function kernels, C, and gamma, were optimized in the training set, leveraging a cross-validated grid search. For the figures of merit, we used the average kappa score metric on the predicted labels from the test set for each of the 100 bootstrap trials.

2.5. Eigendecomposition-Based Ranking of Binary Classifications

In this study, we examined the interactions between mental tasks via a series of 21 binary classification tasks, aiming to critically assess the influence of amplitude modulation-based features. However, this approach posed a challenge as the insights derived from individual binary classification could only offer a partial understanding of the mental task patterns. To address this, we employed a methodology utilizing eigendecomposition to derive a ranking score, hereafter referred to as the 'prestige score' based on terminology from the established literature [65]. The detailed process of extracting the prestige score is depicted in Figure 2.

This approach provides insights into the relative importance of mental tasks by considering the entire set of classification performances. The prestige score not only reflects the immediate kappa scores of mental tasks in discriminating against other tasks but also synthesizes the entire set of classification performances represented by kappa scores for task pairs. In particular, when a task is inherently difficult to identify and consequently yields a low discriminative score, having one or two binary classifications with high kappa scores can lead to the overestimation of its true discriminative capacity. With this methodology, we could analytically discern the genuine discriminative scores of the tasks that secured high kappa scores, revealing the relative ease of achieving such scores and thereby justifying the adjustment of these scores to accurately reflect the task's true discriminative nature.

Figure 2 depicts an illustration of the process of extracting the prestige scores. On the right side of the figure, the step-by-step procedure for applying eigendecomposition to a 7×7 interaction matrix is demonstrated. This matrix was constructed by incorporating the 21 kappa or Fisher score measurements in its upper triangle, each element of which represented a measurement from a pair of tasks. The analytical process yielded the prestige score, associated with the eigenvector with the highest eigenvalue. Utilizing the Perron–Frobenius theorem, these prestige scores were identified as normalized positive vectors that revealed the latent influences that each class held as dictated by the chosen metric [66]. To maintain the relative magnitudes of the interactions captured in the different dataset folds during data amalgamation, each prestige score was scaled by its corresponding eigenvalue. This approach avoided the empirical analysis that is traditionally conducted on 7×7 interaction matrices, allowing for a more robust and succinct interpretation of the influences of mental tasks through 7-element vectors. Consequently, plotting the feature distribution of individual mental tasks became a more streamlined alternative to the conventional method of illustrating all 21 pairs of mental task combinations. Subsequently, the left side of Figure 2 clarifies the iterative procedure employed to amalgamate the prestige scores.

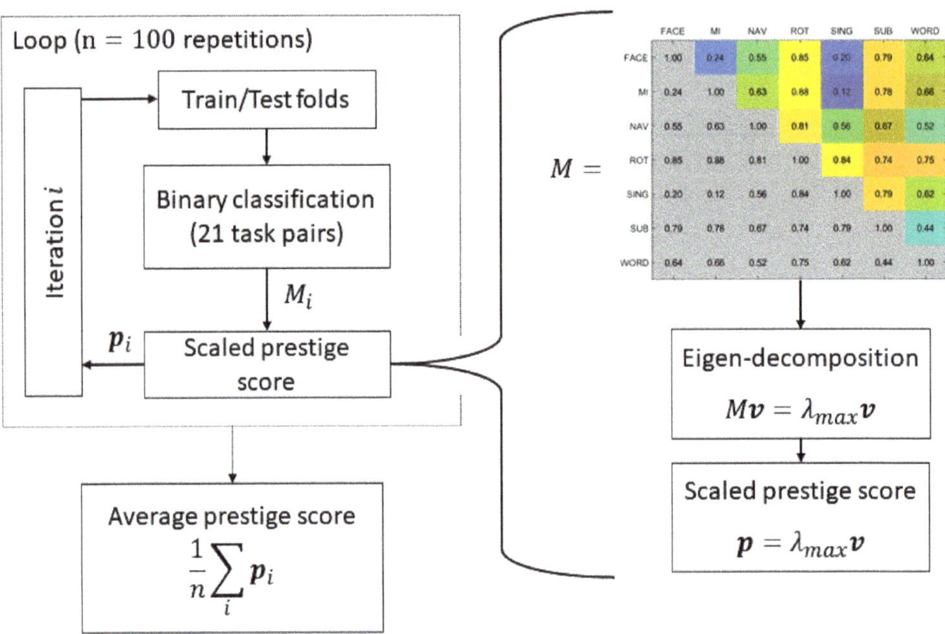

Figure 2. Flow chart illustrating the computation of the scaled prestige scores, depicting how the 21 kappa score measurements were generated in each iteration and arranged in the 7×7 interaction matrix 'M', followed by the eigendecomposition and final averaging of the prestige scores. Kappa scores are used as an illustrative example.

3. Experimental Results

In this section, we describe the experimental results in terms of the impact of the proposed features, the ranking of the mental tasks, and a discussion on the top-selected features.

3.1. Estimation of Optimal Feature Set Size

To estimate the most suitable balance between reducing overfitting risk and ensuring high classifier accuracy, we proceeded to identify the optimal number of features by systematically testing various feature set sizes. Figure 3 depicts the relationship between the quantity of the top-selected features, ranked using the Fisher linear discriminant scores, and the performance of a vanilla SVM classifier. The x-axis represents the number of selected features and the y-axis represents the grand average kappa score derived from all 21 task pair classifications. In total, three combinations of feature types were tested and the model performance using PSD, AMP+PSD, and AMP+CCORAM+PSD are shown in yellow, orange, and blue, respectively. We used the kappa scores to analyze the curve and determine the exact number of features for when the accuracy started to plateau. As can be seen, this occurred after 2000 features for all feature type combinations. Above this threshold, the performance gain was negligible and potential overfitting issues could appear. Henceforth, only experiments with classifiers trained using this number of top features will be reported.

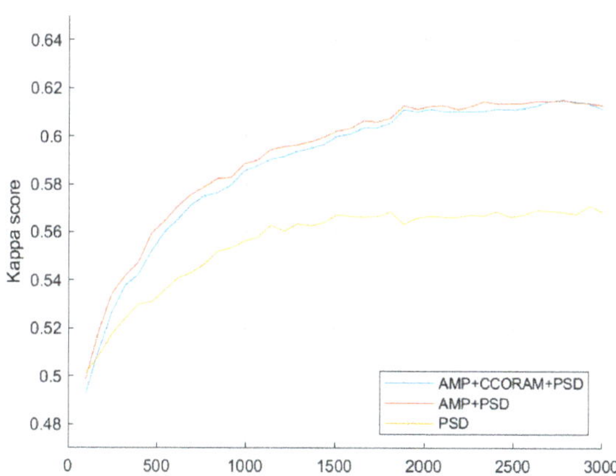

Figure 3. Grand average of task pair kappa scores versus the number of features. The top 2000 features were chosen for subsequent experiments.

Following feature selection, we conducted an in-depth analysis of the distribution of all selected features, categorizing them by type and analyzing them within each top feature subset based on both appearance frequency and FLD score. The AMP features were predominant, representing 74.14% of the selected pool with a cumulative score of 70.02%. PSD features followed, accounting for 22.72% with a cumulative score of 27.11%. The CCORAM features were the least represented at 3.14%, contributing 2.86% to the cumulative score.

3.2. Impact of Proposed Features on Classification Performance

Figure 4 shows the kappa scores obtained for each of the 21 possible task pairs for PSD features alone (orange), PSD features combined with the proposed AMP features (red), and all three feature sets together (blue). As can be seen, the incorporation of AMP features significantly enhanced performance across 17 of the 21 task pairs. For the remaining four task pairs, the accuracy still increased in three, just not significantly (i.e., ROT-WORD, NAV-WORD, and FACE-SING), while the accuracy actually dropped relative to using PSD features alone in one (ROT-SUB). Task pairs associated with mental activities, such as SUB, WORD, FACE, and SING, generally showed negligible enhancements. Notably, the MI-SING and FACE-SING pairs registered the lowest kappa values. These tasks, which are intricately linked to cognitive functions, such as working memory, long-term affective retrieval, and both auditory and visual memory processing, could inherently possess greater inter-subject variability due to their abstract nature, while factors such as individual cognitive approach, mental state, and problem-solving strategy could influence the outcome, leading to subjective disparities.

Overall, the average kappa scores across all 21 task pairs were 0.6221, 0.6237, and 0.5761 for the AMP+CCORAM+PSD, AMP+PSD, and PSD models, respectively. These findings suggest that in the majority of cases, AMP features capture exclusive information that is not obtained through PSD alone.

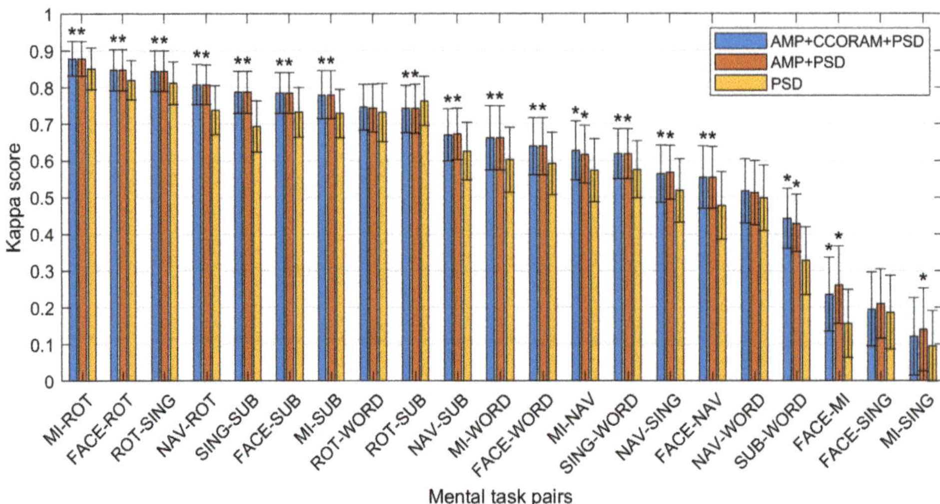

Figure 4. Comparison of the kappa scores of the AMP+CCORAM+PSD (blue), AMP+PSD (red), and PSD (orange) models, derived from the selection of the 2000 optimal features. Columns AMP+CCORAM+PSD and/or AMP+PSD with a symbol * indicate a significant difference relative to PSD results, with a significance level alpha of 0.05.

3.3. Ranking of Mental Task Kappa Scores

By employing the eigendecomposition ranking technique, we estimated the discriminative power of the individual mental tasks. Figure 5 illustrates the estimated individual kappa scores for each of the seven mental tasks, which were derived from the averaged prestige score vectors across the 100 dataset folds. These vectors were procured through the ranking method outlined previously, employing the kappa score interaction matrix. In essence, each iteration of the classification pipeline produced a single interaction matrix, from which a prestige score was extracted. This procedure was repeated across all folds to ultimately average the prestige scores, thereby providing an estimate of the "true" kappa score for each mental task. As a result, mental tasks with higher relative scores were suggestive of a consistently superior discriminating pattern quality, revealing exclusive patterns that distinctly characterized such tasks. The outcomes of this analysis are shown in Figure 5, where the discriminatory power of each task is displayed in decreasing order of prestige score.

As can be seen, the rotation (ROT) task achieved the highest score. Interestingly, in [48], this task was shown to be preferred amongst the participants, despite its demanding nature. This capacity for differentiation could be attributed to its primarily visual nature, which reduces the subjectivity typically present in tasks involving complex processing, such as memory or emotion, thereby exhibiting more discernable neuronal patterns. Conversely, our findings showed that the FACE and SING tasks were the least accurately identified, a result that was in alignment with feedback given by subjects in [48], suggesting that these were their least favorite tasks.

The navigation (NAV) and subtraction (SUB) tasks placed second and third in this analysis, both of which were liked by subjects in [48]. Interestingly, while the motor imagery (MI) task demonstrated a similar kappa score to the FACE and WORD tasks in our study, its use is widespread within the active BCI community. It has been hypothesized that MI-induced patterns may be overshadowed when paired with brain teaser tasks. This could be observed in task pairs such as FACE–MI and SING–MI. The intricate dynamics of MI, involving both promising classifications in certain configurations and limitations in others, underscores a complex landscape that necessitates a careful and balanced consideration when selecting tasks for BCI applications.

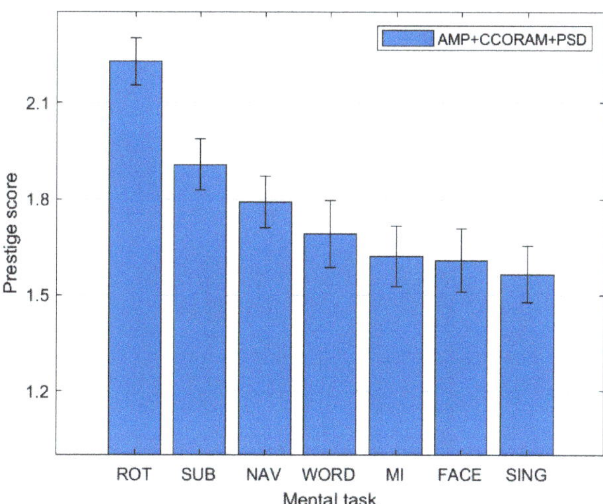

Figure 5. Ranking of individual mental tasks using PSD+AMP+CCORAM features, derived from pairwise classification kappa scores. A higher rank indicates a more distinct pattern, facilitating the model's ability to discriminate one mental task from others.

3.4. Mental Task Feature Analysis

In a similar vein to the analysis performed in the previous section and depicted by Figure 2, we re-conducted this analysis with the goal of understanding which feature patterns most efficiently captured neural patterns for different tasks. We started by constructing a 7×7 matrix for each feature. Each cell in these matrices corresponded to the FLD score for a pair of mental tasks when differentiated using that particular feature. Consequently, each matrix provided a snapshot of the discriminative power of a single feature across different task pairs. From these matrices, prestige scores were derived for each feature, reflecting its distributed discriminative power across the mental tasks.

This analysis was repeated for all features and then aggregated per feature type group, i.e., AMP, CCORAM, or PSD. Figure 6 summarizes our findings. As can be seen, the AMP features were the most efficient in capturing discerning patterns for each individual mental task, followed by the traditional PSD features. Particularly in the ROT task, AMP demonstrated a marked effectiveness over PSD, whereas for the SUB task, both AMP and PSD showed similar discriminative power. Meanwhile, the contribution of the CCORAM feature type was notably lower, showing no clear leaning towards any specific task, indicating its limited utility in this context. It is hoped that these findings will provide insights for researchers developing active BCIs regarding what features to use depending on the mental task in question.

3.5. Multidimensional Analysis of Relevant Features

For each individual feature, which were associated with unique combinations of channels, bands, and feature types, we computed a distinct prestige score, representing their ranking across mental tasks. Utilizing these scores, we constructed topoplots to visually represent the distribution of feature rankings across the EEG scalp. For the AMP feature type, the corresponding topoplots for each band are illustrated in Figures 7–11. Similarly, for the PSD feature type, Figures 12–14 display the topoplots of the prestige score distributions for each band.

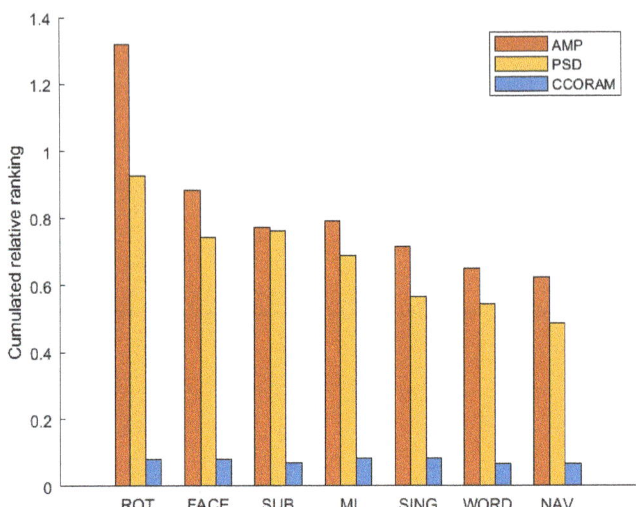

Figure 6. Average Fisher-based prestige scores of mental tasks per feature type (AMP, PSD, and CCORAM).

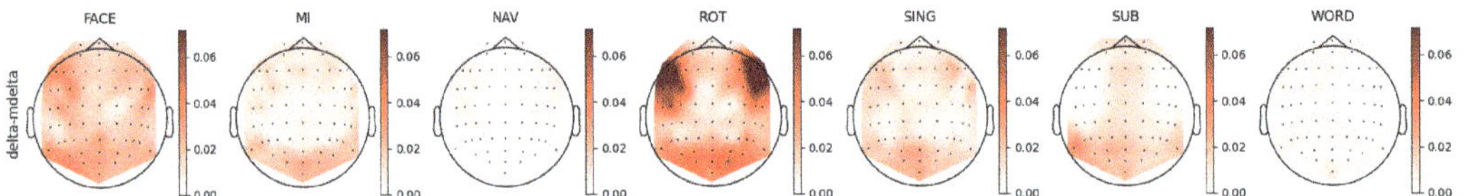

Figure 7. Ranking distribution of AMP features related to the delta band.

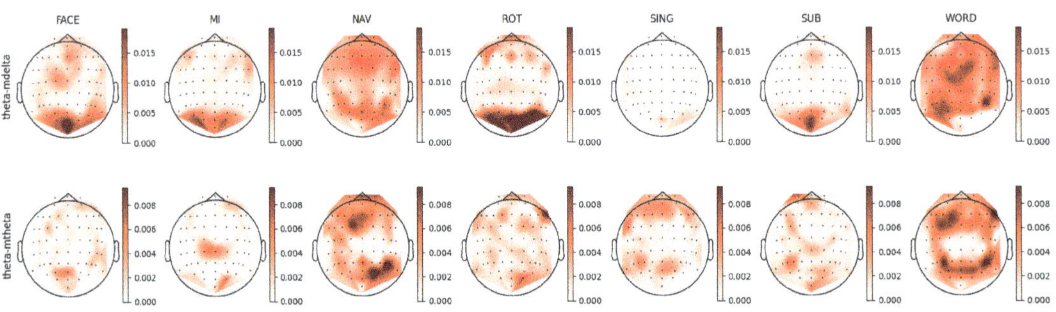

Figure 8. Ranking distribution of AMP features related to the theta band.

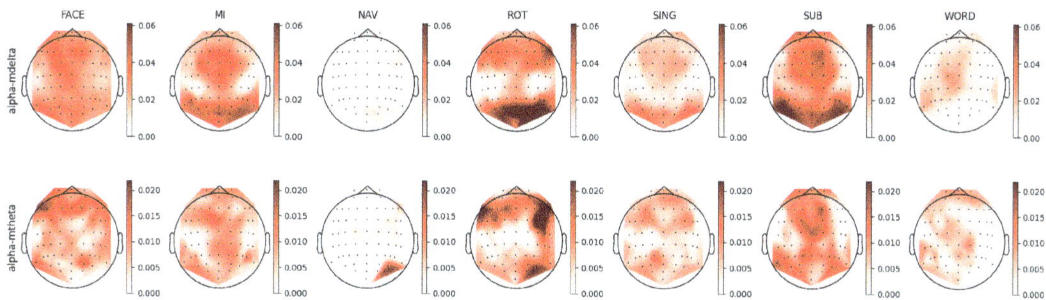

Figure 9. Ranking distribution of AMP features related to the alpha band.

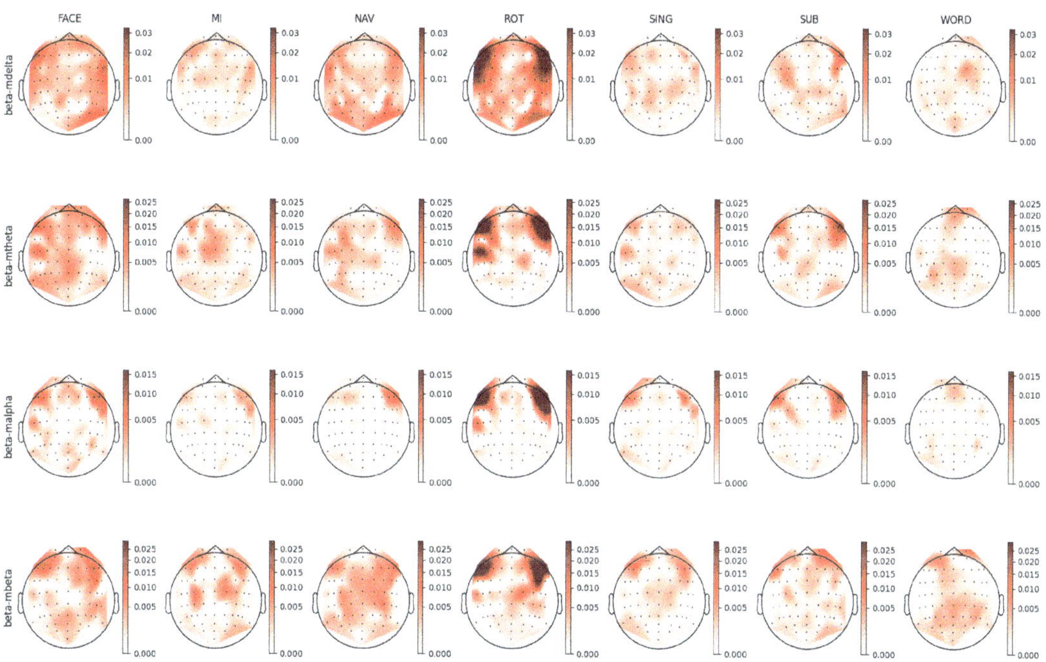

Figure 10. Ranking distribution of AMP features related to the beta band.

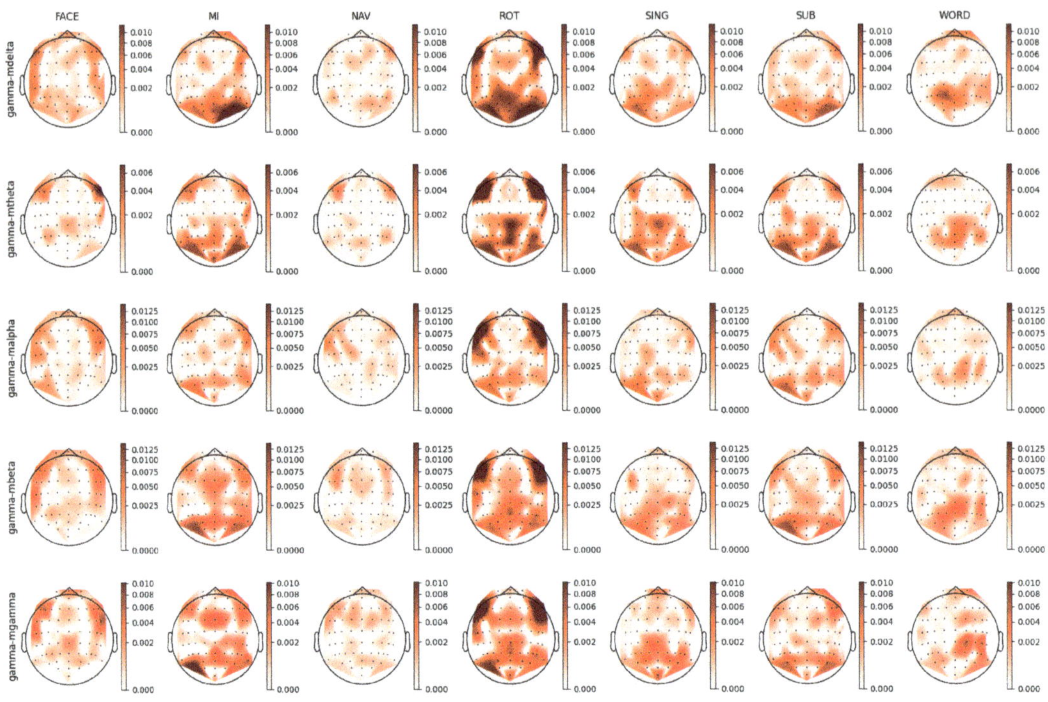

Figure 11. Ranking distribution of AMP features related to the gamma band.

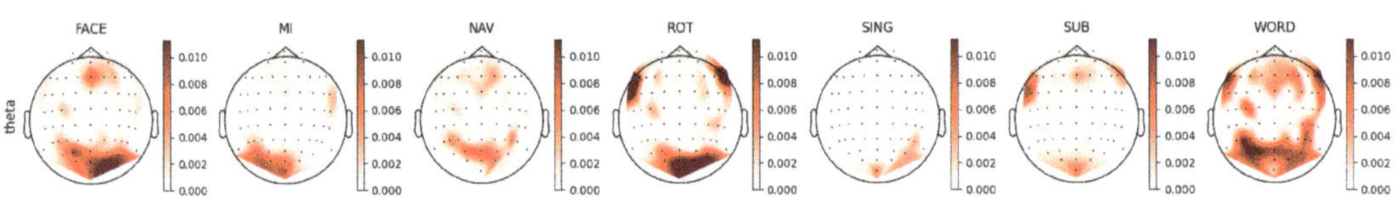

Figure 12. Ranking distribution of PSD features for the theta band.

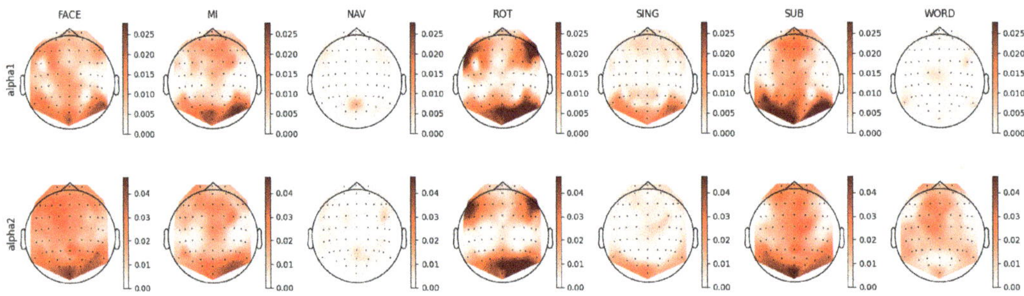

Figure 13. Ranking distribution of PSD features for the alpha band.

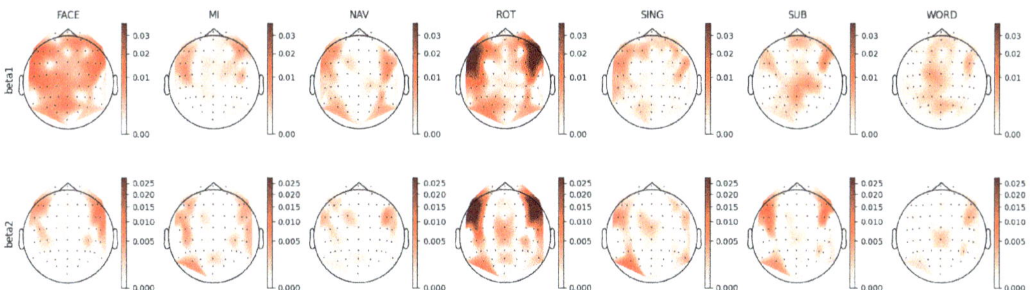

Figure 14. Ranking distribution of PSD features for the beta band.

4. Discussion

The proposed amplitude modulation features are postulated to mirror core biological mechanisms within the brain that are potentially engaged in processing stimuli and directing behavioral responses [42,67–69]. Numerous studies have suggested that amplitude modulation patterns in EEG signals could be indicative of mechanisms for controlled inhibition [69–72] and cognitive process integration [73–75], and correlate with energy consumption in cortical tissues [76]. Based on our experiments, such features offer a supplemental layer of information that could enrich traditional PSD features in classifying mental states and tasks, thus improving the robustness of active BCI systems. In the following subsections, we postulate the underpinnings of these new features and their complementary roles to PSDs.

4.1. Beta Band Analysis

The beta frequency sub-band, which serves numerous roles depending on its location and context in the brain [73,77], is dominantly involved in long-term memory and stimuli processes [78,79]. It facilitates the binding of temporally separated information into meaningful entities and aids in maintaining task-specific cognitive processes over extended periods [73]. Figure 14 illustrates topoplots that depict the spatial distribution of the prestige scores for relevant PSD features associated with the beta band. A comparative examination of the topoplots for the FACE and ROT tasks revealed analogous patterns, with both demonstrating bilateral frontotemporal patterns in the beta band.

A similar bilateral frontal pattern is seen in the beta–mbeta band in the AMP topoplots in Figure 10. Additionally, the AMP topoplots unveiled noteworthy patterns in the parieto-occipital-temporal region for the beta–mdelta and beta–mtheta bands. Beta activity modulated by the lower delta and theta bands was disseminated across the scalp, while the distribution of faster beta dynamics, specifically beta–malpha and beta–mbeta, was inclined towards the frontal area. The observation of slow temporal beta activity dynamics in the occipital-parietal area could suggest top-down control via the slow wave dynamics of specialized visual processing brain structures [73,80–82]. This activity is congruent with the visual nature of the FACE task. Conversely, fast beta activity dynamics were localized in the frontal area, implying the potential association with long-term memory and recall processes. It is well established that slow waves play a pivotal role in facilitating long-distance communication between distinct brain structures, with the interactions between the beta and theta bands contributing to the binding of neuronal information across both time and space [73,83].

In both the motor imagery (MI) and navigation (NAV) tasks, there was a notable presence of activity in the motor cortex, emphasizing their shared reliance on motor functions. For the MI task, the literature has typically reported amplitude changes in the mu band (7.5 Hz to 12.5 Hz) and beta band oscillations in the motor cortex [84,85], although this is not evidenced in the PSD topoplots presented in Figure 14. Nevertheless, this typical MI beta activity can be observed in the AMP topoplots in Figure 10. Here, beta activity in the

MI task could be delineated into beta–mtheta and beta–mbeta AM bands, suggesting a role for beta in maintaining long-term motor action.

For the NAV task, a slight similarity was discerned between the distributions of the beta1 PSD and corresponding AMP topoplots. This resemblance unveiled insights into the multifarious nature of beta band activity, hinting at processes such as the recall of familiar memories and the long-term binding of information. In particular, the beta–mdelta and beta–mbeta bands in the AMP topoplot exhibited relevant bilateral frontotemporal and occipital patterns. We hypothesized that beta–mdelta activity in frontotemporal areas was indicative of processes related to recalling familiar memories, possibly due to the involvement of the limbic system [73,81,83,86,87]. This system is a well-known generator of delta band oscillations, with delta band activity from the medial temporal cortex being notably associated with assessing the familiarity of sensory stimuli [76,88,89].

Further, beta–mbeta activity may be associated with the long-term temporal binding of information, a crucial aspect considering that the NAV task involved navigating through familiar terrain over extended durations. This band, depicting the fast temporal dynamics of beta oscillations in the prefrontal area, could reflect communication from the limbic system to the cortex, a mechanism that is integral to working and long-term memory processes [73,82,83]. In addition, beta–mtheta activity revealed an exclusive pattern in the left cortical hemisphere, potentially related to memory restitution.

In both the ROT and SING tasks, a general agreement between AMP and PSD was observed, with the presence of beta–mbeta activity in the motor cortex aligning with the necessity for advanced visuomotor functions in ROT and the long-term maintenance of specific processes in SING [73,81]. These observations collectively highlighted the consistent binding role of beta–mbeta activity across varied tasks.

In the PSD topoplots in Figure 14, relevant activity in the beta1 band was observed in the parietal and motor cortices during the WORD task. This pattern can be similarly observed in the AMP topoplots in Figure 10, within the beta–mdelta, beta–mtheta, and beta–mbeta bands. Notably, there was a diminished contribution from the anterior frontal region in the case of the beta–malpha band.

Following these observations, the AMP topoplots illustrated in Figure 10 underscored the significance of the motor cortex during the WORD task. This suggests the pivotal roles of motor functions in linguistic tasks [73,90]. The presence of the beta–mbeta and beta–mtheta bands further accentuated the potential of fast beta band dynamics in binding large neuronal assemblies [73,80,91].

4.2. Theta Band Analysis

As shown in Figure 12, the FACE task presented a relevant feature distribution in the occipital-parietal region and a partial presence in the frontal cortex. This pattern could be divided into two complementary patterns observed in the theta–mdelta and theta–mtheta bands in Figure 8. While the theta–mdelta band clearly represented most of the relevant activity located in the prefrontal, motor, and occipital cortices, the theta–mtheta band exhibited precise relevant activity in the occipital-parietal area. The theta band is known to be involved in memory retrieval [76–78,82,86,87,92] and the process of visual information encoding. This activity aligned with the requirements of the FACE task, which was a visual task necessitating memory retrieval.

For the MI task, the slow temporal dynamics of the theta band, represented by theta–mdelta, suggested modulated attention across time by the PFC. In contrast, faster theta power changes, represented by the theta–mtheta band, seemed mainly to contribute to the decoding of recalling visual information. In the MI task, the patterns observed across PSD and AMP, particularly in theta–mtheta band, supported the hypothesis of sensory feedback processing, providing a nuanced understanding of the reactivation of specialized sensory brain structures.

In the NAV task, both the PSD and AMP topoplots in Figures 8 and 12, respectively, showcase parietal and prefrontal activity, hinting at the activation of visual processing regions and working memory involvement, with theta oscillations likely reflecting a decoding mechanism for previously learned information [75,80]. Similarly, the ROT task presented theta patterns in the occipital-parietal regions that are associated with visual stimuli encoding, while frontal patterns in theta–mdelta band could be interpreted as reflecting problem-solving processes.

Despite the absence of prominent patterns in PSD for the SING task, the theta–mtheta activity in AMP revealed the involvement of the frontal and parietal regions, potentially indicating memorized information retrieval, maintaining the consistent hypothetical role of theta's involvement in information encoding [76,78,93].

Regarding the WORD task, the AMP topoplots in Figure 8 provide nuanced insights into the theta band's temporal dynamics and reveal a left prefrontal cortex lateralization, aligning with the literature on hemisphere association with linguistic functions. Further, the importance of the motor cortex, as highlighted in the AMP topoplots, aligned with the literature regarding its involvement in linguistic tasks [94].

4.3. Alpha Band Analysis

By examining the FACE task in Figure 13, it can be seen that both the alpha1 and alpha2 bands exhibited relevant occipital-parietal spatial distributions, as well as left frontal distribution. In the AMP topoplots presented in Figure 9, such patterns are also identifiable; however, the alpha–mdelta band showed diffused relevant activity across the entire scalp, while the alpha–mtheta band distribution was focused in the frontal and left occipital areas. These findings suggest two levels of speed in the alpha power temporal dynamics: slow variation in alpha activity seemed to be related to global scope function, while faster temporal dynamics seemed to be related to specific functions, such as executive and visual processing. Alpha oscillations, known for their relation with inhibition [70,81,95], show modulation by slow waves in the delta band, indicating the possible involvement of the limbic system as a modulated attention process. We could hypothesize that remembering a familiar face requires the sustained inhibition of new stimuli processing, as well as coordination between visually specialized brain structures (occipital) and specialized executive/emotion-related brain structures (frontal) [95].

Regarding the MI task, the mu band (7.5 to 12.5 Hz) is typically considered to be a relevant oscillation frequency within the motor cortex during motor imagery [96,97]. This pattern was also discernible in our feature analysis of the PSD topoplots in Figure 13, where we noted relevant activity for both the alpha1 (8 to 10 Hz) and alpha2 (10 to 12 Hz) bands in the motor cortex, coupled with observations of parietal-occipital activity for both bands. Upon contrasting this observed activity with that in the AMP topoplots, similar distributions of the relevant features were evident in the alpha–mdelta and alpha–mtheta bands. Specifically, the topoplot of alpha–mdelta band in Figure 9 illustrates pronounced and widespread activity in the motor cortex, suggesting that mu activity becomes more discernible when considering the slow temporal dynamics of alpha oscillations. We further hypothesized that the MI task necessitated the partial deactivation of the motor region as the movement was imagined and not executed, explaining the presence of slow modulated inhibition in the motor area. Furthermore, the alpha–mdelta band exhibited highly relevant activity in the parietal-occipital regions, potentially indicating the inhibition of visual stimuli [77], analogous to observations made in the FACE task. The alpha–mtheta band displayed a subdued pattern in the parietal-occipital cortex, with the distribution being more centralized in the somatosensory cortex. This activity was noteworthy as it suggested the presence of some form of virtual sensory feedback accompanying the imagined motor activity.

For the ROT task, the PSD topoplots depicted in Figure 13 exhibit consistent patterns for the alpha1 and alpha2 bands, showcasing activity in the bilateral frontal and right occipital regions. Delving deeper into the AMP topoplots, as seen in Figure 9, a clear distinction arises between two brain areas: the alpha–mdelta band was concentrated in the right occipital area, while alpha–mtheta band was prominently active in the bilateral frontal regions. This differentiation implied that the alpha–mdelta band was notably associated with controlled attention in the visual cortex, a nuanced insight provided by the comprehensive temporal analysis inherent to AMP features. This discerned alpha activity in both frontal and occipital areas potentially reflected a dynamic shift in attention and modulated inhibition across time. This interpretation aligned coherently with the demands of the ROT task, which necessitated the processing of visual stimuli coupled with logical evaluation.

For the SING task, the alpha–mtheta band was notably relevant in the prefrontal cortex and left parietal cortex, as evidenced more clearly by AMP than PSD. The alpha–mdelta band presented a comparable, albeit more diffuse, cortical-frontal activity. Unlike the other tasks, SING did not primarily involve visual processing, explaining the absence of any significant occipital activity. However, the task could necessitate imagined visual constructs or advanced sensorial representation, implicating the somatosensory, motor, and prefrontal cortices. The pronounced presence of alpha–mdelta activity in the prefrontal cortex could signify the greater influence of sympathetic structures during song recollection.

Intriguingly, the SUB task exhibited distribution patterns analogous to those of the SING task. This similarity was characterized by pronounced prefrontal activity in both the alpha–mdelta and alpha–mtheta bands, along with a bilateral occipital-parietal pattern. This pattern is observable in the alpha1 and alpha2 bands in the PSD topoplots represented in Figure 13 and the alpha–mdelta and alpha–mtheta bands in the AMP topoplots depicted in Figure 9. These observed distributions suggested the potential reliance on working memory and visual cues in the SUB task [76,90,98]. Lastly, the AMP topoplots for the WORD task revealed a left asymmetry in the cortical distribution of relevant features, a characteristic only present in the alpha2 band in the PSD topoplots. This asymmetry aligned with reports in the existing scientific literature [99].

4.4. Delta Band Analysis

As shown in Figure 7, the delta band exhibited a uniform distribution of relevant features across the scalp in all tasks, illustrating its role in modulating various cognitive processes throughout the brain. Notably, in the FACE and ROT tasks, the increased relevance of delta activity in the bilateral frontotemporal and parietal-occipital regions was observed, emphasizing its pivotal role in modulating attention [71,72] and facilitating the integration and coordination of sensory information [79]. This temporal synchronization is especially crucial in tasks requiring coherent communication among different brain regions for structured cognitive task execution. In the MI, SING, and SUB tasks, a subtle yet noticeable relevance in the parietal-occipital regions underscored the delta band's versatility in contributing to functions like motor imagery, sensory processing, and working memory. The insights derived from analyzing the delta band AMP topoplots across tasks (Figure 7) underscored the delta band's fundamental role in attention modulation, the temporal binding of neurological activity, and sustaining cognitive processing.

4.5. Gamma Band Analysis

As can be seen in Figure 11, for the FACE task, the relevant gamma band activity in the somatosensory area and bilateral frontotemporal regions emphasized its role in the processing of sensory information associated with recalling previously learned stimuli. The rapid oscillations of the gamma band were indicative of localized neuronal activity [70,73,80], making neurons more receptive to specific stimuli and thereby aiding in the retrieval of specific visual information [76,77,88,93]. In the MI task, the prominence of the gamma band in the motor and premotor cortices was consistent with the localized activation of neuronal structures associated with imagined motor activity. This suggests that the gamma band is closely linked with tasks that involve motor planning without actual movement. In the ROT task, all modulated gamma bands depicted in Figure 11 exhibited consistent distributions of relevant activity across a variety of cortical areas. The observed patterns of activity aligned with the multifaceted demands of the ROT task, which necessitated the concurrent engagement of abstract visual representation, advanced sensory processing, working memory, and executive functions. In both the SING and SUB tasks, the primary patterns in the somatosensory and parietal cortices marked the gamma band's significance in sensory processing, which is required in tasks in which sensorial patterns are reactivated, such as mental singing. Lastly, the WORD task displayed lateral patterns in the gamma–mdelta and gamma–mbeta bands in the left parietal cortex, with a shift towards the right parietal area in the gamma–mtheta, gamma–malpha, and gamma–mgamma bands. Such hemispheric specialization echoed with the previously observed patterns in the theta and alpha bands and further aligned with the well-established concept of hemisphere dominance in language processing [99].

4.6. Performance Interpretation

In light of the distinctive patterns observed across the different bands and mental tasks, we propose additional interpretations for some of the performance results. For the ROT task, unique distribution patterns were apparent in both the alpha and beta bands, depending on the temporal dynamics, alongside noticeable activity in the right motor cortex in the beta–mbeta band. These patterns, revealed by the AMP topoplots, offered additional insights beyond what PSDs could provide, thereby potentially explaining the clear differences in the accumulated prestige scores observed in Figure 6.

When comparing the MI task to the SING task, several similarities emerged, including the involvement of the alpha band in the motor cortex and parietal-occipital regions and the presence of beta band activity in the motor and sensorimotor cortices, as well as frontotemporal areas. These similarities could account for the low kappa score observed for the MI-SING mental task pair in Figure 4. Lastly, the SUB task demonstrated clear similarities in both the PSD and AMP topoplots for the alpha and beta bands and comparable occipital activity in the theta band. These similarities in the distributions of both feature types could account for the small differences in prestige scores observed in Figure 6.

In conclusion, distinguishing between the band amplitude dynamics in slow and fast components significantly enhanced our understanding of the roles of such bands. This was particularly crucial as slow oscillations are typically associated with long-distance communication between distinct brain regions, while fast oscillations correspond to intensive neurological activity. While our analysis provided an initial step towards understanding the neural oscillations and regional distributions underlying different tasks, a complete picture of brain dynamics is yet to emerge. Further investigations into functional connectivity and synchronization between different brain areas during tasks will provide more insights into the complex brain networks involved in mental processes.

Overall, our findings highlight the potential of AM features as promising tools for uncovering task-related changes across different frequencies and brain regions. By reflecting fundamental neurological mechanisms, such as inter-areal communication and top-down control, AM features could potentially enrich our understanding of the complexities of brain activity. These findings underscore the need to incorporate and investigate AM features in future BCI research.

4.7. Limitations

This work was not without its limitations. First, the number of participants, while in line with both clinical [37,100] and non-clinical BCI studies [101–103], was limited and future work could explore larger population sizes. Moreover, the proposed classification tasks relied on discriminating between two different mental tasks. In the future, a third resting class could also be used. Additionally, while the present study was interested in gauging the benefits of the proposed features and their existence across participants and session dates, the data from all participants were first combined and then partitioned into training and test sets. To further improve classifier accuracy, future work could explore a more common per-subject analysis where classifiers are fine-tuned to each individual user [104] or use data from certain days for training and those from other days for testing [105].

The extraction of AMP and CCORAM features in our study employed a 4 s time interval, introducing some latency. While the impact was minimized with a sliding window with 3 s overlap, the initial latency was a limiting factor compared to, e.g., PSD-based measures, which can rely on windows of 1 s and overlaps of milliseconds. As such, active BCIs based on the proposed features may have lower information transfer rates than other methods. While this may not be an issue for many applications, including neurorehabilitation, future studies could investigate the use of shorter epoch sizes to optimize the balance between latency and accuracy.

Moreover, while using CCORAM provided a more robust estimation of synchronized regions relative to the traditional coherence and phase locked value metrics, it did not reveal the direction of information flow. This information, in turn, could lead to additional insights, such as differentiation between motor control and sensory feedback information [106]. Additionally, phase-to-phase metrics are believed to measure cross-frequency interactions in the brain more accurately compared to phase–amplitude coupling-related features [74]; thus, future work could also explore such measures. Regarding frequency bands, here, we relied on the five conventional bands to allow for comparisons to previous works. However, some active BCIs may achieve improved performance with other bands (e.g., the mu band for MI-based BCIs). Future work could also explore the benefits of using alternative frequency band representations.

Lastly, we chose to use Fisher linear discriminant-based feature selection due to its computational effectiveness and ease-of-use for binary tasks. However, such simple selection methods do not remove potentially redundant features, which could be the cause of the small fluctuations seen in Figure 3. In the future, other more advanced feature selection algorithms, such as the mRMR algorithm [62], may provide a more concise feature pool for analysis.

5. Conclusions

In this paper, we proposed a new feature set for active BCIs based on the amplitude modulation dynamics of different EEG sub-bands. Via extensive experimentation, we showed the benefits of the proposed features, as well as their complementarity to conventional power spectral features. An in-depth discussion was provided to explore the complex cognitive mechanisms being measured by the new features and conjecture their roles in improving BCI accuracy for different mental tasks. As for future work, we wish to explore the use of the proposed features for adaptive BCIs, where the mental states of users can be tracked in real time (e.g., fatigue, frustration) to adjust active BCI classification. Such cognitive states are known to affect active BCI accuracy and may be a major limiting factor in BCI use in neurorehabilitation applications [14,107,108].

Author Contributions: Conceptualization, T.H.F.; Methodology, T.H.F.; Software, O.R.; Formal analysis, O.R.; Investigation, O.R.; Resources, T.H.F.; Writing—original draft, O.R.; Writing—review & editing, T.H.F., A.A.d.O.; Supervision, A.A.d.O. and T.H.F. All authors have read and agreed to the published version of the manuscript.

Funding: This research received no external funding.

Institutional Review Board Statement: The study was approved by the INRS Ethics Review Board.

Informed Consent Statement: Informed consent was obtained from all subjects involved in the study (Certificate CER 2014-354).

Data Availability Statement: Data was collected in 2014, as described in [48]. Data can be made available upon request to T.H.F.

Conflicts of Interest: The authors declare no conflict of interest.

Abbreviations

The following abbreviations are used in this manuscript:

AM	Amplitude Modulation
AMP	Amplitude Modulation Power
CCORAM	Circular Correlation of Amplitude Modulation
PSD	Power Spectral Density
BCI	Brain–Computer Interface
ROT	Rotation Imagery Task
SING	Sing Imagery Task
FACE	Face Imagery Task
MI	Motor Imagery Task
NAV	Navigational Imagery Task
SUB	Arithmetic Task
WORD	Word Completion Task
EOG	Electrooculography
EEG	Electroencephalogram

References

1. Rashid, M.; Sulaiman, N.; Abdul Majeed, A.P.P.; Musa, R.M.; Ab Nasir, A.F.; Bari, B.S.; Khatun, S. Current status, challenges, and possible solutions of EEG-based brain-computer interface: A comprehensive review. *Front. Neurorobot.* **2020**, *14*, 25. [CrossRef]
2. He, C.; Chen, Y.Y.; Phang, C.R.; Stevenson, C.; Chen, I.P.; Jung, T.P.; Ko, L.W. Diversity and Suitability of the State-of-the-Art Wearable and Wireless EEG Systems Review. *IEEE J. Biomed. Health Inform.* **2023**, *27*, 3830–3843. [CrossRef]
3. Vasiljevic, G.A.M.; De Miranda, L.C. Brain–computer interface games based on consumer-grade EEG Devices: A systematic literature review. *Int. J.-Hum.-Comput. Interact.* **2020**, *36*, 105–142. [CrossRef]
4. LaRocco, J.; Le, M.D.; Paeng, D.G. A systemic review of available low-cost EEG headsets used for drowsiness detection. *Front. Neuroinform.* **2020**, *14*, 42. [CrossRef]
5. Said, R.R.; Heyat, M.B.B.; Song, K.; Tian, C.; Wu, Z. A Systematic Review of Virtual Reality and Robot Therapy as Recent Rehabilitation Technologies Using EEG-Brain—Computer Interface Based on Movement-Related Cortical Potentials. *Biosensors* **2022**, *12*, 1134. [CrossRef] [PubMed]
6. Värbu, K.; Muhammad, N.; Muhammad, Y. Past, present, and future of EEG-based BCI applications. *Sensors* **2022**, *22*, 3331. [CrossRef] [PubMed]
7. Milani, G.; Antonioni, A.; Baroni, A.; Malerba, P.; Straudi, S. Relation Between EEG Measures and Upper Limb Motor Recovery in Stroke Patients: A Scoping Review. *Brain Topogr.* **2022**, *35*, 651–666. [CrossRef]
8. Vatinno, A.A.; Simpson, A.; Ramakrishnan, V.; Bonilha, H.S.; Bonilha, L.; Seo, N.J. The prognostic utility of electroencephalography in stroke recovery: A systematic review and meta-analysis. *Neurorehabilit. Neural Repair* **2022**, *36*, 255–268. [CrossRef]
9. Mansour, S.; Ang, K.K.; Nair, K.P.; Phua, K.S.; Arvaneh, M. Efficacy of brain–computer interface and the impact of its design characteristics on poststroke upper-limb rehabilitation: A systematic review and meta-analysis of randomized controlled trials. *Clin. EEG Neurosci.* **2022**, *53*, 79–90. [CrossRef]
10. Sebastián-Romagosa, M.; Cho, W.; Ortner, R.; Sieghartsleitner, S.; Von Oertzen, T.J.; Kamada, K.; Laureys, S.; Allison, B.Z.; Guger, C. Brain–computer interface treatment for gait rehabilitation in stroke patients. *Front. Neurosci.* **2023**, *17*, 1256077. [CrossRef] [PubMed]
11. Chaudhary, U.; Mrachacz-Kersting, N.; Birbaumer, N. Neuropsychological and neurophysiological aspects of brain-computer-interface (BCI) control in paralysis. *J. Physiol.* **2021**, *599*, 2351–2359. [CrossRef] [PubMed]
12. Vanutelli, M.E.; Salvadore, M.; Lucchiari, C. BCI Applications to Creativity: Review and Future Directions, from little-c to C2. *Brain Sci.* **2023**, *13*, 665. [CrossRef]
13. Ding, Q.; Lin, T.; Wu, M.; Yang, W.; Li, W.; Jing, Y.; Ren, X.; Gong, Y.; Xu, G.; Lan, Y. Influence of iTBS on the acute neuroplastic change after BCI training. *Front. Cell. Neurosci.* **2021**, *15*, 653487. [CrossRef] [PubMed]
14. Myrden, A.; Chau, T. Effects of user mental state on EEG-BCI performance. *Front. Hum. Neurosci.* **2015**, *9*, 308. [CrossRef]
15. Nijboer, F.; Birbaumer, N.; Kübler, A. The influence of psychological state and motivation on brain–computer interface performance in patients with amyotrophic lateral sclerosis—A longitudinal study. *Front. Neuropharmacol.* **2010**, *4*, 55. [CrossRef]
16. Kleih, S.C.; Kübler, A. Psychological factors influencing brain-computer interface (BCI) performance. In Proceedings of the 2015 IEEE International Conference on Systems, Man, and Cybernetics, Hong Kong, China, 9–12 October 2015; pp. 3192–3196.
17. Jeunet, C.; N'Kaoua, B.; Lotte, F. Advances in user-training for mental-imagery-based BCI control: Psychological and cognitive factors and their neural correlates. *Prog. Brain Res.* **2016**, *228*, 3–35. [PubMed]
18. Xu, M.; He, F.; Jung, T.P.; Gu, X.; Ming, D. Current challenges for the practical application of electroencephalography-based brain–computer interfaces. *Engineering* **2021**, *7*, 1710–1712. [CrossRef]
19. Lee, Y.E.; Kwak, N.S.; Lee, S.W. A real-time movement artifact removal method for ambulatory brain-computer interfaces. *IEEE Trans. Neural Syst. Rehabil. Eng.* **2020**, *28*, 2660–2670. [CrossRef]

20. Naser, M.Y.; Bhattacharya, S. Towards Practical BCI-Driven Wheelchairs: A Systematic Review Study. *IEEE Trans. Neural Syst. Rehabil. Eng.* **2023**, *31*, 1030–1044. [CrossRef]

21. Liu, Z.; Shore, J.; Wang, M.; Yuan, F.; Buss, A.; Zhao, X. A systematic review on hybrid EEG/fNIRS in brain-computer interface. *Biomed. Signal Process. Control* **2021**, *68*, 102595. [CrossRef]

22. Li, Z.; Zhang, S.; Pan, J. Advances in hybrid brain-computer interfaces: Principles, design, and applications. *Comput. Intell. Neurosci.* **2019**, *2019*, 3807670. [CrossRef]

23. Varshney, A.; Ghosh, S.K.; Padhy, S.; Tripathy, R.K.; Acharya, U.R. Automated classification of mental arithmetic tasks using recurrent neural network and entropy features obtained from multi-channel EEG signals. *Electronics* **2021**, *10*, 1079. [CrossRef]

24. Angsuwatanakul, T.; O'Reilly, J.; Ounjai, K.; Kaewkamnerdpong, B.; Iramina, K. Multiscale entropy as a new feature for EEG and fNIRS analysis. *Entropy* **2020**, *22*, 189. [CrossRef] [PubMed]

25. Agarwal, P.; Kumar, S. EEG-based imagined words classification using Hilbert transform and deep networks. *Multimed. Tools Appl.* **2023**, *20*, 026040. [CrossRef]

26. Dzianok, P.; Kołodziej, M.; Kublik, E. Detecting attention in Hilbert-transformed EEG brain signals from simple-reaction and choice-reaction cognitive tasks. In Proceedings of the 2021 IEEE 21st International Conference on Bioinformatics and Bioengineering (BIBE), Kragujevac, Serbia, 25–27 October 2021; pp. 1–4.

27. Ko, W.; Jeon, E.; Jeong, S.; Suk, H.I. Multi-scale neural network for EEG representation learning in BCI. *IEEE Comput. Intell. Mag.* **2021**, *16*, 31–45. [CrossRef]

28. Bascil, M.S.; Tesneli, A.Y.; Temurtas, F. Spectral feature extraction of EEG signals and pattern recognition during mental tasks of 2-D cursor movements for BCI using SVM and ANN. *Australas. Phys. Eng. Sci. Med.* **2016**, *39*, 665–676. [CrossRef] [PubMed]

29. Tiwari, S.; Goel, S.; Bhardwaj, A. MIDNN—A classification approach for the EEG based motor imagery tasks using deep neural network. *Appl. Intell.* **2022**, *52*, 4824–4843. [CrossRef]

30. Cattai, T.; Colonnese, S.; Corsi, M.C.; Bassett, D.S.; Scarano, G.; Fallani, F.D.V. Phase/amplitude synchronization of brain signals during motor imagery BCI tasks. *IEEE Trans. Neural Syst. Rehabil. Eng.* **2021**, *29*, 1168–1177. [CrossRef]

31. Tao, X.; Yi, W.; Wang, K.; He, F.; Qi, H. Inter-stimulus phase coherence in steady-state somatosensory evoked potentials and its application in improving the performance of single-channel MI-BCI. *J. Neural Eng.* **2021**, *18*, 046088. [CrossRef]

32. Nisar, H.; Boon, K.W.; Ho, Y.K.; Khang, T.S. Brain-Computer Interface: Feature Extraction and Classification of Motor Imagery-Based Cognitive Tasks. In Proceedings of the 2022 IEEE International Conference on Automatic Control and Intelligent Systems (I2CACIS), Shah Alam, Malaysia, 25–25 June 2022; pp. 42–47.

33. Chakladar, D.D.; Dey, S.; Roy, P.P.; Dogra, D.P. EEG-based mental workload estimation using deep BLSTM-LSTM network and evolutionary algorithm. *Biomed. Signal Process. Control* **2020**, *60*, 101989. [CrossRef]

34. Wang, L.; Huang, W.; Yang, Z.; Zhang, C. Temporal-spatial-frequency depth extraction of brain-computer interface based on mental tasks. *Biomed. Signal Process. Control* **2020**, *58*, 101845. [CrossRef]

35. He, C.; Liu, J.; Zhu, Y.; Du, W. Data augmentation for deep neural networks model in EEG classification task: A review. *Front. Hum. Neurosci.* **2021**, *15*, 765525. [CrossRef] [PubMed]

36. Mai, N.D.; Long, N.M.H.; Chung, W.Y. 1D-CNN-based BCI system for detecting Emotional states using a Wireless and Wearable 8-channel Custom-designed EEG Headset. In Proceedings of the 2021 IEEE International Conference on Flexible and Printable Sensors and Systems (FLEPS), Manchester, UK, 20–23 June 2021; pp. 1–4.

37. Roy, Y.; Banville, H.; Albuquerque, I.; Gramfort, A.; Falk, T.H.; Faubert, J. Deep learning-based electroencephalography analysis: A systematic review. *J. Neural Eng.* **2019**, *16*, 051001. [CrossRef]

38. Meng, L.; Jiang, X.; Wu, D. Adversarial robustness benchmark for EEG-based brain–computer interfaces. *Future Gener. Comput. Syst.* **2023**, *143*, 231–247. [CrossRef]

39. Zhang, X.; Wu, D.; Ding, L.; Luo, H.; Lin, C.T.; Jung, T.P.; Chavarriaga, R. Tiny noise, big mistakes: Adversarial perturbations induce errors in brain–computer interface spellers. *Natl. Sci. Rev.* **2021**, *8*, nwaa233. [CrossRef] [PubMed]

40. Cassani, R.; Falk, T.H. Alzheimer's disease diagnosis and severity level detection based on electroencephalography modulation spectral "patch" features. *IEEE J. Biomed. Health Inform.* **2019**, *24*, 1982–1993. [CrossRef]

41. Jesus, B., Jr.; Cassani, R.; McGeown, W.J.; Cecchi, M.; Fadem, K.; Falk, T.H. Multimodal prediction of Alzheimer's disease severity level based on resting-state EEG and structural MRI. *Front. Hum. Neurosci.* **2021**, *15*, 700627. [CrossRef] [PubMed]

42. Clerico, A.; Tiwari, A.; Gupta, R.; Jayaraman, S.; Falk, T.H. Electroencephalography amplitude modulation analysis for automated affective tagging of music video clips. *Front. Comput. Neurosci.* **2018**, *11*, 115. [CrossRef]

43. Grass, A.M.; Gibbs, F.A. A Fourier transform of the electroencephalogram. *J. Neurophysiol.* **1938**, *1*, 521–526. [CrossRef]

44. Buzsáki, G.; Anastassiou, C.A.; Koch, C. The origin of extracellular fields and currents—EEG, ECoG, LFP and spikes. *Nat. Rev. Neurosci.* **2012**, *13*, 407–420. [CrossRef]

45. Barlow, J.S. Methods of analysis of nonstationary EEGs, with emphasis on segmentation techniques: A comparative review. *J. Clin. Neurophysiol.* **1985**, *2*, 267–304. [CrossRef] [PubMed]

46. Cassani, R.; Falk, T.H. Spectrotemporal modeling of biomedical signals: Theoretical foundation and applications. In *Reference Module in Biomedical Sciences*; University of Quebec: Montreal, QC, Canada, 2018.

47. Trambaiolli, L.R.; Cassani, R.; Falk, T.H. EEG spectro-temporal amplitude modulation as a measurement of cortical hemodynamics: An EEG-fNIRS study. In Proceedings of the 2020 42nd Annual International Conference of the IEEE Engineering in Medicine & Biology Society (EMBC), Virtual, 20–24 July 2020; pp. 3481–3484.

48. Banville, H.; Gupta, R.; Falk, T.H. Mental task evaluation for hybrid NIRS-EEG brain-computer interfaces. *Comput. Intell. Neurosci.* **2017**, *2017*, 3524208. [CrossRef] [PubMed]

49. Herrmann, C.S.; Senkowski, D.; Röttger, S. Phase-locking and amplitude modulations of EEG alpha: Two measures reflect different cognitive processes in a working memory task. *Exp. Psychol.* **2004**, *51*, 311–318. [CrossRef]

50. Thatcher, R.W. Coherence, phase differences, phase shift, and phase lock in EEG/ERP analyses. *Dev. Neuropsychol.* **2012**, *37*, 476–496. [CrossRef] [PubMed]

51. French, C.C.; Beaumont, J.G. A critical review of EEG coherence studies of hemisphere function. *Int. J. Psychophysiol.* **1984**, *1*, 241–254. [CrossRef]

52. Aydore, S.; Pantazis, D.; Leahy, R.M. A note on the phase locking value and its properties. *Neuroimage* **2013**, *74*, 231–244. [CrossRef]

53. Delorme, A.; Makeig, S. EEGLAB: An open source toolbox for analysis of single-trial EEG dynamics including independent component analysis. *J. Neurosci. Methods* **2004**, *134*, 9–21. [CrossRef]

54. Hyvarinen, A. Fast and robust fixed-point algorithms for independent component analysis. *IEEE Trans. Neural Netw.* **1999**, *10*, 626–634. [CrossRef]

55. Mognon, A.; Jovicich, J.; Bruzzone, L.; Buiatti, M. ADJUST: An automatic EEG artifact detector based on the joint use of spatial and temporal features. *Psychophysiology* **2011**, *48*, 229–240. [CrossRef]

56. Gramfort, A.; Luessi, M.; Larson, E.; Engemann, D.A.; Strohmeier, D.; Brodbeck, C.; Goj, R.; Jas, M.; Brooks, T.; Parkkonen, L.; et al. MEG and EEG data analysis with MNE-Python. *Front. Neurosci.* **2013**, *7*, 267. [CrossRef]

57. Newson, J.J.; Thiagarajan, T.C. EEG frequency bands in psychiatric disorders: A review of resting state studies. *Front. Hum. Neurosci.* **2019**, *12*, 521. [CrossRef] [PubMed]

58. Trajin, B.; Chabert, M.; Régnier, J.; Faucher, J. Space vector analysis for the diagnosis of high frequency amplitude and phase modulations in induction motor stator current. In Proceedings of the 5th International Conference on Condition Monitoring and Machinery Failure Prevention Technologies—CM/MFPT 200810, Stratford-upon-Avon, UK, 22–24 June 2008; pp. 1–8.

59. Fraga, F.J.; Falk, T.H.; Kanda, P.A.; Anghinah, R. Characterizing Alzheimer's disease severity via resting-awake EEG amplitude modulation analysis. *PLoS ONE* **2013**, *8*, e72240. [CrossRef] [PubMed]

60. Burgess, A.P. On the interpretation of synchronization in EEG hyperscanning studies: A cautionary note. *Front. Hum. Neurosci.* **2013**, *7*, 881. [CrossRef] [PubMed]

61. Murias, M.; Webb, S.J.; Greenson, J.; Dawson, G. Resting state cortical connectivity reflected in EEG coherence in individuals with autism. *Biol. Psychiatry* **2007**, *62*, 270–273. [CrossRef]

62. Ding, C.; Peng, H. Minimum redundancy feature selection from microarray gene expression data. *J. Bioinform. Comput. Biol.* **2005**, *3*, 185–205. [CrossRef]

63. Guyon, I.; Weston, J.; Barnhill, S.; Vapnik, V. Gene selection for cancer classification using support vector machines. *Mach. Learn.* **2002**, *46*, 389–422. [CrossRef]

64. Fu, R.; Tian, Y.; Shi, P.; Bao, T. Automatic detection of epileptic seizures in EEG using sparse CSP and fisher linear discrimination analysis algorithm. *J. Med. Syst.* **2020**, *44*, 1–13. [CrossRef]

65. Sikarwar, R.; Shakya, H.K.; Singh, S.S. A Review on Social Network Analysis Methods and Algorithms. In Proceedings of the 2021 13th International Conference on Computational Intelligence and Communication Networks (CICN), Lima, Peru, 22–23 September 2021; pp. 1–5.

66. Keener, J.P. The Perron–Frobenius theorem and the ranking of football teams. *SIAM Rev.* **1993**, *35*, 80–93. [CrossRef]

67. Trambaiolli, L.; Cassani, R.; Biazoli, C.; Cravo, A.; Sato, J.; Falk, T. Resting-awake EEG amplitude modulation can predict performance of an fNIRS-based neurofeedback task. In Proceedings of the 2018 IEEE International Conference on Systems, Man, and Cybernetics (SMC), Miyazaki, Japan, 7–10 October 2018; pp. 1128–1132.

68. Autthasan, P.; Du, X.; Arnin, J.; Lamyai, S.; Perera, M.; Itthipuripat, S.; Yagi, T.; Manoonpong, P.; Wilaiprasitporn, T. A single-channel consumer-grade EEG device for brain–computer interface: Enhancing detection of SSVEP and its amplitude modulation. *IEEE Sens. J.* **2019**, *20*, 3366–3378. [CrossRef]

69. Hilla, Y.; Von Mankowski, J.; Föcker, J.; Sauseng, P. Faster visual information processing in video gamers is associated with EEG alpha amplitude modulation. *Front. Psychol.* **2020**, *11*, 599788. [CrossRef]

70. Jensen, O.; Mazaheri, A. Shaping functional architecture by oscillatory alpha activity: Gating by inhibition. *Front. Hum. Neurosci.* **2010**, *4*, 186. [CrossRef] [PubMed]

71. Prada, L.; Barcelo, F.; Herrmann, C.S.; Escera, C. EEG delta oscillations index inhibitory control of contextual novelty to both irrelevant distracters and relevant task-switch cues. *Psychophysiology* **2014**, *51*, 658–672. [CrossRef] [PubMed]

72. Zarjam, P.; Epps, J.; Lovell, N.H.; Chen, F. Characterization of memory load in an arithmetic task using non-linear analysis of EEG signals. In Proceedings of the 2012 Annual International Conference of the IEEE Engineering in Medicine and Biology Society, San Diego, CA, USA, 28 August–1 September 2012; pp. 3519–3522.

73. Weiss, S.; Mueller, H.M. "Too many betas do not spoil the broth": The role of beta brain oscillations in language processing. *Front. Psychol.* **2012**, *3*, 201. [CrossRef]

74. Palva, J.M.; Palva, S. Functional integration across oscillation frequencies by cross-frequency phase synchronization. *Eur. J. Neurosci.* **2018**, *48*, 2399–2406. [CrossRef]

75. Daume, J.; Gruber, T.; Engel, A.K.; Friese, U. Phase-amplitude coupling and long-range phase synchronization reveal frontotemporal interactions during visual working memory. *J. Neurosci.* **2017**, *37*, 313–322. [CrossRef]
76. Herweg, N.A.; Apitz, T.; Leicht, G.; Mulert, C.; Fuentemilla, L.; Bunzeck, N. Theta-alpha oscillations bind the hippocampus, prefrontal cortex, and striatum during recollection: Evidence from simultaneous EEG–fMRI. *J. Neurosci.* **2016**, *36*, 3579–3587. [CrossRef] [PubMed]
77. Freeman, W.J.; Vitello, G. Matter and mind are entangled in EEG amplitude modulation and its double. *Soc. Mind Matter Res.* **2016**, *14*, 7–24.
78. Klimesch, W. EEG alpha and theta oscillations reflect cognitive and memory performance: A review and analysis. *Brain Res. Rev.* **1999**, *29*, 169–195. [CrossRef]
79. Harmony, T.; Fernández, T.; Silva, J.; Bernal, J.; Díaz-Comas, L.; Reyes, A.; Marosi, E.; Rodríguez, M.; Rodríguez, M. EEG delta activity: An indicator of attention to internal processing during performance of mental tasks. *Int. J. Psychophysiol.* **1996**, *24*, 161–171. [CrossRef]
80. Silberstein, R.B. Dynamic sculpting of brain functional connectivity and mental rotation aptitude. *Prog. Brain Res.* **2006**, *159*, 63–76.
81. Mizuhara, H.; Wang, L.Q.; Kobayashi, K.; Yamaguchi, Y. Long-range EEG phase synchronization during an arithmetic task indexes a coherent cortical network simultaneously measured by fMRI. *Neuroimage* **2005**, *27*, 553–563. [CrossRef] [PubMed]
82. Plank, M.; Müller, H.J.; Onton, J.; Makeig, S.; Gramann, K. Human EEG correlates of spatial navigation within egocentric and allocentric reference frames. In *Proceedings of the Spatial Cognition VII: International Conference, Spatial Cognition 2010, Mt. Hood/Portland, OR, USA, 15–19 August 2010*; Proceedings 7; Springer: Berlin/Heidelberg, Germany, 2010; pp. 191–206.
83. Mizuhara, H.; Yamaguchi, Y. Human cortical circuits for central executive function emerge by theta phase synchronization. *Neuroimage* **2007**, *36*, 232–244. [CrossRef] [PubMed]
84. Han, Y.; Wang, B.; Luo, J.; Li, L.; Li, X. A classification method for EEG motor imagery signals based on parallel convolutional neural network. *Biomed. Signal Process. Control* **2022**, *71*, 103190. [CrossRef]
85. Xu, K.; Huang, Y.Y.; Duann, J.R. The sensitivity of single-trial mu-suppression detection for motor imagery performance as compared to motor execution and motor observation performance. *Front. Hum. Neurosci.* **2019**, *13*, 302. [CrossRef]
86. Alahmari, A. Neuroimaging Role in Mental Illnesses. *Int. J. Neural Plast.* **2021**, *4*. [CrossRef]
87. Mizuhara, H.; Wang, L.Q.; Kobayashi, K.; Yamaguchi, Y. A long-range cortical network emerging with theta oscillation in a mental task. *Neuroreport* **2004**, *15*, 1233–1238. [CrossRef]
88. Norman, K.A.; O'Reilly, R.C. Modeling hippocampal and neocortical contributions to recognition memory: A complementary-learning-systems approach. *Psychol. Rev.* **2003**, *110*, 611. [CrossRef] [PubMed]
89. Collins, E.; Robinson, A.K.; Behrmann, M. Distinct neural processes for the perception of familiar versus unfamiliar faces along the visual hierarchy revealed by EEG. *NeuroImage* **2018**, *181*, 120–131. [CrossRef] [PubMed]
90. Soltanlou, M.; Artemenko, C.; Dresler, T.; Haeussinger, F.B.; Fallgatter, A.J.; Ehlis, A.C.; Nuerk, H.C. Increased arithmetic complexity is associated with domain-general but not domain-specific magnitude processing in children: A simultaneous fNIRS-EEG study. *Cogn. Affect. Behav. Neurosci.* **2017**, *17*, 724–736. [CrossRef]
91. Anwar, A.R.; Muthalib, M.; Perrey, S.; Galka, A.; Granert, O.; Wolff, S.; Heute, U.; Deuschl, G.; Raethjen, J.; Muthuraman, M. Effective connectivity of cortical sensorimotor networks during finger movement tasks: A simultaneous fNIRS, fMRI, EEG study. *Brain Topogr.* **2016**, *29*, 645–660. [CrossRef]
92. Cebolla, A.M.; Palmero-Soler, E.; Leroy, A.; Cheron, G. EEG spectral generators involved in motor imagery: A swLORETA study. *Front. Psychol.* **2017**, *8*, 2133. [CrossRef] [PubMed]
93. Mukundan, C.; Sumit, S.; Chetan, S. Brain Electrical Oscillations Signature profiling (BEOS) for measuring the process of remembrance. *EC Neurol.* **2017**, *8*, 217–230.
94. Vukovic, N.; Feurra, M.; Shpektor, A.; Myachykov, A.; Shtyrov, Y. Primary motor cortex functionally contributes to language comprehension: An online rTMS study. *Neuropsychologia* **2017**, *96*, 222–229. [CrossRef] [PubMed]
95. Sauseng, P.; Klimesch, W.; Stadler, W.; Schabus, M.; Doppelmayr, M.; Hanslmayr, S.; Gruber, W.R.; Birbaumer, N. A shift of visual spatial attention is selectively associated with human EEG alpha activity. *Eur. J. Neurosci.* **2005**, *22*, 2917–2926. [CrossRef]
96. Yu, H.; Ba, S.; Guo, Y.; Guo, L.; Xu, G. Effects of motor imagery tasks on brain functional networks based on EEG Mu/Beta rhythm. *Brain Sci.* **2022**, *12*, 194. [CrossRef]
97. Molla, M.K.I.; Al Shiam, A.; Islam, M.R.; Tanaka, T. Discriminative feature selection-based motor imagery classification using EEG signal. *IEEE Access* **2020**, *8*, 98255–98265. [CrossRef]
98. Spüler, M.; Walter, C.; Rosenstiel, W.; Gerjets, P.; Moeller, K.; Klein, E. EEG-based prediction of cognitive workload induced by arithmetic: A step towards online adaptation in numerical learning. *ZDM* **2016**, *48*, 267–278. [CrossRef]
99. Riès, S.K.; Dronkers, N.F.; Knight, R.T. Choosing words: Left hemisphere, right hemisphere, or both? Perspective on the lateralization of word retrieval. *Ann. N. Y. Acad. Sci.* **2016**, *1369*, 111–131. [CrossRef] [PubMed]
100. Lennon, O.; Tonellato, M.; Del Felice, A.; Di Marco, R.; Fingleton, C.; Korik, A.; Guanziroli, E.; Molteni, F.; Guger, C.; Otner, R.; et al. A systematic review establishing the current state-of-the-art, the limitations, and the DESIRED checklist in studies of direct neural interfacing with robotic gait devices in stroke rehabilitation. *Front. Neurosci.* **2020**, *14*, 578. [CrossRef]
101. Al-Saegh, A.; Dawwd, S.A.; Abdul-Jabbar, J.M. Deep learning for motor imagery EEG-based classification: A review. *Biomed. Signal Process. Control* **2021**, *63*, 102172. [CrossRef]

102. Arpaia, P.; Esposito, A.; Natalizio, A.; Parvis, M. How to successfully classify EEG in motor imagery BCI: A metrological analysis of the state of the art. *J. Neural Eng.* **2022**, *19*, 031002. [CrossRef] [PubMed]

103. Kaya, M.; Binli, M.; Ozbay, E.; Yanar, H.; Mishchenko, Y. A large electroencephalographic motor imagery dataset for electroencephalographic brain computer interfaces. *Sci. Data* **2018**, *5*, 180211. [CrossRef] [PubMed]

104. Dos Santos, E.M.; San-Martin, R.; Fraga, F.J. Comparison of subject-independent and subject-specific EEG-based BCI using LDA and SVM classifiers. *Med. Biol. Eng. Comput.* **2023**, *61*, 835–845. [CrossRef] [PubMed]

105. Śliwowski, M.; Martin, M.; Souloumiac, A.; Blanchart, P.; Aksenova, T. Impact of dataset size and long-term ECoG-based BCI usage on deep learning decoders performance. *Front. Hum. Neurosci.* **2023**, *17*, 1111645. [CrossRef]

106. Liang, T.; Zhang, Q.; Liu, X.; Dong, B.; Liu, X.; Wang, H. Identifying bidirectional total and non-linear information flow in functional corticomuscular coupling during a dorsiflexion task: A pilot study. *J. Neuroeng. Rehabil.* **2021**, *18*, 1–15. [CrossRef]

107. Hougaard, B.I.; Rossau, I.G.; Czapla, J.J.; Miko, M.A.; Skammelsen, R.B.; Knoche, H.; Jochumsen, M. Who willed it? decreasing frustration by manipulating perceived control through fabricated input for stroke rehabilitation BCI games. *Proc. ACM-Hum.-Comput. Interact.* **2021**, *5*, 1–19. [CrossRef]

108. Évain, A.; Argelaguet, F.; Strock, A.; Roussel, N.; Casiez, G.; Lécuyer, A. Influence of error rate on frustration of BCI users. In Proceedings of the International Working Conference on Advanced Visual Interfaces, Bari, Italy, 7–10 June 2016; pp. 248–251.

Article

A Symbols Based BCI Paradigm for Intelligent Home Control Using P300 Event-Related Potentials

Faraz Akram [1,*], Ahmed Alwakeel [2,3], Mohammed Alwakeel [3], Mohammad Hijji [3] and Usman Masud [4,5,*]

1   Department of Biomedical Engineering, Riphah International University, Islamabad 46000, Pakistan
2   Sensor Networks and Cellular Systems Research Center, University of Tabuk, Tabuk 71491, Saudi Arabia
3   Faculty of Computers & Information Technology, University of Tabuk, Tabuk 71491, Saudi Arabia
4   Faculty of Electrical and Electronics Engineering, University of Engineering and Technology, Taxila 47050, Pakistan
5   Department of Electrical Communication Engineering, University of Kassel, 34127 Kassel, Germany
*   Correspondence: faraz.akram@riphah.edu.pk (F.A.); usmanmasud123@hotmail.com (U.M.)

Abstract: Brain-Computer Interface (BCI) is a technique that allows the disabled to interact with a computer directly from their brain. P300 Event-Related Potentials (ERP) of the brain have widely been used in several applications of the BCIs such as character spelling, word typing, wheelchair control for the disabled, neurorehabilitation, and smart home control. Most of the work done for smart home control relies on an image flashing paradigm where six images are flashed randomly, and the users can select one of the images to control an object of interest. The shortcoming of such a scheme is that the users have only six commands available in a smart home to control. This article presents a symbol-based P300-BCI paradigm for controlling home appliances. The proposed paradigm comprises of a 12-symbols, from which users can choose one to represent their desired command in a smart home. The proposed paradigm allows users to control multiple home appliances from signals generated by the brain. The proposed paradigm also allows the users to make phone calls in a smart home environment. We put our smart home control system to the test with ten healthy volunteers, and the findings show that the proposed system can effectively operate home appliances through BCI. Using the random forest classifier, our participants had an average accuracy of 92.25 percent in controlling the home devices. As compared to the previous studies on the smart home control BCIs, the proposed paradigm gives the users more degree of freedom, and the users are not only able to control several home appliances but also have an option to dial a phone number and make a call inside the smart home. The proposed symbols-based smart home paradigm, along with the option of making a phone call, can effectively be used for controlling home through signals of the brain, as demonstrated by the results.

Keywords: brain-computer interface; smart home; phone control; event-related potentials; EEG; P300

Citation: Akram, F.; Alwakeel, A.; Alwakeel, M.; Hijji, M.; Masud, U. A Symbols Based BCI Paradigm for Intelligent Home Control Using P300 Event-Related Potentials. *Sensors* **2022**, *22*, 10000. https://doi.org/10.3390/s222410000

Academic Editors: Imran Khan Niazi, Hendrik Santosa and Noman Naseer

Received: 28 October 2022
Accepted: 15 December 2022
Published: 19 December 2022

1. Introduction

A neurological disease known as a locked-in syndrome is caused by the total paralysis of all voluntary muscles throughout the body. It could be caused by brain or spinal cord damage, amyotrophic lateral sclerosis, brainstem stroke, diseases of the circulatory system, damage to nerve cells, and several other neuromuscular diseases. People who have locked-in syndrome are completely cognizant and able to reason and think, but they are unable to talk or move anything other than their eyes. All other voluntary muscle movement is blocked, making it impossible for them to speak or move [1–3]. Locked-in syndrome patients may generally move their eyes and can occasionally blink to communicate, but some seriously ill patients can also lose control of their eye movements and may become completely paralyzed. It is extremely difficult to interact with individuals who have locked-in syndrome and other forms of paralysis since they are unable to speak or express their needs or sentiments to those around them due to their lack of muscle-based modes of communication. For such patients, a direct brain-computer interface (BCI), which sends messages and instructions to the outside world via a new, non-muscular communication and control channel, may be an effective option. These people may be able to communicate once again because of brain-computer interfaces that open a new line of communication between their brain signals and computers. Recent research has demonstrated that BCI technology makes it viable for the brain to directly interact with the outside world, enabling paralyzed people to interact and gain control. The BCI technology enables users to interact with computers and operate appliances without using their muscles. BCI research aims to improve the overall quality of life for the disabled by giving them access to technology that allows them to interact with their environment. To attain this objective, several BCI applications have been presented in the literature. including character spelling [4,5], word typing [6], controlling a wheelchair [7,8], controlling a robotic/prosthetic limb [9], virtual reality control [10,11], neurorehabilitation [12], controlling a car [13], web browser control [14], Unmanned Aerial Vehicle (UAV) control [15,16], and games [17–19]. All these BCIs take commands from the brain and transform them to control signals for the desired application.

There are numerous ways of reading the brain's activity such as functional Magnetic Resonance Imaging (fMRI), Magnetoencephalography (MEG), Positron Emission Tomography (PET), Computer Tomography (CT), Electrocorticography (ECoG), and functional Near-Infrared Spectroscopy (fNIRS). However, Electroencephalography (EEG) has been the most popular type for BCIs because it is non-invasive, relatively cheap, and easier to use. The brain's EEG signal is made up of numerous signals with varied characteristics, each of which corresponds to a different mental activity. P300 Event-Related Potential (ERP), one of the signals, has been used in various BCI applications. When the user is exposed to an uncommon stimulus in a sequence of common stimuli, about 300 milliseconds after the start of the target/rare stimulus, a positive wave called P300 appears. It is induced when a person recognizes an occasional "target" stimulus among a regular stream of common "non-target" stimuli. When the user concentrates on a specific stimulus while simultaneously being exposed to several stimuli, P300 can be induced. In 1998, Farwell and Donchin [20] were the first to demonstrate a character spelling application of the P300-ERP. They proposed displaying a flashing matrix, having 6 rows and 6 columns of characters and numbers, to the users, and taking EEG signals of the users. They could get the P300 wave only after the target character was flashed, which could be detected using a classifier. The P300 signal has been used for numerous other applications of BCI; for instance, directing the movement of a cursor or a screen item [21,22], browsing the internet [23], file explorer [24], Playing games [25], and navigating a wheelchair [26].

There have been few studies on using the BCI to operate a smart-home environment. Hoffmann et al. [27] proposed an interface consisting of 6 images of home appliances for disabled subjects, and the users could select one of the images. Their proposed image flashing paradigm had been the most popular paradigm for smart-home control and several researchers used the same scheme. Achanccaray et al. [28] and Cortez et al. [29] used the same paradigm to control six home appliances. The problem with this image flashing paradigm is that the users have only six commands available, therefore they can perform only six tasks in a smart home. Carabalona et al. [30] developed an icon-based smart home control system. They proposed to use icons instead of characters in a smart home control system. However, the accuracy achieved using their proposed system was very low i.e., 50%. Park et al. [31] used Steady-State Visual Evoked Potentials (SSVEP) to operate three household appliances: a robotic vacuum cleaner, an air cleaner, and a humidifier. Kais et al. [32] proposed an implementation of home devices control using the motor imagery BCI. They used the BCI competition dataset where subjects performed four types of motor imagery (Left hand, Right hand, Foot, and Tongue). These four motor imagery states can be used to turn on/off two devices. Edlinger et al. [33] combined P300 and SSVEP for controlling a virtual home environment and tested their system with three subjects. Katyal et al. [34] also proposed a hybrid paradigm containing SSVEP and P300 to increase the number of decision options. Similarly, Chai et al. [35] proposed combining SSVEP with electromyography (EMG) signals. Uyanik et al. [36] proposed an SSVEP-based BCI to control a wheelchair along with a smart home. Kim et al. [37] combined ERPs with speech imagery. Usman et al. [38], proposed a symbols-based smart home control system using the P300. The suggested method used a 6×4 matrix, akin to the character spelling paradigm. The proposed system was tested on three subjects and achieved an accuracy of 87.5%. Taejun Lee et al. [39] utilized a BCI based on the P300 to control three home applications: the door lock, the electric light, and the speaker. Vega et al. [40] used six symbols of commonly used home appliances and employed a deep learning model for the classification of EEG. Shukla et al. [41,42] used a 2 × 3 matrix containing six symbols (TV, Mobile, Door, Fan, Light bulb, and Heater) and used P300-ERP to control these six appliances. They tested their system with 14 participants and achieved an average accuracy of 92.44%. The problem with such systems is the limited number of control options available to the users.

This paper proposes a P300-based intelligent home control system that allows users to control multiple home appliances and allows them to dial a phone call. Our interface comprises a matrix of twelve symbols. Each symbol indicates a smart home activity that can be executed. Users can choose any symbol to perform their desired action, such as turning on or off lights. To detect P300, we employed a supervised machine learning algorithm, Random Forest (RF), and evaluated our proposed system with 10 healthy participants for smart home control and making phone calls. The accuracy rate was 92.25 percent on average.

The rest of the paper is organized as follows. Section 2 presents a detailed methodology of the proposed home control system. The experimental results are presented in Section 3. Finally, Section 4 presents a conclusion and discussion.

2. Materials and Methods

Our methodology comprises displaying a flashing paradigm to the user as shown in Figure 1. While the user is observing the flashing paradigm, EEG waves are recorded using a 32-channel EEG cap. After preprocessing, a random forest classifier is then used to classify the P300 signal of the brain. The classification of the P300 signal leads to choosing the target symbol. Ideally, the P300 should be found after 300 ms of the flashing of the symbol the user is focusing on.

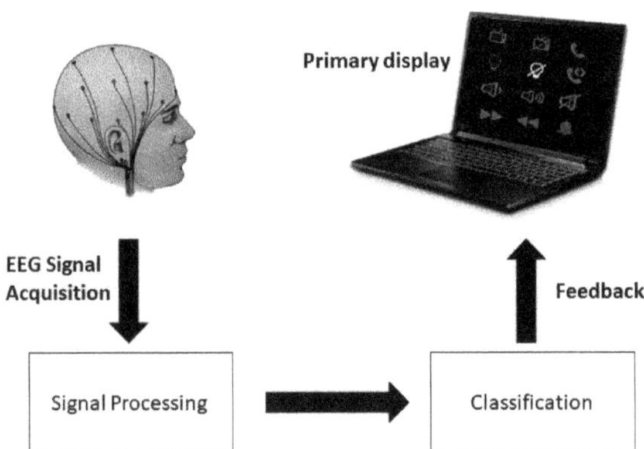

Figure 1. Block diagram of the proposed system.

2.1. Participants

Ten healthy volunteers (7 males and 3 females; ages 18 to 35 years) participated in the experiments. The participants' vision was normal or corrected and had no record of any neurological brain illnesses. None of the participants had any prior knowledge or experience with BCIs of any kind. The experimental protocol was explained to the participants and all participants signed a written informed consent form before the start of the experiment.

2.2. Primary Display for Controlling Home Appliances

In this paper, we propose an interface that consists of a 4×3 matrix containing 12 symbols. Each symbol is randomly intensified as shown in Figure 2, and the users must concentrate on the symbol they want to choose. The description of those symbols is shown in Figure 3. Rather than intensifying rows and columns of the matrix as in traditional P300 displays, we chose to intensify each symbol individually because previous research has shown that the prior probability of the target is inversely related to the amplitude of the P300 [43]. Higher probabilities of the appearance of the target lead to lower amplitudes of the P300-ERP, which in turn makes the classification difficult and hence reduces accuracy. In the case of our proposed smart home control, we have only 12 symbols in the main display making it a 4×3 matrix. Using the row and column-wise intensification will lead us to have a target probability of 2/7 because out of seven intensifications (4 rows and 3 columns) two should contain the target symbol (one row and one column). Whereas intensifying each symbol gives us the probability of the appearance of the target symbol to be 1/12 which is far less than the row/column intensification.

Figure 2. Primary Display to control the smart home.

The subjects were told to randomly choose a symbol and focus on the chosen symbol and count how many times it flashed quietly. After 300 ms of the target symbol flashing, P300 appears. Intensifications are randomized in blocks. Each symbol is intensified exactly once in random order in each block of twelve intensifications. This intensification block is repeated fifteen times in total. We employed a 100 ms intensification time with a 75 ms blank interval in between.

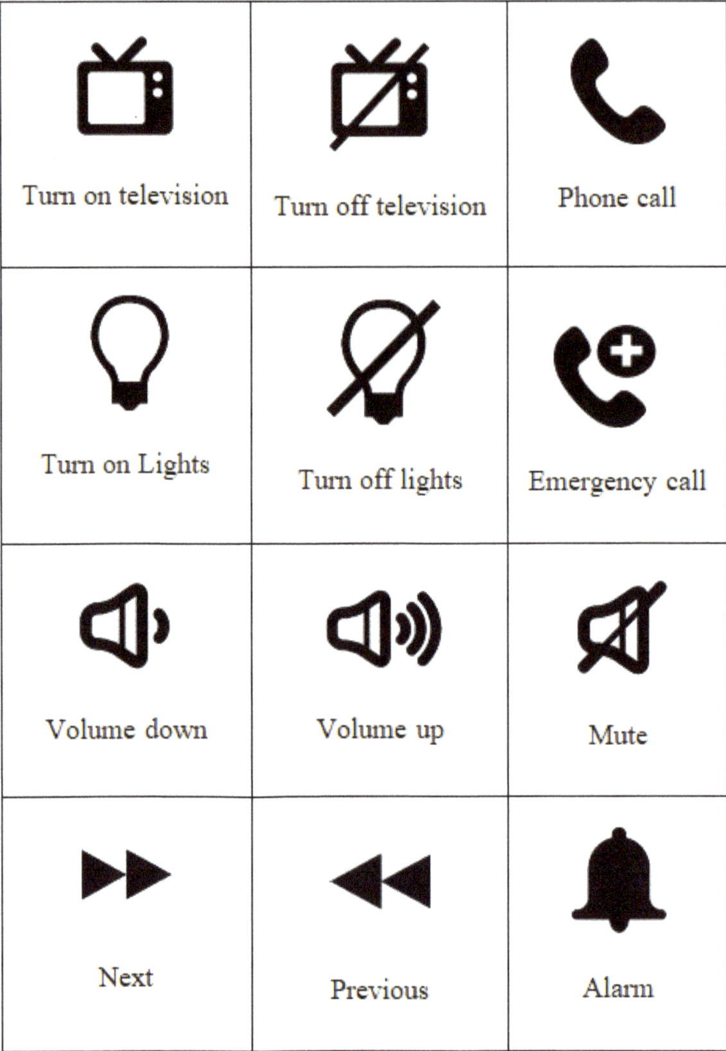

Figure 3. Functional symbols on the main Interface and their description.

On the primary display, the users have twelve symbols to choose from, and each symbol represents an action to be taken in the smart home such as controlling the TV, lights, volume, and phone. The description of all those symbols is shown in Figure 3.

2.3. Secondary Display for Making Phone Calls

In our proposed system, the users have the option of making phone calls along with controlling the home appliances. On the main display, there is a symbol for a phone call as shown in Figures 2 and 3. If the user selects that phone call symbol, the display gets changed to a secondary display containing a 4 × 3 matrix of numbers that are shown in Figure 4. The secondary display contains numbers (0–9) for typing the phone number to be called, a call symbol for connecting the phone call, and a return symbol to disconnect the phone call and return to the primary display.

Figure 4. Secondary display for making phone calls.

2.4. Experimental Setup

A 32-channel EEG data was acquired, as per the 10–20 international electrode system. The placement of electrodes is shown in Figure 5. Out of those 32 channels, only 8 were used for the classification of ERPs. The subjects were seated on a cozy chair in front of a display screen and were instructed to focus on the desired symbol from the displayed symbols (as shown in Figure 2). Each participant was required to attend two sessions: training and testing. Each participant was instructed to select a random symbol/number during the training session, and this experiment was repeated 10 times. During the test session, each participant chose 40 symbols/numbers at random (20 symbols on the main display and 20 numbers on the secondary display).

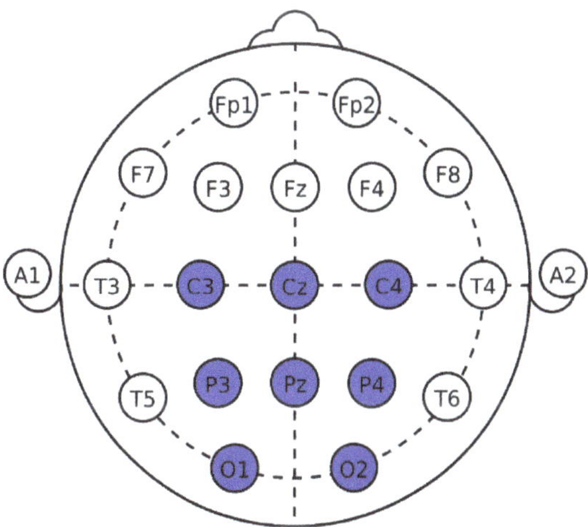

Figure 5. Electrode placement (highlighted electrodes are used in this study).

2.5. Signal Processing and Classification

In this study, we used 8 channels for signal processing and classification. The locations of the used electrodes are shown in Figure 5. It is known from the previous studies [44,45] that the P300-ERP is most dominant in the Pz area and the nearby electrodes. Therefore, we have chosen only those locations. The acquired EEG signals are bandpass filtered using a third-order Butterworth filter between 0.1 Hz to 25 Hz to remove the noise. The filtered signal is then segmented, and segments of 800 ms are extracted after the onset of each stimulus, i.e., flashing of each symbol from each of the used channels. To construct a single feature vector, the data segments are concatenated over the eight channels.

For the classification of P300-ERPs, we employed the random forest classifier because RF had shown superior performance in P300 classification in our previous studies [38,46]. We also compared the results of the random forest classifier with commonly used classification methods such as Support Vector Machine (SVM), Linear Discriminant Analysis (LDA), and k-nearest neighbors (kNN).

Random Forest is an ensemble classification technique. The general idea of the random forest is to build multiple decision tree classifiers and combine their results for better accuracy. Using multiple classifiers in an ensemble gives a more stable prediction. Random forest utilizes a random subset of the data to train each of the decision trees in the forest and then combines the result of each decision tree by majority voting. In this way, the weak classifiers are combined to form a strong classifier.

Random Forest was used to detect the presence of P300 ERPs in the recorded EEG signal. The system selects the symbol that elicits P300-ERP and displays the result as feedback to the user.

3. Results

We conducted experiments using our proposed paradigm on ten healthy volunteers. The 32 channels' EEG data was acquired at a sampling frequency of 250 Hz. The participants were instructed to pay close attention to the target symbol and silently count how many times it flashed. This silent counting helps users in maintaining their attention. Each participant attended two sessions, namely training, and testing. During training, each participant selected 10 randomly selected symbols and during the test session, 40 symbols/numbers were selected, one at a time, as per the choice of the participants. The users were shown the main display, as shown in Figure 2. The symbols on the display were randomly flashing. The flashing time is shown in Table 1.

Table 1. Flashing time of the proposed paradigm.

Intensification time	100 ms
Inter-stimulus blank time	75 ms
Total Symbols	12
Number of repetitions for each symbol	15

Flashing a single symbol takes 175 ms: 100 ms for flashing and 75 ms of blank time in between two flashes. The proposed paradigm has a total of 12 symbols on the screen. Therefore, to flash each symbol once, the system requires 2.1 s. We had chosen to flash each symbol 15 times. Therefore, if we calculate the time required to select a symbol by each user, it comes out to be 31.5 s (2.1 × 15). The same timings were used in the second (numbers-based) paradigm to make a phone call. The users can select each number in 31.5 s. Figure 6 shows the images of the data collection.

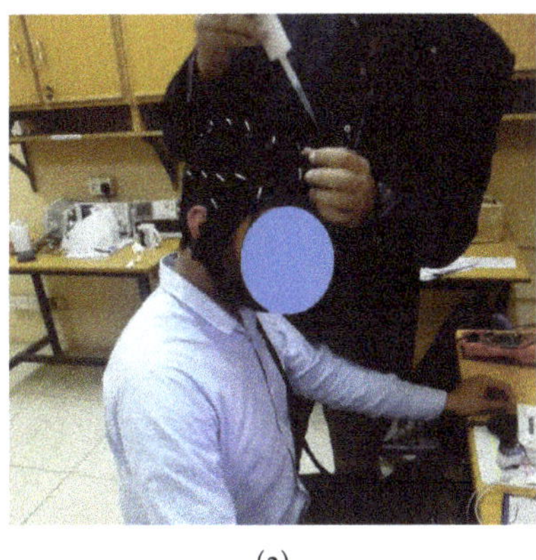 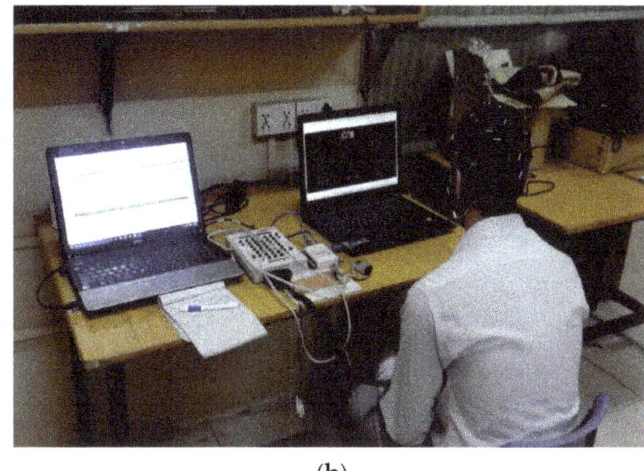

(a)   (b)

Figure 6. Data Collection. (**a**) Preparation for EEG data collection; (**b**) A participant during data collection.

3.1. Experimental Results

During the test session, each user was asked to select twenty random symbols on the primary symbols-based display, and twenty random numbers on the phone call interface. Table 2 presents the accuracy of classification for each subject on both displays. The random forest classifier achieved an overall accuracy of 92.25 percent.

Table 2. Classification accuracies on the proposed displays using random forest classifier.

Subjects	Accuracy on Primary Display (%)	Accuracy on Secondary Display (%)	Average Accuracy (%)
S1	95	100	97.5
S2	85	85	85
S3	100	95	97.5
S4	90	95	92.5
S5	80	85	82.5
S6	95	100	97.5
S7	85	90	87.5
S8	95	95	95
S9	90	95	92.5
S10	95	95	95
Mean	**91**	**93.5**	**92.25**

We also compared the results of the random forest classifier with other commonly used classifiers such as SVM, LDA, kNN. The comparison of the accuracies obtained by these classifiers is presented in Table 3. All these classifiers performed well with minor differences in accuracies. RF classifier performed the best with an average accuracy of 92.25, whereas the SVM, LDA, and kNN achieved average classification accuracies of 91, 89.25, and 90.25 percent, respectively.

Table 3. Comparison of classification accuracies using RF, SVM, LDA, and kNN classifiers.

Subjects	Random Forest	SVM	LDA	kNN
S1	97.5	95	92.5	95
S2	85	85	82.5	85
S3	97.5	97.5	95	92.5
S4	92.5	95	90	92.5
S5	82.5	85	82.5	80
S6	97.5	92.5	90	92.5

Table 3. *Cont.*

Subjects	Random Forest	SVM	LDA	kNN
S7	87.5	85	87.5	87.5
S8	95	95	92.5	90
S9	92.5	92.5	90	95
S10	95	87.5	90	92.5
Mean	**92.25**	**91**	**89.25**	**90.25**

3.2. Waveform Morphologies

The averaged ERPs for target and non-target stimuli are shown in Figure 7 for both the proposed displays. The P300 event-related potential is visible for the target cases in both paradigms. However, it was noted that the amplitude of the P300 was smaller in the case of the symbols-based paradigm. The accuracy was also lesser in the case of the symbols-based paradigm as shown in Table 2.

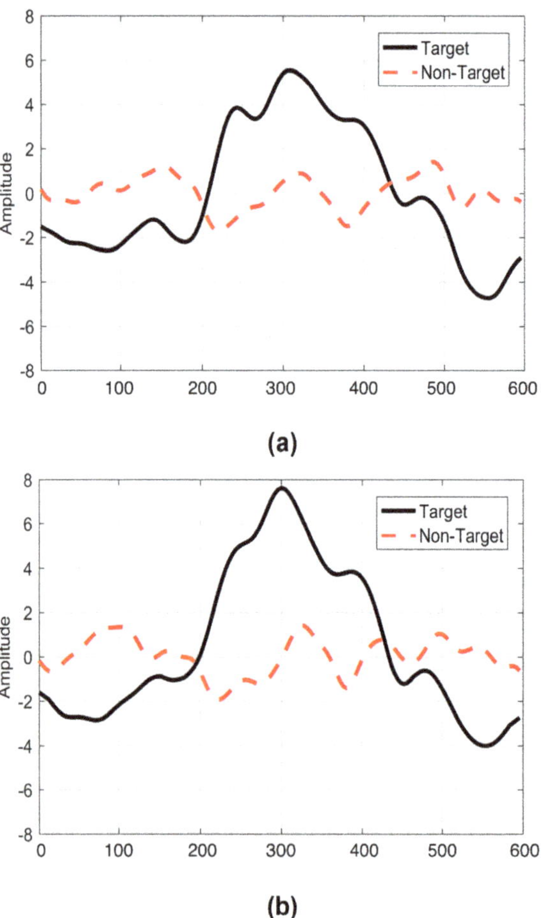

(a)

(b)

Figure 7. Waveforms for target and non−target stimuli. (**a**) Primary symbols−based display for controlling home appliances. (**b**) Secondary display (Numbers−based) for making phone calls.

A comparison of the P300 waveforms for the primary and secondary paradigms is shown in Figure 8. The amplitude of the numbers-based paradigm is larger than the symbols-based paradigm. The reason for this difference in P300 amplitude can be the familiarity of the users with numbers. The users are more familiar with the numbers, as compared to the symbols. This familiarity improves the user's P300 response, which leads to improved classification accuracy. The comparison of accuracies on both the paradigms is presented in Figure 9.

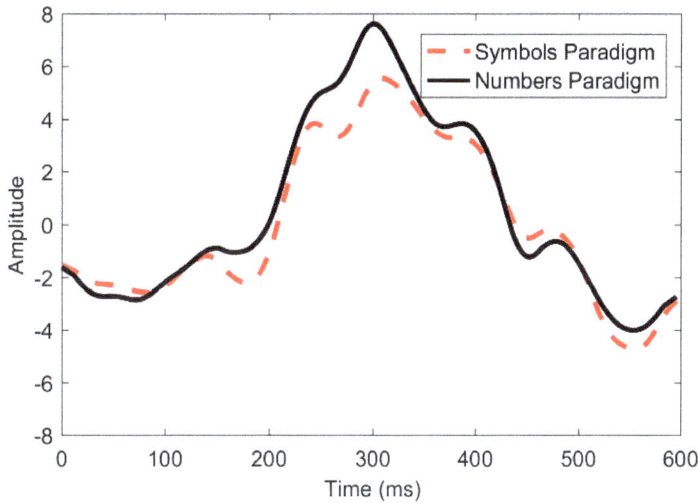

Figure 8. Comparison of the ERPs for both the displays.

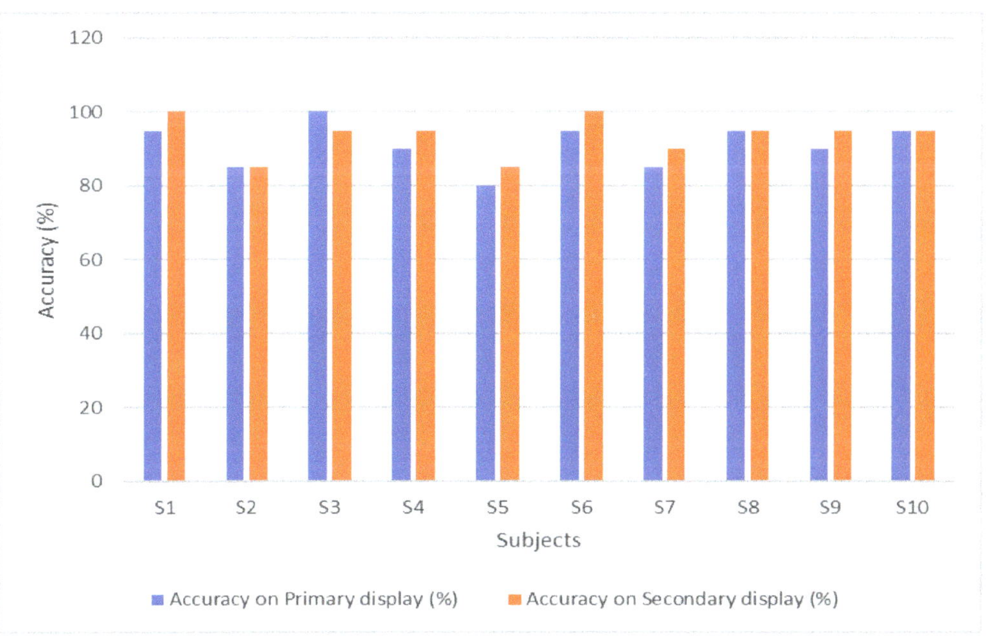

Figure 9. Comparison of the classification accuracies for both the displays.

A slight difference in accuracies of the primary and secondary display can be seen in Figure 9, because the amplitude of P300-ERPs was smaller in the case of the symbols-based paradigm. The numbers-based paradigm produced better P300 however, the difference is not that significant. On average, the numbers-based paradigm has 2.5% better accuracy than the symbols-based paradigm.

4. Discussion and Conclusions

This paper presented a novel BCI paradigm to control home appliances using the P300-ERP of the brain. The proposed paradigm included common household appliances such as televisions, lights, and music. We also included an option to make a phone call in our smart home. As evident from the results, the proposed P300-BCI system is capable of both controlling appliances and making phone calls in a smart home setting.

Compared to our previously proposed smart home system [38], we have achieved better accuracy in this paper. The reason for this improvement in accuracy is a single symbol flashing instead of flashing rows and columns. When we do rows/columns flashing in a 4 × 3 matrix, there are a total of 7 flashes (4 rows and 3 columns). Out of these seven flashes, two flashes correspond to the target symbol/number (one row and one column). This leads the prior probability of the target to be 2/7 or 0.286. Whereas flashing individual numbers/symbols in the same 4 × 3 matrix gives us a prior probability of target as 1/12 or 0.083 (one symbol out of a total twelve). It is known from previous studies [43] that in an oddball paradigm, the P300's magnitude is negatively correlated with the target's prior probability. The probability of flashing the target symbol is very low in this proposed paradigm as compared to the previous studies. Therefore, it leads to better P300 s, which in turn gives us better classification accuracy. Moreover, we have included a secondary paradigm for making phone calls and the users were able to dial phone numbers with an accuracy of 93.5%.

We can see from the results that the accuracy of the secondary display (numbers-based) is slightly better than the symbols-based display, one reason for this can be the familiarity of the users with the numbers as compared to the symbols. These results are consistent with Kaufmann et al. [47] where they had replaced the characters with the popular faces in a P300-based paradigm and achieved better accuracies.

Compared with the other smart home applications proposed in the literature, the proposed paradigm offers a more degree of freedom with an increased number of devices to control. The users can control 12 devices at a time using a single interface. The users also have the freedom to dial a phone number of their choice and make a phone call, which is an important aspect in improving the quality of life of the disabled. The number of symbols can further be increased and can be changed depending on the application and requirements of the users.

We have calculated the time required by the users to perform each command in the smart home. The users require 31.5 s to perform each command (selecting a symbol or a number). The time can be reduced by reducing the number of repetitions in flashing. However, if we reduce the number of flashes, the classification of P300 s gets difficult and may decrease the classification accuracy. In the future, better classification models can be employed to solve this problem of P300 recognition in a lesser number of repetitions.

Author Contributions: Conceptualization, F.A. and U.M.; Methodology, F.A. and U.M.; Software, F.A., A.A. and M.A.; Validation, A.A., M.A. and M.H.; Formal analysis, F.A. and U.M.; Data curation, F.A., A.A., M.A. and M.H.; Investigation, F.A. and U.M.; Resources, A.A., M.A. and M.H.; Writing—original draft preparation, F.A. and U.M.; writing—review and editing, F.A., A.A., M.A., M.H. and U.M.; Supervision, F.A.; Project Administration, U.M.; Funding Acquisition, A.A., M.A. and M.H. All authors have read and agreed to the published version of the manuscript.

Funding: This research received no external funding.

Institutional Review Board Statement: The study was conducted in accordance with the Declaration of Helsinki and approved by the Research Ethics Committee of Riphah International University.

Informed Consent Statement: Informed consent was obtained from all subjects involved in the study.

Conflicts of Interest: The authors declare no conflict of interest.

Abbreviations

The following abbreviations are used in this manuscript:

BCI	Brain-Computer Interface
CT	Computer Tomography
ECoG	Electrocorticography
EEG	Electroencephalography
EMG	Electromyography
ERP	Event Related Potential
fMRI	functional Magnetic Resonance Imaging
fNIRS	functional near-infrared spectroscopy
kNN	k-Nearest Neighbors
LDA	Linear Discriminant Analysis
MEG	Magnetoencephalography
PET	Positron Emission Tomography
RF	Random Forest
SSVEP	Steady-State Visual Evoked Potentials
SVM	Support Vector Machine

References

1. Laureys, S.; Pellas, F.; van Eeckhout, P.; Ghorbel, S.; Schnakers, C.; Perrin, F.; Berré, J.; Faymonville, M.E.; Pantke, K.H.; Damas, F.; et al. The Locked-in Syndrome: What Is It like to Be Conscious but Paralyzed and Voiceless? *Prog. Brain Res.* **2005**, *150*, 495–611. [CrossRef] [PubMed]
2. Smith, E.; Delargy, M. Locked-in Syndrome. *BMJ* **2005**, *330*, 406–409. [CrossRef] [PubMed]
3. Lulé, D.; Zickler, C.; Häcker, S.; Bruno, M.A.; Demertzi, A.; Pellas, F.; Laureys, S.; Kübler, A. Life Can Be Worth Living in Locked-in Syndrome. *Prog. Brain Res.* **2009**, *177*, 339–351. [CrossRef] [PubMed]
4. Kaufmann, T.; Völker, S.; Gunesch, L.; Kübler, A. Spelling Is Just a Click Away—A User-Centered Brain-Computer Interface Including Auto-Calibration and Predictive Text Entry. *Front. Neurosci.* **2012**, *6*, 72. [CrossRef]
5. Kübler, A.; Furdea, A.; Halder, S.; Hammer, E.M.; Nijboer, F.; Kotchoubey, B. A Brain-Computer Interface Controlled Auditory Event-Related Potential (P300) Spelling System for Locked-In Patients. *Ann. N. Y. Acad. Sci.* **2009**, *1157*, 90–100. [CrossRef]
6. Akram, F.; Han, H.S.; Kim, T.S. A P300-Based Brain Computer Interface System for Words Typing. *Comput. Biol. Med.* **2014**, *45*, 118–125. [CrossRef]
7. Zgallai, W.; Brown, J.T.; Ibrahim, A.; Mahmood, F.; Mohammad, K.; Khalfan, M.; Mohammed, M.; Salem, M.; Hamood, N. Deep Learning AI Application to an EEG Driven BCI Smart Wheelchair. In Proceedings of the 2019 Advances in Science and Engineering Technology International Conferences, ASET 2019, Dubai, United Arab Emirates, 26 March–10 April 2019; Institute of Electrical and Electronics Engineers Inc.: Piscataway, NJ, USA, 2019.
8. Sakkalis, V.; Krana, M.; Farmaki, C.; Bourazanis, C.; Gaitatzis, D.; Pediaditis, M. Augmented Reality Driven Steady-State Visual Evoked Potentials for Wheelchair Navigation. *IEEE Trans. Neural Syst. Rehabil. Eng.* **2022**, *30*, 2960–2969. [CrossRef]
9. McFarland, D.J.; Wolpaw, J.R. Brain-Computer Interface Operation of Robotic and Prosthetic Devices. *Computer* **2008**, *41*, 52–56. [CrossRef]
10. Heo, D.; Kim, M.; Kim, J.; Choi, Y.J.; Kim, S.-P. The Uses of Brain-Computer Interface in Different Postures to Application in Real Life. In Proceedings of the 10th International Winter Conference on Brain-Computer Interface (BCI), Gangwon-do, Republic of Korea, 21–23 February 2022; Institute of Electrical and Electronics Engineers (IEEE): Piscataway, NJ, USA, 2022; pp. 1–5.
11. Coogan, C.G.; He, B. Brain-Computer Interface Control in a Virtual Reality Environment and Applications for the Internet of Things. *IEEE Access* **2018**, *6*, 10840–10849. [CrossRef]
12. Khan, M.A.; Das, R.; Iversen, H.K.; Puthusserypady, S. Review on Motor Imagery Based BCI Systems for Upper Limb Post-Stroke Neurorehabilitation: From Designing to Application. *Comput. Biol. Med.* **2020**, *123*, 103843. [CrossRef]
13. Yu, Y.; Zhou, Z.; Yin, E.; Jiang, J.; Tang, J.; Liu, Y.; Hu, D. Toward Brain-Actuated Car Applications: Self-Paced Control with a Motor Imagery-Based Brain-Computer Interface. *Comput. Biol. Med.* **2016**, *77*, 148–155. [CrossRef] [PubMed]
14. Gannouni, S.; Alangari, N.; Mathkour, H.; Aboalsamh, H.; Belwafi, K. BCWB: A P300 Brain-Controlled Web Browser. *Int. J. Semant. Web Inf. Syst.* **2017**, *13*, 55–73. [CrossRef]
15. Prasath, M.S.; Naveen, R.; Sivaraj, G. Mind-Controlled Unmanned Aerial Vehicle (UAV) Using Brain–Computer Interface (BCI). *Unmanned Aer. Veh. Internet Things* **2021**, *13*, 231–246. [CrossRef]

16. Masud, U.; Saeed, T.; Akram, F.; Malaikah, H.; Akbar, A. Unmanned Aerial Vehicle for Laser Based Biomedical Sensor Development and Examination of Device Trajectory. *Sensors* **2022**, *22*, 3413. [CrossRef] [PubMed]
17. Li, M.; Li, F.; Pan, J.; Zhang, D.; Zhao, S.; Li, J.; Wang, F. The MindGomoku: An Online P300 BCI Game Based on Bayesian Deep Learning. *Sensors* **2021**, *21*, 1613. [CrossRef] [PubMed]
18. Dutta, S.; Banerjee, T.; Roy, N.D.; Chowdhury, B.; Biswas, A. Development of a BCI-Based Gaming Application to Enhance Cognitive Control in Psychiatric Disorders. *Innov. Syst. Softw. Eng.* **2021**, *17*, 99–107. [CrossRef]
19. Louis, J.D.; Alikhademi, K.; Joseph, R.; Gilbert, J.E. Mind Games: A Web-Based Multiplayer Brain-Computer Interface Game. *Proc. Hum. Factors Ergon. Soc. Annu. Meet.* **2022**, *66*, 2234–2238. [CrossRef]
20. Farwell, L.A.; Donchin, E. Talking off the Top of Your Head: Toward a Mental Prosthesis Utilizing Event-Related Brain Potentials. *Electroencephalogr. Clin. Neurophysiol.* **1988**, *70*, 510–523. [CrossRef]
21. Piccione, F.; Giorgi, F.; Tonin, P.; Priftis, K.; Giove, S.; Silvoni, S.; Palmas, G.; Beverina, F. P300-Based Brain Computer Interface: Reliability and Performance in Healthy and Paralysed Participants. *Clin. Neurophysiol.* **2006**, *117*, 531–537. [CrossRef]
22. Citi, L.; Poli, R.; Cinel, C.; Sepulveda, F. P300-Based BCI Mouse with Genetically-Optimized Analogue Control. *IEEE Trans. Neural Syst. Rehabil. Eng.* **2008**, *16*, 51–61. [CrossRef]
23. Martínez-Cagigal, V.; Santamaría-Vázquez, E.; Gomez-Pilar, J.; Hornero, R. A Brain–Computer Interface Web Browser for Multiple Sclerosis Patients. *Neurol. Disord. Imaging Phys. Vol. 2 Eng. Clin. Perspect. Mult. Scler.* **2019**, *2*, 327–357. [CrossRef]
24. Bai, L.; Yu, T.; Li, Y. A Brain Computer Interface-Based Explorer. *J. Neurosci. Methods* **2015**, *244*, 2–7. [CrossRef] [PubMed]
25. Finke, A.; Lenhardt, A.; Ritter, H. The MindGame: A P300-Based Brain-Computer Interface Game. *Neural. Netw.* **2009**, *22*, 1329–1333. [CrossRef] [PubMed]
26. Eidel, M.; Kübler, A. Wheelchair Control in a Virtual Environment by Healthy Participants Using a P300-BCI Based on Tactile Stimulation: Training Effects and Usability. *Front. Hum. Neurosci.* **2020**, *14*, 265. [CrossRef] [PubMed]
27. Hoffmann, U.; Vesin, J.-M.; Ebrahimi, T.; Diserens, K. An Efficient P300-Based Brain-Computer Interface for Disabled Subjects. *J. Neurosci. Methods* **2008**, *167*, 115–125. [CrossRef] [PubMed]
28. Achanccaray, D.; Flores, C.; Fonseca, C.; Andreu-Perez, J. A P300-Based Brain Computer Interface for Smart Home Interaction through an ANFIS Ensemble. In Proceedings of the IEEE International Conference on Fuzzy Systems, Naples, Italy, 9–12 July 2017; Institute of Electrical and Electronics Engineers Inc.: Piscataway, NJ, USA, 2017.
29. Cortez, S.A.; Flores, C.; Andreu-Perez, J. A Smart Home Control Prototype Using a P300-Based Brain–Computer Interface for Post-Stroke Patients. In *Proceedings of the 5th Brazilian Technology Symposium. Smart Innovation, Systems and Technologies*; Springer Science and Business Media Deutschland GmbH: Cham, Switzerland, 2021; Volume 202, pp. 131–139. [CrossRef]
30. Carabalona, R.; Grossi, F.; Tessadri, A.; Castiglioni, P.; Caracciolo, A.; de Munari, I. Light on! Real World Evaluation of a P300-Based Brain-Computer Interface (BCI) for Environment Control in a Smart Home. *Ergonomics* **2012**, *55*, 552–563. [CrossRef]
31. Park, S.; Cha, H.S.; Kwon, J.; Kim, H.; Im, C.H. Development of an Online Home Appliance Control System Using Augmented Reality and an SSVEP-Based Brain-Computer Interface. In Proceedings of the 8th International Winter Conference on Brain-Computer Interface, BCI 2020, Gangwon-do, Republic of Korea, 26–28 February 2020; Institute of Electrical and Electronics Engineers Inc.: Piscataway, NJ, USA, 2020.
32. Kais, B.; Ghaffari, F.; Romain, O.; Djemal, R. An Embedded Implementation of Home Devices Control System Based on Brain Computer Interface. *Proc. Int. Conf. Microelectron. ICM* **2014**, *2015*, 140–143. [CrossRef]
33. Edlinger, G.; Holzner, C.; Guger, C. A Hybrid Brain-Computer Interface for Smart Home Control. In Proceedings of the International Conference on Human-Computer Interaction, Orlando, FL, USA, 9–14 July 2011; Springer: Berlin/Heidelberg, Germany, 2011; pp. 417–426.
34. Katyal, A.; Singla, R. A Novel Hybrid Paradigm Based on Steady State Visually Evoked Potential & P300 to Enhance Information Transfer Rate. *Biomed. Signal Process. Control* **2020**, *59*, 101884. [CrossRef]
35. Chai, X.; Zhang, Z.; Guan, K.; Lu, Y.; Liu, G.; Zhang, T.; Niu, H. A Hybrid BCI-Controlled Smart Home System Combining SSVEP and EMG for Individuals with Paralysis. *Biomed. Signal Process. Control* **2020**, *56*, 101687. [CrossRef]
36. Uyanik, C.; Khan, M.A.; Das, R.; Hansen, J.P.; Puthusserypady, S. Brainy Home: A Virtual Smart Home and Wheelchair Control Application Powered by Brain Computer Interface. *Biodevices* **2022**, *1*, 134–141. [CrossRef]
37. Kim, H.J.; Lee, M.H.; Lee, M. A BCI Based Smart Home System Combined with Event-Related Potentials and Speech Imagery Task. In Proceedings of the 2020 8th International Winter Conference on Brain-Computer Interface (BCI), Gangwon-do, Republic of Korea, 26–28 February 2020. [CrossRef]
38. Masud, U.; Baig, M.I.; Akram, F.; Kim, T.-S. A P300 Brain Computer Interface Based Intelligent Home Control System Using a Random Forest Classifier. In Proceedings of the 2017 IEEE Symposium Series on Computational Intelligence, Honolulu, HI, USA, 27 November–1 December 2017; Volume 2018.
39. Lee, T.; Kim, M.; Kim, S.-P. Improvement of P300-Based Brain–Computer Interfaces for Home Appliances Control by Data Balancing Techniques. *Sensors* **2020**, *20*, 5576. [CrossRef] [PubMed]
40. Vega, C.F.; Quevedo, J.; Escandón, E.; Kiani, M.; Ding, W.; Andreu-Perez, J. Fuzzy Temporal Convolutional Neural Networks in P300-Based Brain–Computer Interface for Smart Home Interaction. *Appl. Soft Comput.* **2022**, *117*, 108359. [CrossRef]
41. Shukla, P.K.; Chaurasiya, R.K.; Verma, S.; Sinha, G.R. A Thresholding-Free State Detection Approach for Home Appliance Control Using P300-Based BCI. *IEEE Sens. J.* **2021**, *21*, 16927–16936. [CrossRef]

42. Shukla, P.K.; Chaurasiya, R.K.; Verma, S. Performance Improvement of P300-Based Home Appliances Control Classification Using Convolution Neural Network. *Biomed. Signal Process. Control* **2021**, *63*, 102220. [CrossRef]
43. Allison, B.Z.; Pineda, J.A. ERPs Evoked by Different Matrix Sizes: Implications for a Brain Computer Interface (BCI) System. *IEEE Trans. Neural Syst. Rehabil. Eng.* **2003**, *11*, 110–113. [CrossRef]
44. Linden, D.E.J. The P300: Where in the Brain Is It Produced and What Does It Tell Us? *Neuroscientist* **2016**, *11*, 563–576. [CrossRef]
45. Bledowski, C.; Prvulovic, D.; Hoechstetter, K.; Scherg, M.; Wibral, M.; Goebel, R.; Linden, D.E.J. Localizing P300 Generators in Visual Target and Distractor Processing: A Combined Event-Related Potential and Functional Magnetic Resonance Imaging Study. *J. Neurosci.* **2004**, *24*, 9353–9360. [CrossRef]
46. Akram, F.; Han, S.M.; Kim, T.-S. An Efficient Word Typing P300-BCI System Using a Modified T9 Interface and Random Forest Classifier. *Comput. Biol. Med.* **2015**, *56*, 30–36. [CrossRef]
47. Kaufmann, T.; Schulz, S.M.; Grünzinger, C.; Kübler, A. Flashing Characters with Famous Faces Improves ERP-Based Brain-Computer Interface Performance. *J. Neural. Eng.* **2011**, *8*, 56016. [CrossRef]

Article

Multi-Kernel Temporal and Spatial Convolution for EEG-Based Emotion Classification

Taweesak Emsawas [1,*], Takashi Morita [2], Tsukasa Kimura [2], Ken-ichi Fukui [2] and Masayuki Numao [2]

[1] Graduate School of Information Science and Technology, Osaka University, Osaka 565-0871, Japan
[2] The Institute of Scientific and Industrial Research (ISIR), Osaka University, Osaka 567-0047, Japan
* Correspondence: taweesak@ai.sanken.osaka-u.ac.jp

Abstract: Deep learning using an end-to-end convolutional neural network (ConvNet) has been applied to several electroencephalography (EEG)-based brain–computer interface tasks to extract feature maps and classify the target output. However, the EEG analysis remains challenging since it requires consideration of various architectural design components that influence the representational ability of extracted features. This study proposes an EEG-based emotion classification model called the multi-kernel temporal and spatial convolution network (MultiT-S ConvNet). The multi-scale kernel is used in the model to learn various time resolutions, and separable convolutions are applied to find related spatial patterns. In addition, we enhanced both the temporal and spatial filters with a lightweight gating mechanism. To validate the performance and classification accuracy of MultiT-S ConvNet, we conduct subject-dependent and subject-independent experiments on EEG-based emotion datasets: DEAP and SEED. Compared with existing methods, MultiT-S ConvNet outperforms with higher accuracy results and a few trainable parameters. Moreover, the proposed multi-scale module in temporal filtering enables extracting a wide range of EEG representations, covering short- to long-wavelength components. This module could be further implemented in any model of EEG-based convolution networks, and its ability potentially improves the model's learning capacity.

Keywords: brain–computer interface (BCI); electroencephalography (EEG); emotion classification; machine learning; convolutional neural network (ConvNet)

Citation: Emsawas, T.; Morita, T.; Kimura, T.; Fukui, K.-i.; Numao, M. Multi-Kernel Temporal and Spatial Convolution for EEG-Based Emotion Classification. *Sensors* **2022**, *22*, 8250. https://doi.org/10.3390/s22218250

Academic Editors: Imran Khan Niazi, Hendrik Santosa and Noman Naseer

Received: 29 September 2022
Accepted: 23 October 2022
Published: 27 October 2022

1. Introduction

1.1. Background

The brain—computer interface (BCI) technology is a technology that acquires brain signals and interprets neuronal information into the desired action. BCI has been used in various medical and non-medical applications [1], such as assistive technology [2,3], game playing [4,5], and mental state recognition [6,7]. There are several ways to record brain activity. One of the most popular modalities of BCI is electroencephalography (EEG). The EEG method has been used to record electrical signals in the human brain by measuring tiny voltage fluctuations using electrode sensors. The EEG recording can be performed by attaching electrodes to the scalp surface without surgery and implantation. This non-invasive EEG system holds the promise of real-world BCI applications and is currently entering the mass market. For example, EEG has been used to diagnose abnormalities of the human brain by detecting the presence of aberrant electrical activity [8,9]. In non-medical fields, attempts have been made to transduce EEG states from human players to control video games [10]. Moreover, the acquisition of EEG signals is also exploited to recognize human psychological states [11]. Distinct patterns of EEG signals can describe users' emotions and feelings in response to specific circumstances with less human bias. Despite its several potential applications, the non-stationarity and high-dimensionality of EEG signals pose a challenge to reliable implementation. EEG recordings could be easily affected by various sources of noise, including eye movement, muscle contraction, and environmental settings, which present significant difficulties in EEG interpretation. Because the brain is a complex organ with different parts that function and respond differently, to evaluate spatial brain activity, EEG data are generally recorded in more than one and up to 64 electrode positions increasing space dimensionality and complexity during feature extraction and analysis.

Therefore, the capability of developed EEG methods should consider several aspects, including interpretability, performance, and usability for real-world applications. Ascertaining methods aim to understand complex EEG signals derived from the human brain extracting their informative features and classifying them. Conventionally, beneficial knowledge is extracted from raw EEG signals by manually defined features and reported through statistical reports. For instance, frequency bands in slow to hyperactivity brain waves reflect the distinction in the intensity of emotions [12]. Positive emotions, such as joy or happiness, are relatively associated with the left frontal hemisphere. In contrast, negative emotions, such as sadness or fear, are relatively associated with the right frontal hemisphere [13,14]. Nevertheless, these defined features are easily sensitive to noise, and their computation requires appropriate data cleansing, signal preprocessing, and expertise.

1.2. Research Gap

Recently, many studies have applied deep learning (DL) to extract and interpret EEG signals. An end-to-end convolutional neural network (ConvNet or CNN) is constructed on the basis of a shared-weight architecture that can reasonably learn joint optimization of feature extraction and classification. The appropriate design, such as the layer's depth and width, kernel size, and optimizer, becomes an essential consideration that significantly affects the representational ability and classification performance. It has been demonstrated that deep and shallow convolution networks can automatically extract essential temporal and spatial information from raw EEG signals [15]. Nevertheless, as the feature extraction of these models is based on a single kernel size, their learning ability and EEG representational performance are limited. Larger ranges of signal transformations are needed to represent differences in slow and hyperactive EEG frequencies. Parameter tuning and optimization are also required for a particular task. In addition to the kernel, parameter dependence and computational cost in training and testing processes have to be taken into account [16]. A separation of network layers was suggested to help reduce the computational cost and model complexity. To further improve model efficiencies, the network architectures, including kernel sizes and parameter sets, should be rationally designed for each specific EEG analysis task.

1.3. Motivation and Contribution

Accordingly, achieving good EEG data analysis performance using DL-based techniques requires a wide range of useful representations while preserving optimized trainable parameters. This study proposes an architecture called multi-kernel temporal and spatial convolution for EEG-based emotion classification. We present (1) multi-kernel learning for temporal convolution and (2) filter recalibration with a lightweight gating mechanism. The proposed model makes the classification process more efficient by factoring the kernel into a series of operations to capture various short and long patterns. Then, separable convolution is applied for spatial learning by considering each channel separately and convolving over electrode channels. Moreover, we recalibrate weights after temporal and spatial convolving to utilize the limited data available. We compare our model with existing techniques and investigate task accuracy by conducting classification experiments on BCI datasets.

2. Related Works

2.1. Feature Extraction and Classification

In the classification of emotions learned from EEG signals, feature extraction is an essential step in which these emotions are represented and categorized into the desired labels. Conventionally, EEG features can be categorized into three domains: time, frequency, and time–frequency domains [17]. The time domain observes time-series characteristics and variations using conventional techniques, such as linear prediction and component analysis. The frequency or spectral domain is the standard method used for quantifying EEG signals by adapting the Fourier transform. Fourier analysis reflects the frequency content using sums of trigonometric functions and then distributes the average power into the PSD. As shown in Figure 1a, the PSD in a human EEG signal is approximately divided into several ranges, including theta (4–7 Hz), alpha (8–12 Hz), beta (13–32 Hz), and gamma (>32 Hz) bands. These frequency bands reflect brain activities through the strength of variation. High-frequency bands, such as gamma or beta waves, indicate hyperactive brain activity and alertness, and low-frequency bands, such as delta or theta waves, indicate deep meditation and relaxation. Theta (4–7 Hz) is a slow wave associated with the subconscious mind, deep relaxation, and meditation. Changes in alpha (8–12 Hz) and beta (13–32 Hz) waves are the most discriminative for emotional states [12,18]. Gamma (>33 Hz) is a hyperactivity wave associated with problem-solving and concentration and is related to positive and negative emotions but on different sides; left for negative and right for positive [19].

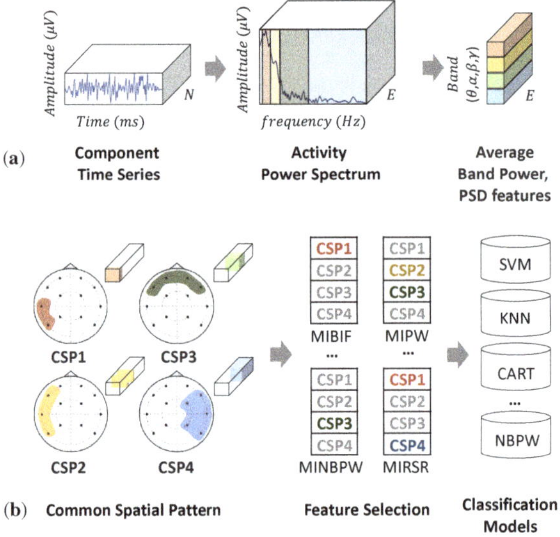

Figure 1. (**a**) Power spectral density (PSD) features, extracted from raw EEG signals. (**b**) Filter bank common spatial pattern (FBCSP).

Moreover, in addition to frequency or temporal features, spatial features can be extracted using multi-channel electrodes. Each part of the human brain can convey different information through the asymmetric hemisphere. For example, the frontal lobe is associated with reasoning, parts of speech, and emotion. The temporal lobe is associated with the perception and recognition of auditory stimuli, whereas the occipital lobe is responsible for vision. For this reason, EEG signals require a method that can manipulate both temporal and spatial information for feature learning and classification. A common spatial pattern (CSP) [20] is a mathematical method for computing the variance of features to discriminate window patterns or emotional classes. This method uses the simultaneous diagonalization of two covariance matrices to construct optimal spatial filters [21,22]. CSP patterns can be used as features for machine learning (ML) to classify emotions [23]. However, the classification of CSP features requires a specific frequency range, which significantly depends on the subject or task. To address this problem, filter bank common spatial pattern (FBCSP) [21] has been proposed to perform an autonomous feature selection through temporal-spatial filtering. Compared with CSP methods, FBCSP as shown in Figure 1b, adds two more processes to perform feature selection and classification, respectively. The first two stages perform temporal and spatial filtering to construct a filter bank of discriminative CSP features. Then, features are selected independently depending on the classifiers in the third stage. Popular techniques with good EEG classification accuracy include random forest (RF), K-nearest neighbor (KNN), support vector machine (SVM), and fully connected network (FCN) [24,25]. However, these techniques employ manual parameterization to classify EEG features, and the appropriate selection for a particular task relies on the complexity and cleanliness of data recording.

2.2. End-to-End Convolutional Neural Networks

In recent years, the use of end-to-end CNNs (ConvNets) has been introduced for effective and reasonable EEG analysis in feature extraction and classification tasks. First, the raw EEG signals are measured in microvolts (μV) and recorded in a 2D space, with temporal (T, time) and spatial dimensions (E, the number of electrode channels). Then, the convolutional operation is applied to extract informative features through three types of filtering, i.e., convolving across time, space, and both time and space, respectively, as shown in Figure 2. Temporal filtering convolves information across the time—space domain with a $1 \times t$ kernel size. The temporal content of each electrode channel is extracted by shared weighting over the time—space input. Spatial filtering proceeds with the filter matrix $E \times 1$ across all electrode channels, learning the variance of features. The temporal—spatial kernel with a $k_1 \times k_2$ 2D kernel size, where k represents a specific kernel size, is applied to both time and channel domains. In addition, there is a need to determine the suitable size for learning various representations and distinctions. Typically, the kernel size is manually defined and used in feature extractors in CNNs. A single-scale kernel might be trapped with a limited amount of representations. Especially for EEG feature extraction, the learning of representations needs to be diverse and capable of effectively extracting temporal and spatial features.

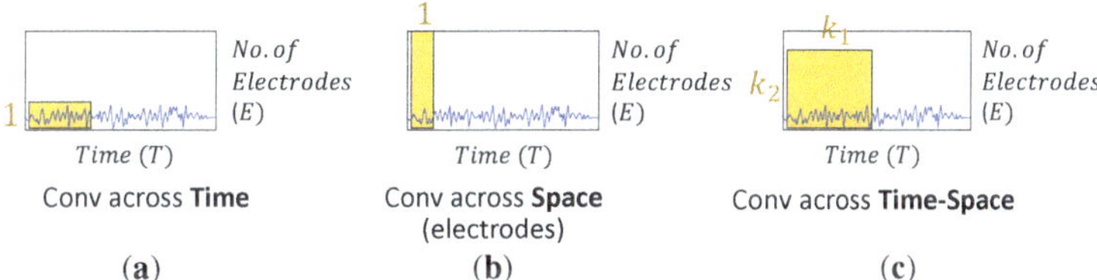

Figure 2. EEG convolution filter types include (**a**) temporal filter, (**b**) spatial filter, and (**c**) temporal-–spatial filter. The yellow blocks denote the two-dimensional (2D) kernel for performing feature extraction.

Apart from kernel designs, the network depth significantly influences the learning strategy, particularly low–high-level feature decoding. For example, a deep convolutional network (Deep ConvNet) has been used for EEG decoding with a single-layer temporal filter, and then, the output is fed into multi-layer spatial convolution and pooling layers, as shown in Figure 3a. The EEG classification results show that the Deep ConvNet outperforms the widely used FBCSP (mean decoding accuracy of 84.0% and 82.1%, respectively) [15]. Deep ConvNets can achieve competitive accuracy and can be applied to general EEG decoding tasks. On the other hand, according to the FBCSP pipeline, a shallow convolutional network (Shallow ConvNet) is designed for tailoring decoding band power features, as shown in Figure 3b. The first layer performs a temporal convolution to simulate bandpass extraction. Then, the second layer performs a spatial convolution that analogizes the CSP spatial filter in FBCSP. The extracted features of Shallow ConvNet are related to log band power and are designed explicitly for oscillatory signal classification.

Moreover, another critical aspect that affects informative features and learning ability is reducing the network size while maintaining good performance. One of the simpler networks is a separable convolution that enables networks to construct informative features by fusing both spatial and channel information. The operation consists of spatial mapping independently performed over each input channel and feed-forward mapping to project the output onto a new feature space. These operations enable networks to construct informative features by fusing spatial and temporal information within local receptive fields at each layer. A compact CNN for EEG-based BCI, called EEGNet, is introduced to construct an EEG-specific model with separable convolution [16]. The number of trainable parameters in EEGNet is significantly less than that of Deep ConvNet and Shallow ConvNet (170 and 100 times, respectively), but EEGNet still achieves performance comparable to that of Deep ConvNet and Shallow ConvNet.

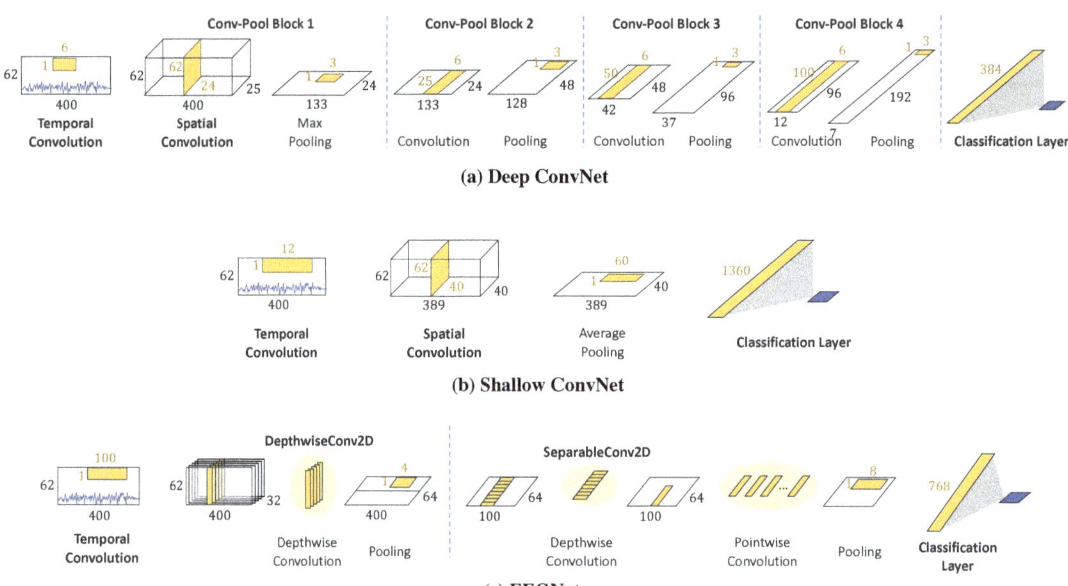

Figure 3. Model architectures of (**a**–**c**).

3. Proposed Method

This study proposes an end-to-end CNN called multi-kernel temporal and spatial convolution (MultiT-S ConvNet) for EEG-based emotion recognition. The proposed model enables multi-scale representation learning and improves classification performance. The model consists of three parts, namely, a temporal learner, spatial learner, and classifier, which simultaneously learn discriminative representations in the time and electrode channel dimensions. The model architecture is depicted in Figure 4, and the details are described in the following sections.

3.1. Multi-Kernel Temporal Processing

In this study, we hypothesize that multi-kernel filters can improve temporal representations learned from raw EEG data. Considering the kernel design of temporal convolution, the larger size can learn higher resolutions in the time domain but necessitates high computational costs. On the other hand, a smaller kernel size essentially learns low-level features or shorter temporal patterns in the time domain and can reduce computational costs. Multi-scale convolution kernels combining long and short patterns are applied by factoring a kernel into various kernel sizes that would independently convolve and map them in order. Learning a wide range of representations while avoiding high computational costs can improve the performance of EEG classification. The main advantage of this architecture is significant quality gain at a modest increase in computational requirements compared with shallower and narrower architectures. Accordingly, we specifically adopted four kernels of length 25 ms (=5 samples under the sampling rate of 200 Hz), 50 ms (=10 samples), 100 ms (=20 samples), and 200 ms (=40 samples). These choices were based on four frequency bands, namely, theta (4–7 Hz), alpha (8–12 Hz), beta (13–32 Hz), and gamma (>33 Hz), extensively used to characterize brain states/activities. The sampling rate is also considered because wavelengths can be captured with the same time scale, even from different datasets. These multi-kernels can capture an extensive range of representations in EEG signals. The larger kernel size can learn various informative features, along longer temporal patterns. Meanwhile, the smaller kernel size specifically extracts shorter temporal patterns.

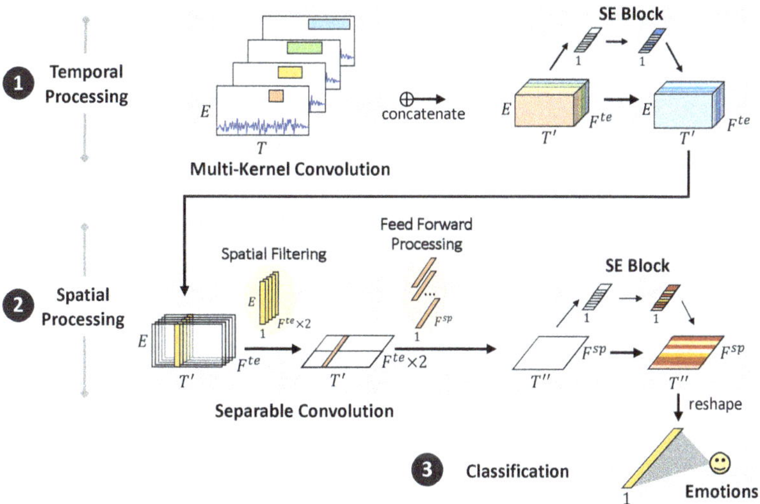

Figure 4. MultiT-S ConvNet architecture. The model comprises temporal filtering, spatial filtering, and classification layers. The first layer employs four types of temporal convolutions to learn multiple temporal features from raw EEG signals. Then, the separable convolution is used to learn temporal-specific spatial filters across electrode channels. The SE block is used to recalibrate the features of both filters. The final classification layer is used to discriminate the emotional labels.

The raw EEG signals in Figure 4 can be represented as 2D time series whose dimensions are time (T) and electrode channels (E). The window size is set as 2 s (T is 400 data points for the SEED dataset) based on the change in emotion over time. We duplicate the input into four modules and convolve them using four different kernel sizes: (1×5), (1×10), (1×20), and (1×40). Each module is padded with zeros to retain the same size. Along the multi-kernel temporal convolution, a tensor of size $(T' \times E \times F^{te})$ is produced, where $T' \times$ denotes the temporal dimensionality after convolution and F^{te} denotes the total number of temporal features after concatenation. All convolutions, including those modules, apply rectified linear activation (ReLU) and average pooling to data before sending them to the next layer.

3.2. Remaining Modules

3.2.1. Spatial Processing

For each time step (T') and temporal feature (F^{te}), we collected information across electrodes using two full-length filters. A separable convolution was applied for spatial feature learning and to reduce computational costs. Figure 5 depicts the separable convolution and the virtualization of topoplots. It consists of two steps: spatial filtering and feed-forward processing. Spatial filtering connects each temporal feature map individually to learn specific spatial filters across electrodes. We manually set the feature multiplier to 2 to increase the number of parameters in the neural network to learn more traits better. The network gathers data with the $F^{te} \times 2$ of the respective filter $(1 \times E)$ and stacks the outputs into feature maps $(T' \times 1 \times (F^{te} \times 2))$. The interpretation of this step is to build multiple filters that can learn informative features across all electrode channels individually. Then, feed-forward processing continues to learn and optimizes a temporal—spatial summary for each feature map with independent filters $(1 \times 1 \times (F^{te} \times 2))$ across the spatial feature maps. Here we learn how to weigh the useful set of filters from the previous extraction. After this spatial processing, temporal—spatial feature maps $(T'' \times 1 \times F^{sp})$ are extracted, where T'' represents the time after temporal—spatial processing and pooling.

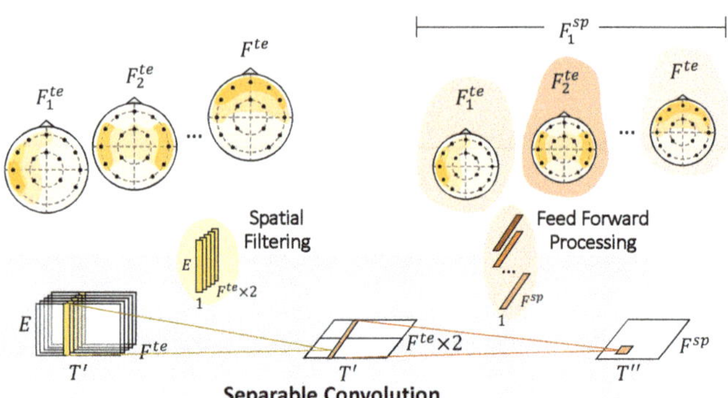

Figure 5. Spatial Convolution. Spatial processing separately convolves across the electrodes on the temporal feature maps. Feedforward processing subsequently weights the feature maps to extract temporal—spatial features. Shading represents how the network emphasizes or understates the corresponding feature maps.

3.2.2. Channel Recalibration

From the outputs, after each temporal and spatial processing, we need to utilize the limited dataset as much as possible regarding the time-cost consumption of EEG recording. The squeeze-and-excitation block (SE block) is further applied to adaptively recalibrate the gathering of informative feature responses by explicitly modeling interdependencies between feature maps [26]. Adding SE blocks after the convolutional layers can help model weighting that learns to emphasize and dismiss informative features. The weights of each feature map are equally informative after performing temporal convolution in Section 3.1. The temporal SE block is briefly described in two steps. First, in the squeezing step, global information is compressed into a feature descriptor $(1 \times 1 \times F^{te})$ using average pooling. Next, a dense layer is added with ReLU activation to reduce the channel complexity by a ratio (r) while r is set to 16. Then, another dense layer with sigmoid activation is added to give each channel a smooth gating function. Finally, the excitation step obtains the aggregated information and fully captures the channel-wise dependencies. According to the temporal feature maps, the features are multiplied by the weights of the temporal SE block and output $(T' \times E \times F^{te})$.

Moreover, after spatial processing in Section 3.2.1, the spatial SE is applied to adaptively recalibrate temporal-spatial feature responses. The squeeze step is to compress global temporal-spatial information into a feature descriptor $(1 \times 1 \times F^{sp})$. Next, two dense layers reduce the channel complexity $(r = 16)$. In the squeezing step, the temporal-spatial feature maps are multiplied by the weights of the spatial SE block. After performing these convolutions, the extracted features are generated and transmitted to the classification layer.

3.2.3. Classification

The outputs from the previous module are flattened into one vector representing all extracted features and linearly transformed into classification logits with an FCN. For binary classification, the final layer unit is 1. The loss function is binary cross-entropy, and the activation function is a sigmoid function. While 3 classes classification, the final layer unit is 3. The loss function is categorical cross-entropy, and the activation function is a softmax function. The final output is a discrete emotion label with probability.

4. Experiments and Results

To validate the method's performance, feature extractability, and classification accuracy, we examined and compared it with existing methods. The experiments and results consist of subject-dependent and independent classification on the SEED and DEAP datasets.

4.1. Experiment Setting

4.1.1. Datasets

This study conducted experiments on two open-access datasets: SEED (SJTU emotion EEG dataset) [22] and DEAP (dataset for emotion analysis using physiological signals) [27] datasets. These datasets have been widely used in emotion recognition using multimodal physiological signals. The SEED dataset [22] consists of 64 channels of EEG signals recorded from 15 subjects. All subjects were asked to watch 15 excerpts of movie clips for 3 trials. Each clip was approximately 4 min long, and the time interval between trials was one week or longer. The emotional labels corresponded to three types of movies to stimulate emotional states: positive, neutral, and negative. The recorded EEG signals were downsampled from 1000 to 200 Hz and applied a bandpass filter from 0–75 Hz. The DEAP dataset [27] is a multimodal dataset that consists of EEG and peripheral physiological signals. The EEG signals of 32 subjects were recorded by 32 electrodes using the BioSemi ActiveTwo system; 40 one-min excerpts of music videos were used to stimulate emotional states. All subjects were asked to rate the emotion score (1–9) in the valence-arousal space. The EEG signal was recorded with a 512 sample rate. The signals were downsampled to 128 Hz, a bandpass filter was applied to them from 4 to 45 Hz, and EOG artifacts were removed. According to the review paper [25], the average accuracy of the SEED dataset was 90.0%, significantly higher than 83.6% of the DEAP dataset within-subject dependencies.

This study conducted SEED experiments in a three-class classification, including positive, neutral, and negative labels, and a binary classification with less complexity and fewer data considered positive and negative labels. On the other hand, DEAP employed 2D models rated by subjects caused discrepancies between movie types and real subject emotions. To simplify the problem, we investigate a binary classification of valence and arousal scores with positive and negative labels.

4.1.2. Baseline and Existing Models

In baseline experiments, the PSD with four bands was used as extracted features, along with ML classifiers. The baseline classifiers related to EEG analysis comprise the KNN, RF, and FCN. These models were tuned by grid search with the optimal hyperparameters selected, resulting in the most accurate performance. Accordingly, the number of neighbors for the KNN classification is set to 21. The number of RF trees is set to 100. The FCN has 2 hidden layers with 100 nodes, which are then forwarded to the output layer. The activation function is the logistic sigmoid function. Adam optimization is applied while training the FCN model.

Moreover, we compare the performance of MultiT-S ConvNet against existing models, including the Deep ConvNet, Shallow ConvNet, and EEGNet models. The Deep ConvNet architecture [15] consists of one temporal convolution layer, four spatial convolution and pooling layers, and a classification layer in order. The Shallow ConvNet architecture [15] consists of single temporal and spatial convolution layers. Then, it sequentially passes data to a pooling layer and a classification layer. For the EEGNet architecture, we refer to the original network [16], which consists of temporal convolution and a separable convolution layer followed by a classification layer. The MultiT-S ConvNet architecture is depicted in Figure 6. The number of filters and trainable parameters are shown in Figure 2. We applied the Adam optimizer to all DL models in the training process. The loss function for binary classification is binary cross-entropy, whereas three-class classification is categorical cross-entropy. Each model was trained for 200 epochs with 100 batch sizes. These were trained on GPU, NVIDIA Quadro P6000 (Santa Clara, United States), using Tensorflow [28] and Keras [29].

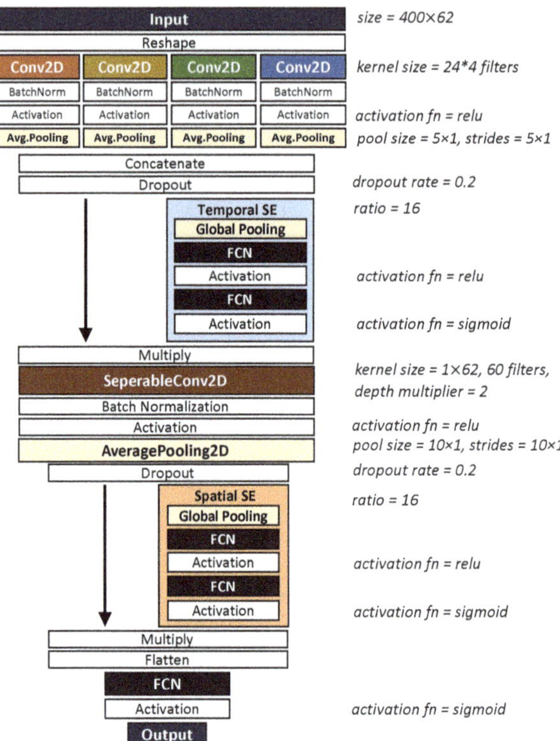

Figure 6. MultiT-S ConvNet model setting.

4.1.3. Comparison Approaches

All models were examined on the subject-dependent and independent classification of the SEED and DEAP datasets. In the subject-dependent experiments, 5-fold cross-validation was applied, experimenting with different partitions of 80% training data and 20% test data; 30% of the training data were held out for validation of free parameters, and the model parameters were optimized on the basis of the remaining 70%. Moreover, the validation set was randomly picked on the desired subject, and the 5-fold cross-validation was applied to all subjects. For the subject-independent experiment, we applied the leave-one-out cross-validation to access model performance. The model was trained using data from all but one subject and tested on the held-out subject; 30% of the training data were randomly selected for free-parameter validation without balancing among subjects.

The performance metric was classification accuracy. Moreover, we show the chance level, which is the obtained accuracy when consistently predicting the majority class. To verify multiple comparisons, analysis of variance (ANOVA) was applied to determine whether the means in the desired group were significantly different. Then, Dunnett's test was used to observe the many-to-one comparisons with our proposed method.

4.2. SEED Experiments and Results

For the SEED dataset experiments, we performed two-class and three-class classification to study both subject-dependent and subject-independent results. The two-class experiment uses two-thirds of the three-class dataset to simplify the problem's complexity and additional investigation. Their accuracy is shown in Figure 7 and Table 1. For the performance results, all models can achieve the 50% and 33.3% chance levels in the subject-dependent and subject-independent experiments, respectively. We used one-way ANOVA to compare differences in the means of all models, and we found that the SEED dataset's results are significant. The post hoc Dunnett's test is then conducted to compare every result with that of MultiT-S ConvNet for observing the significant difference. RF, Deep ConvNet, Shallow ConvNet, EEGNet, and MultiT-S ConvNet significantly outperform the chance levels ($p < 0.01$). Moreover, MultiT-S ConvNet achieves the highest accuracy in all experiments. For two-class and three-class classifications, the subject-dependent results are 95.2% ± 3.2% and 86.0% ± 5.3% respectively, whereas the subject-independent results are 75.1% ± 8.4% and 54.6% ± 6.8% respectively. RF outperforms other baseline techniques by a significant level ($p < 0.05$). The accuracy of all DL models is significantly higher than KNN ($p < 0.01$) and FCN ($p < 0.05$) models. However, there is no statistical difference among deep learning models ($p > 0.05$), except in three-class and subject-independent experiments, where the accuracy of MultiT-S ConvNet is significantly higher than that of EEGNet ($p < 0.05$).

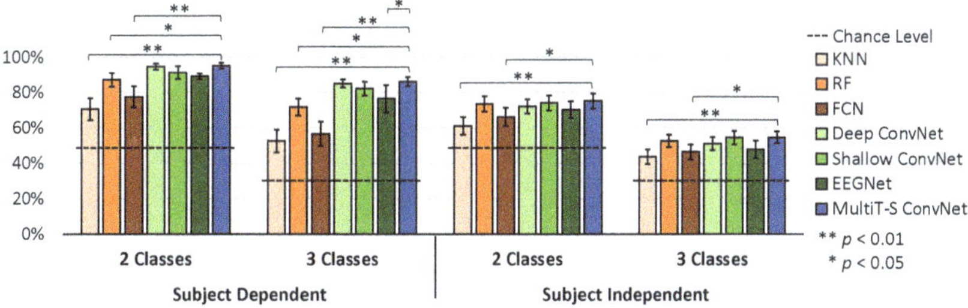

Figure 7. Average classification accuracy for all subjects on the SEED dataset. Error bars denote a 95% confidence interval (CI) computed from all subject's means. The horizontal dashed lines indicate the chance level. The stars indicate significant differences compared with MultiT-S ConvNet.

Table 1. The accuracy results of subject-dependent and subject-independent classification on SEED and DEAP datasets. The bottom row indicates our proposed MultiT-S ConvNet results. Bold indicates the highest accuracy.

Models	SEED-2 Classes		SEED-3 Classes		DEAP-Valence		DEAP-Arousal	
	Subj. Dep.	Subj. Indep.	Subj. Dep.	Subj. Indep.	Subj. Dep.	Subj. Indep.	Subj. Dep.	Subj. Indep.
Chance Level	50.0 ± 0.0	50.0 ± 0.0	33.3 ± 0.0	33.3 ± 0.0	56.6 ± 9.2	**56.6 ± 9.2**	58.9 ± 15.5	**58.9 ± 15.5**
KNN	70.5 ± 12.3	72.0 ± 7.8	52.5 ± 12.7	43.7 ± 8.0	55.9 ± 6.9	52.9 ± 5.2	61.6 ± 11.0	53.0 ± 10.7
RF	87.0 ± 7.7	73.4 ± 8.5	71.7 ± 9.5	52.7 ± 6.9	55.6 ± 7.3	51.1 ± 3.6	61.8 ± 10.6	50.0 ± 7.5
FCN	77.4 ± 11.9	66.1 ± 10.1	56.5 ± 13.6	46.4 ± 8.4	54.8 ± 6.4	53.9 ± 6.6	60.8 ± 10.9	50.6 ± 13.2
Deep ConvNet	94.8 ± 3.4	72.0 ± 7.8	85.0 ± 4.5	51.1 ± 7.4	76.6 ± 3.4	47.5 ± 7.0	78.6 ± 4.6	51.8 ± 13.0
Shallow ConvNet	91.2 ± 7.4	74.0 ± 8.5	82.1 ± 7.7	54.4 ± 7.4	76.0 ± 6.9	49.1 ± 7.7	78.9 ± 6.7	51.0 ± 14.7
EEGNet	89.1 ± 3.6	70.1 ± 9.5	76.3 ± 15.1	47.8 ± 9.7	**79.2 ± 6.3**	49.3 ± 7.2	82.1 ± 6.9	51.4 ± 10.3
MultiT-S ConvNet	**95.2 ± 3.2**	**75.1 ± 8.4**	**86.0 ± 5.3**	**54.6 ± 6.8**	77.6 ± 5.8	52.6 ± 8.8	**82.2 ± 6.5**	51.5 ± 10.4

4.3. DEAP Experiments and Results

Using the DEAP dataset, we conducted four experiments of valence and arousal binary classification on both subject dependence and subject independence. The classes were defined by emotion scores, where 1 to 4 is a negative class, and 5 to 9 is a positive class. DEAP dataset's results are shown in Figure 8 and Table 1. Similar to the SEED dataset, one-way ANOVA with Dunnett's post hoc is applied to investigate significant differences among all models. The chance level of valence is 56.6% ± 9.2%, whereas the chance level of arousal is 58.9% ± 15.5%. For subject-dependent experiments, EEGNet outperforms the other models with a valence classification accuracy of 79.2% ± 5.8%, and MultiT-S ConvNet outperforms the other models with an arousal classification accuracy of 82.2% ± 6.5%. In both valence and arousal experiments, all CNN models explicitly and significantly outperform the baseline models ($p < 0.01$). Nevertheless, these chance levels could not be achieved in the subject-independent experiments.

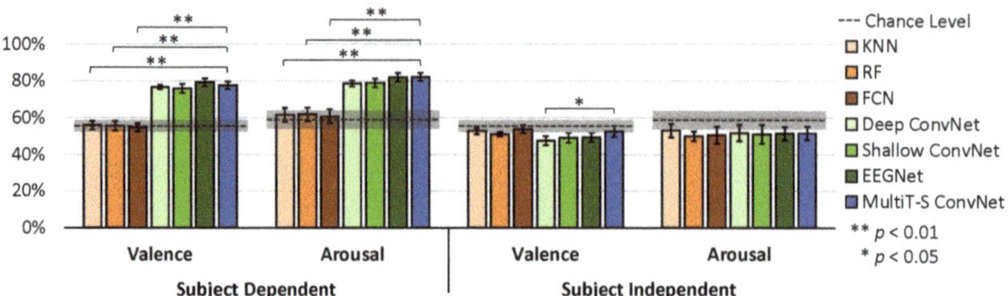

Figure 8. Average classification accuracy for all subjects on the DEAP dataset. Error bars denote a 95% CI computed from all subject's means. The horizontal dashed lines indicate the chance level, and the grey areas indicate the CI of chance levels. The stars indicate significant differences compared with MultiT-S ConvNet.

4.4. Ablation Study

To verify the effectiveness of our temporal filtering, we further investigate the effect of temporal convolution on the other CNN models. Based on the first blocks of each model in Figure 3, we reform a temporal convolution with multi-kernel convolution. Meanwhile, the spatial filters were halved to preserve the number of parameters and computational cost. Here, we consider a two-class classification with a subject-independent setting on the SEED dataset, and the results are shown in Table 2. In addition, a paired t-test is used to observe the difference between the corresponding models. Deep ConvNet with multi-kernel temporal convolution consists of 12×4 temporal (band) filters, 12 spatial filters, and 168 temporal-spatial filters. As a result, the number of trainable parameters decreases from 182,497 to 73,777. It significantly outperforms the original Deep ConvNet with 73.96% accuracy from 72.04% ($p < 0.05$). For EEGNet with multi-kernel temporal convolution, the number of trainable parameters increases from 13,537 to 16,177, but it achieves a significantly better accuracy of 74.00% than the original EEGNet of 70.15% ($p < 0.01$). However, there is no significant difference between Shallow ConvNet with and without multi-kernel convolution ($p > 0.05$). Moreover, we trained the MultiT-S ConvNet model without SE blocks to observe their effects on the learning performance. The accuracy of our model without SE Blocks decreases from 75.13% to 73.95% ($p < 0.05$).

Table 2. The number of filters, trainable parameters, and accuracies before and after applying multi-kernel convolution on the SEED dataset.

	No. of Temp. Filter	No. of Spat. Filter	No. of Temp-Spat Filter	Trainable Parameters	Acc.(%)
Deep ConvNet	24	24	48 + 96 + 192 = 336	182,497	72.04
Deep ConvNet (+)	12 × 4 = 48	12	24 + 48 + 96 = 168	73,777	73.96 *
Shallow ConvNet	40	40		101,281	73.96
Shallow ConvNet (+)	20 × 4 = 80	20		101,721	73.19
EEGNet	32	64	64	13,537	70.15
EEGNet (+)	16 × 4 = 64	32	32	16,177	74.00 **
MultiT-S ConvNet	24 × 4 = 96	64		30,313	75.13

Note: (+) indicates reforming of temporal filters and halving of spatial filters. Stars indicate paired *t*-tests compared to the corresponding model. (** is $p < 0.01$ and * is $p < 0.05$).

5. Discussion

5.1. Classification Performance

In this study, the accuracy performance was examined using SEED and DEAP, which are well-known EEG emotion datasets. In addition, existing models and our proposed model were implemented with similar settings, including data segmentation, pre-processing, parameter tuning, and evaluation metrics, to ensure the reliability of model performance comparison. The experiments and results demonstrate that all ConvNets with appropriate design choices can outperform traditional approaches in terms of accuracy performances in all classification experiments. Moreover, this study compares MultiT-S ConvNet against existing ConvNet models in terms of accuracy performance in EEG-based emotion classification, as shown in Table 3. The comparison shows that the overall performance of MultiT-S ConvNet outperforms Deep ConvNet, Shallow ConvNet, and EEGNet with higher accuracies of 2.2%, 2.3%, and 3.7%, respectively.

Table 3. The comparison of model performance between the proposed MultiT-S ConvNet and existing ConvNets. The accuracy (%) is the average from all experiments in this study. It calculates differences in the accuracy between MultiT-S ConvNet and other models, and the positive mark denotes a higher accuracy of the MultiT-S ConvNet. The minimum frequency (Hz) covered by each model is displayed. The bottom row computes the number of trainable parameters in dependence on the sampling rate of the SEED dataset. The negative and positive marks denote a decrease and increase of trainable parameters required by the MultiT-S ConvNet, compared with indicated models.

	MultiT-S ConvNet	vs. Deep ConvNet	vs. Shallow ConvNet	vs. EEGNet
Accuracy (%)	71.9	+2.2	+2.3	+3.7
Minimum freq. covered	5 Hz	40 Hz	17 Hz	2 Hz
Number of Trainable Parameters	30,313	−152,184	−43,464	+14,136

According to the result, the SEED dataset is more uncomplicated to classify than the DEAP dataset because it contains different sources of emotional labels and the cleanliness of the signals. For the emotion states, the SEED dataset uses the movie types, whereas the DEAP dataset uses a subject questionnaire. This ambiguity directly affects the classification performance and trustworthiness of each dataset. Based on the SEED dataset results, all the obtained accuracy reached the chance level ($p < 0.01$) in both subject-dependent and independent experiments. For the DEAP dataset, only the ConvNet-based techniques reached the chance level ($p < 0.01$) in the subject-dependent experiments. In contrast, for the DEAP dataset, none of the models could surpass the chance level in subject-independent experiments. Many studies using the DEAP dataset have mainly conducted subject-dependent experiments because the data distribution of each subject in this dataset is highly diverse and varied. Observing the topoplots of PSD features in 4 frequency bands by human eyesight shows that the SEED dataset plots can be distinguished, whereas the DEAP dataset plots are very similar. In terms of subject-dependent and subject-independent learning, subject-dependent performance outperforms subject-independent performance significantly in all experiments. Consequently, the individual distribution directly affects learning performance. Moreover, our results are comparable and consistent with existing studies in terms of average accuracy [25].

5.2. Architectures and Design Choices

For several decades, ML has been applied in EEG analysis for individual modules combined with prior knowledge. Most previous studies applied manual parameterization for feature extraction and then transmitted them to an ML classifier [9,11]. These traditional techniques require background knowledge for signal processing, noise reduction, and data manipulation. From the results in terms of the PSD feature, the classifiers are possibly influenced and affected by the noise and complexity of multi-channel EEG signals, resulting in poor performance. However, the ensemble model, RF, performs well overall and significantly outperforms all baseline techniques on the SEED dataset. The RF model potentially learns how to reduce the generalization error of the prediction and deal with various EEG classification tasks.

On the other hand, ConvNet models are a better choice in EEG analysis than manual hand-crafted features. Numerous ConvNets with end-to-end architecture have been proposed to learn informative features and deal with the variation of noises in EEG analysis automatically and efficiently. For instance, attention classification using Shallow ConvNet and long short-term memory (LSTM) network on a three-back task [30], and emotion classification using subject-invariant bilateral variational domain adversarial neural network [31]. However, EEG features need to be extracted into various representations that improve the learning effect, especially for EEG analysis. Here, we proposed the multi-kernel convolution that helps the model achieve better results than the single-kernel convolution. As a result, our module applies to and obtains better accuracy than existing ConvNets while preserving the model size and computation costs. In this study, all ConvNets models outperform the baseline techniques in all experiments, except in subject-independent experiments on the DEAP dataset. The number of parameters verifies that the learned features are useful for classification. It effectively estimates missing data and maintains good accuracy even when a large proportion of the data is missing. With significantly higher accuracy performance, our MultiT-S ConvNet has approximately six times and three times fewer trainable parameters than Deep ConvNet and Shallow ConvNet, respectively. Despite having fewer parameters than our proposed model, EEGNet performance is unstable in terms of the average accuracy and standard deviation of the SEED and DEAP datasets. Moreover, the additional variations of features extracted by multi-kernel reduce overfitting or vanishing gradient problems while training a model.

6. Conclusions

In conclusion, a well-designed end-to-end convolution network is a promising feature extraction and classification tool in EEG-based emotion analysis. We proposed multi-kernel filtering to increase the variation of temporal representations and further recalibrate both temporal and spatial features using the lightweight gating mechanism. The results show that our MultiT-S ConvNet outperforms the traditional and existing models. However, there are some limitations in the current study that could be addressed in future research. This study focused on discrete emotion recognition with the type of movie stimuli. In the future, the applicability of the MultiT-S network will be explored in brain disease recognition, such as epilepsy and Alzheimer's disease. In addition, this proposed model could be further developed to allow detecting and analyzing dynamic changes in EEG signals over time. Ultimately, our module could be added to any model of EEG-based convolution networks and its ability could improve the learning effect of other existing models when there is a limited dataset.

Author Contributions: Conceptualization, T.E., T.M., T.K., K.-i.F. and M.N.; methodology, T.E.; software, T.E.; validation, T.E., T.M., T.K., K.-i.F. and M.N.; formal analysis, T.E.; investigation, T.E.; resources, M.N.; data curation, T.E.; writing—original draft preparation, T.E.; writing—review and editing, T.E., T.M., T.K., K.-i.F. and M.N.; visualization, T.E.; supervision, T.M., T.K., K.-i.F. and M.N.; project administration, M.N.; funding acquisition, M.N. All authors have read and agreed to the published version of the manuscript.

Funding: This research was funded by the Center of Innovation program from Japan Science and Technology Agency, JST, grant number JPMJCE1310, and was supported by the "Crossover Alliance to create the future with people, intelligence and materials" from the Ministry of Education, Culture, Sports, Science, and Technology (MEXT), Japan. T.E. was also supported by the Japan International Cooperation Agency (JICA) under their Innovative Asia scholarship program.

Institutional Review Board Statement: Not applicable.

Data Availability Statement: The code used in this study is available upon request from the corresponding author.

Acknowledgments: We would like to express gratefulness to our labmates at Numao laboratory, Osaka University, Japan, for their guidance and moral support. We would like to express our great appreciation to the secretaries at Numao laboratory, Osaka University, Japan, for their valuable help and support during the development of this research work.

Conflicts of Interest: The authors declare no conflict of interest.

References

1. Al-Nafjan, A.N.; Hosny, M.I.; Al-Ohali, Y.; Al-Wabil, A. Review and Classification of Emotion Recognition Based on EEG Brain-Computer Interface System Research: A Systematic Review. *Appl. Sci.* **2017**, *7*, 1239. [CrossRef]
2. Millan, J.d.R.; Rupp, R.; Müller-Putz, G.; Murray-Smith, R.; Giugliemma, C.; Tangermann, M.; Vidaurre, C.; Cincotti, F.; Kübler, A.; Leeb, R.; et al. Combining Brain–Computer Interfaces and Assistive Technologies: State-of-the-Art and Challenges. *Front. Neurosci.* **2010**, *4*, 161–175. [CrossRef] [PubMed]
3. Jamil, N.; Belkacem, A.N.; Ouhbi, S.; Lakas, A. Noninvasive Electroencephalography Equipment for Assistive, Adaptive, and Rehabilitative Brain–Computer Interfaces: A Systematic Literature Review. *Sensors* **2021**, *21*, 4754. [CrossRef]
4. Tangermann, M.W.; Krauledat, M.; Grzeska, K.; Sagebaum, M.; Vidaurre, C.; Blankertz, B.; Müller, K.R. Playing Pinball with Non-Invasive BCI. In Proceedings of the 21st International Conference on Neural Information Processing Systems, Vancouver, BC, Canada, 8–10 December 2008; Curran Associates Inc.: Red Hook, NY, USA, 2008; NIPS'08; pp. 1641–1648.
5. Singh, A.K.; Wang, Y.K.; King, J.T.; Lin, C.T. Extended Interaction With a BCI Video Game Changes Resting-State Brain Activity. *IEEE Trans. Cogn. Dev. Syst.* **2020**, *12*, 809–823. [CrossRef]
6. Rached, T.S.; Perkusich, A. Emotion Recognition Based on Brain-Computer Interface Systems. In *Brain-Computer Interface Systems*; Fazel-Rezai, R., Ed.; IntechOpen: Rijeka, Croatia, 2013; Chapter 13. [CrossRef]
7. Torres, E.P.; Torres, E.A.; Hernández-Álvarez, M.; Yoo, S.G. EEG-Based BCI Emotion Recognition: A Survey. *Sensors* **2020**, *20*, 5083. [CrossRef]
8. Acharya, U.R.; Molinari, F.; Subbhuraam, V.S.; Chattopadhyay, S.; Kh, N.; Suri, J. Automated diagnosis of epileptic EEG using entropies. *Biomed. Signal Process. Control* **2012**, *7*, 401–408. [CrossRef]
9. Asadzadeh, S.; Yousefi Rezaii, T.; Beheshti, S.; Delpak, A.; Meshgini, S. A systematic review of EEG source localization techniques and their applications on diagnosis of brain abnormalities. *J. Neurosci. Methods* **2020**, *339*, 108740. [CrossRef]
10. van Vliet, M.; Robben, A.; Chumerin, N.; Manyakov, N.V.; Combaz, A.; Van Hulle, M.M. Designing a brain-computer interface controlled video-game using consumer grade EEG hardware. In Proceedings of the 2012 ISSNIP Biosignals and Biorobotics Conference: Biosignals and Robotics for Better and Safer Living (BRC), Manaus, Brazil, 9–11 January 2012; pp. 1–6. [CrossRef]
11. Klimesch, W. EEG alpha and theta oscillations reflect cognitive and memory performance: A review and analysis. *Brain Res. Rev.* **1999**, *29*, 169–195. [CrossRef]
12. Schmidt, L.A.; Trainor, L.J. Frontal brain electrical activity (EEG) distinguishes valence and intensity of musical emotions. *Cogn. Emot.* **2001**, *15*, 487–500. [CrossRef]
13. Huang, D.; Guan, C.; Ang, K.K.; Zhang, H.; Pan, Y. Asymmetric spatial pattern for EEG-based emotion detection. In Proceedings of the International Joint Conference on Neural Networks (IJCNN), San Diego, CA, USA, 17–21 June 1990; IEEE: Piscataway, NJ, USA, pp. 1–7. [CrossRef]
14. Jatupaiboon, N.; Pan-ngum, S.; Israsena, P. Emotion classification using minimal EEG channels and frequency bands. In Proceedings of the 10th international Joint Conference on Computer Science and Software Engineering (JCSSE), Khon Kaen, Thailand, 29–13 May 2013; IEEE: Piscataway, NJ, USA, 2013; pp. 21–24. [CrossRef]
15. Schirrmeister, R.T.; Springenberg, J.T.; Fiederer, L.D.J.; Glasstetter, M.; Eggensperger, K.; Tangermann, M.; Hutter, F.; Burgard, W.; Ball, T. Deep learning with convolutional neural networks for EEG decoding and visualization. *Hum. Brain Mapp.* **2017**, *38*, 5391–5420. [CrossRef]
16. Lawhern, V.J.; Solon, A.J.; Waytowich, N.R.; Gordon, S.M.; Hung, C.P.; Lance, B.J. EEGNet: A compact convolutional neural network for EEG-based brain–computer interfaces. *J. Neural Eng.* **2018**, *15*, 056013. [CrossRef] [PubMed]
17. Jenke, R.; Peer, A.; Buss, M. Feature Extraction and Selection for Emotion Recognition from EEG. *IEEE Trans. Affect. Comput.* **2014**, *5*, 327–339. [CrossRef]
18. Chanel, G.; Kierkels, J.J.; Soleymani, M.; Pun, T. Short-term emotion assessment in a recall paradigm. *Int. J. -Hum.-Comput. Stud.* **2009**, *67*, 607–627. [CrossRef]
19. Bos, D.O. EEG-based emotion recognition. *Influ. Vis. Audit. Stimuli* **2006**, *56*, 1–17.
20. Ramoser, H.; Muller-Gerking, J.; Pfurtscheller, G. Optimal spatial filtering of single trial EEG during imagined hand movement. *IEEE Trans. Rehabil. Eng.* **2000**, *8*, 441–446. [CrossRef] [PubMed]
21. Ang, K.K.; Chin, Z.Y.; Wang, C.; Guan, C.; Zhang, H. Filter Bank Common Spatial Pattern Algorithm on BCI Competition IV Datasets 2a and 2b. *Front. Neurosci.* **2012**, *6*, 39. [CrossRef]
22. Zheng, W.L.; Lu, B.L. Investigating Critical Frequency Bands and Channels for EEG-based Emotion Recognition with Deep Neural Networks. *IEEE Trans. Auton. Ment. Dev.* **2015**, *7*, 162–175. [CrossRef]
23. Li, M.; Lu, B.L. Emotion classification based on gamma-band EEG. In Proceedings of the 2009 Annual International Conference of the IEEE Engineering in Medicine and Biology Society, Minneapolis, MN, USA, 2–6 September 2009; pp. 1223–1226. [CrossRef]
24. Candra, H.; Yuwono, M.; Chai, R.; Handojoseno, A.; Elamvazuthi, I.; Nguyen, H.T.; Su, S. Investigation of window size in classification of EEG-emotion signal with wavelet entropy and support vector machine. In Proceedings of the 2015 37th Annual

International Conference of the IEEE Engineering in Medicine and Biology Society (EMBC), Milan, Italy, 25–19 August 2015; pp. 7250–7253. [CrossRef]

25. Shu, L.; Xie, J.; Yang, M.; Li, Z.; Li, Z.; Liao, D.; Xu, X.; Yang, X. A review of emotion recognition using physiological signals. *Sensors* **2018**, *18*, 2074. [CrossRef]

26. Hu, J.; Shen, L.; Sun, G. Squeeze-and-Excitation Networks. *IEEE Trans. Pattern Anal. Mach. Intell.* **2017**, *42*, 2011–2023. [CrossRef]

27. Koelstra, S.; Mühl, C.; Soleymani, M.; Lee, J.S.; Yazdani, A.; Ebrahimi, T.; Pun, T.; Nijholt, A.; Patras, I. DEAP: A Database for Emotion Analysis Using Physiological Signals. *IEEE Trans. Affect. Comput.* **2011**, *3*, 18–31. [CrossRef]

28. Abadi, M.; Agarwal, A.; Barham, P.; Brevdo, E.; Chen, Z.; Citro, C.; Corrado, G.S.; Davis, A.; Dean, J.; Devin, M.; et al. TensorFlow: Large-Scale Machine Learning on Heterogeneous Systems. 2015. Available online: tensorflow.org (accessed on 28 September 2022).

29. Keras. 2015. Available online: https://github.com/fchollet/keras (accessed on 28 September 2022).

30. Emsawas, T.; Kimura, T.; Fukui, K.i.; Numao, M. Comparative Study of Wet and Dry Systems on EEG-Based Cognitive Tasks. In Proceedings of the International Conference on Brain Informatics, Padua, Italy, 19 September 2020, Springer: Cham, Switzerland, pp. 309–318.

31. Hagad, J.L.; Kimura, T.; Fukui, K.i.; Numao, M. Learning subject-generalized topographical EEG embeddings using deep variational autoencoders and domain-adversarial regularization. *Sensors* **2021**, *21*, 1792. [CrossRef] [PubMed]

Review

Concurrent fNIRS and EEG for Brain Function Investigation: A Systematic, Methodology-Focused Review

Rihui Li [1,2], Dalin Yang [3,4], Feng Fang [2], Keum-Shik Hong [3], Allan L. Reiss [1] and Yingchun Zhang [2,*]

[1] Center for Interdisciplinary Brain Sciences Research, Department of Psychiatry and Behavioral Sciences, Stanford University School of Medicine, Stanford, CA 94305, USA
[2] Department of Biomedical Engineering, University of Houston, Houston, TX 77004, USA
[3] School of Mechanical Engineering, Pusan National University, Pusan 43241, Korea
[4] Mallinckrodt Institute of Radiology, Washington University School of Medicine in St. Louis, 4515 McKinley Avenue, St. Louis, MO 63110, USA
*   Correspondence: yzhang94@uh.edu

Abstract: Electroencephalography (EEG) and functional near-infrared spectroscopy (fNIRS) stand as state-of-the-art techniques for non-invasive functional neuroimaging. On a unimodal basis, EEG has poor spatial resolution while presenting high temporal resolution. In contrast, fNIRS offers better spatial resolution, though it is constrained by its poor temporal resolution. One important merit shared by the EEG and fNIRS is that both modalities have favorable portability and could be integrated into a compatible experimental setup, providing a compelling ground for the development of a multimodal fNIRS–EEG integration analysis approach. Despite a growing number of studies using concurrent fNIRS-EEG designs reported in recent years, the methodological reference of past studies remains unclear. To fill this knowledge gap, this review critically summarizes the status of analysis methods currently used in concurrent fNIRS–EEG studies, providing an up-to-date overview and guideline for future projects to conduct concurrent fNIRS–EEG studies. A literature search was conducted using PubMed and Web of Science through 31 August 2021. After screening and qualification assessment, 92 studies involving concurrent fNIRS–EEG data recordings and analyses were included in the final methodological review. Specifically, three methodological categories of concurrent fNIRS–EEG data analyses, including EEG-informed fNIRS analyses, fNIRS-informed EEG analyses, and parallel fNIRS–EEG analyses, were identified and explained with detailed description. Finally, we highlighted current challenges and potential directions in concurrent fNIRS–EEG data analyses in future research.

Keywords: EEG; functional NIRS; multimodal neuroimaging; concurrent recording; integrated analysis

Citation: Li, R.; Yang, D.; Fang, F.; Hong, K.-S.; Reiss, A.L.; Zhang, Y. Concurrent fNIRS and EEG for Brain Function Investigation: A Systematic, Methodology-Focused Review. *Sensors* **2022**, *22*, 5865. https://doi.org/10.3390/s22155865

Academic Editor: Sung-Phil Kim

Received: 20 June 2022
Accepted: 30 July 2022
Published: 5 August 2022

1. Introduction

The human brain comprises billions of neurons [1]. Each of these forms a number of synapses, establishing a complicated network with quadrillions of connections and thus enabling our brains to function in an adaptive manner [2]. Although our understanding of neurons on a microscopic scale has progressed in recent decades, little is known about how these huge numbers of neurons (and synapses) work collectively to generate macroscopic brain signals and human behaviors. It is believed that human brain functions and associated behaviors are carried out by complex neural activations and networks. These internal activities generally elevate electrical activity (direct effects) accompanied by a hemodynamic and metabolic response (indirect effects), which serve as the basic sources for all noninvasive neuroimaging techniques. Depending on the sources of the signals, these brain imaging techniques can be roughly divided into two categories. The first category refers to imaging techniques that directly capture the neural electrical activities by detecting the induced electrical or magnetic fluctuations over the scalp. The most representative methods in this category are Electroencephalography (EEG) and Magnetoencephalography (MEG).

The second category comprises indirect imaging approaches that rely on hemodynamic (cerebral blood flow, cerebral blood volume) and metabolic (glucose and oxygen utilization) responses induced by neural activity. Commonly available techniques in this category include functional near-infrared spectroscopy (fNIRS), functional magnetic resonance imaging (fMRI), and positron emission tomography (PET). In this perspective, EEG and fNIRS have been gaining popularity in the research community and clinical practice due to their distinct natures, particularly their noninvasiveness, mobility, and flexibility.

1.1. The Fundamental Basis of fNIRS

Functional Near-infrared Spectroscopy (fNIRS), first reported by Jobsis in 1977 [3], is an optical imaging technique for non-invasive investigation of hemodynamic responses in the brain. fNIRS usually utilizes lights with distinct wavelengths (between 600 and 1000 nm) that can penetrate the scalp and reach the cortical surface to measure the concentration changes of oxygenated hemoglobin (HbO) and deoxygenated hemoglobin (HbR) that are coupled with the metabolic activity of neurons in the outer layers of the cortex. This technique is particularly useful for studying the functional activation within the brain due to the inherent relationship between neural activity and hemodynamic responses in the brain [4]. Specifically, fNIRS measures the regional changes of HbO and HbR concentration, which can serve as an indicator of hemodynamic changes associated with neural activity in the brain.

Currently, the continuous wave NIRS (CW-NIRS) is extensively used in the research and clinical settings due to its low cost and simplicity. The measurement of the hemoglobin concentration (HbO and HbR) in CW-NIRS primarily relies on the physical basis that chromophores inside the brain, especially the HbO and HbR, have specific and sensitive absorption characteristics in the near-infrared range (between 600 and 1000 nm). Lights at different wavelengths can then be injected into the brain via the sources (illuminators) placed on the scalp, and the attenuated lights are detected by the optical detectors placed near the illuminators (Figure 1A), from which the concentration changes of HbO and HbR can be computed based on the Modified Beer-Lambert Law [5]. Specifically, CW-NIRS systems typically utilize laser/LED sources to shine two distinct wavelengths into the brain at a constant intensity and use detectors to measure the intensity of diffusely reflected light continuously.

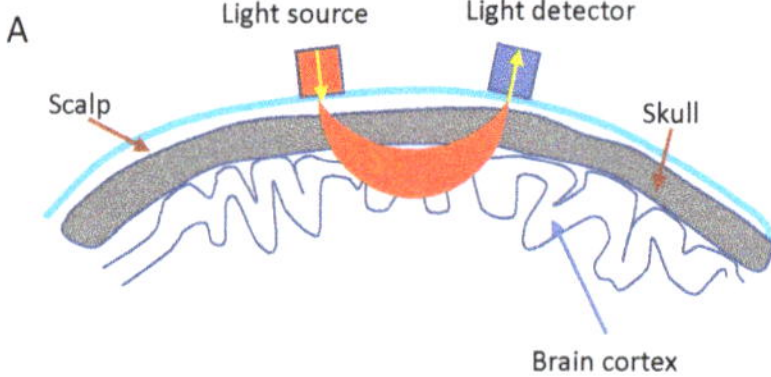

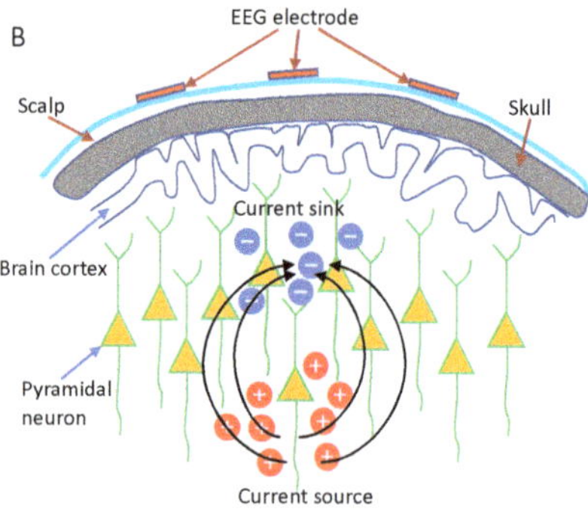

Figure 1. Schematic demonstration: (**A**) fNIRS and (**B**) EEG measurement.

1.2. The Fundamental Basis of EEG

Electroencephalography (EEG), first described by Hans Berger in 1929 [6], is thought to result primarily from the synchronization of post-synaptic potentials at cortical pyramidal neurons [7]. The recorded EEG signal does not represent single neuron depolarization inside the brain. Instead, it is assumed that tens of thousands of synchronized pyramidal neurons within the cortex are firing when the brain is activated, wherein dendritic trunks of the neurons are coherently orientated, parallel with each other and perpendicular to the cortical surface so as to induce sufficient summation and propagation of electrical signals to the scalp (Figure 1B) [8].

Typically, EEG signals are measured through EEG electrodes (including a reference electrode and a ground electrode) placed over a subject's scalp. Voltage differences between the electrodes and the reference electrode are then measured and amplified (Figure 1B). The recorded EEG signals, which represent the large-scale neural oscillatory activity, can be divided into various rhythms depending on characteristic frequency bands, including theta (4–7 Hz), alpha (8–14 Hz), beta (15–25 Hz), and gamma (>25 Hz) [9]. These brain rhythms contain information associated with the ongoing neuronal processing in specific brain areas, which allows EEG to be used as a non-invasive method for the characterization of cortical reorganization, induced by various brain disorders, particularity in the diagnosis of epilepsy and stroke [9–12], and the assessment of brain state alterations [13–15].

1.3. Integration of EEG and fNIRS: Rationale and Advantages

The functional activity of the cerebral cortex can be investigated using various imaging techniques including EEG, fNIRS, fMRI, and their combinations [16–19]. Each of these techniques has its own advantages and disadvantages. However, single-modality imaging techniques can only capture limited information associated with neural activity due to their technical limitations and the inherent complexity of neural processing within the brain. For example, compared to fMRI, fNIRS features higher temporal resolution (<1 s), good portability, lower cost, good resistance to motion artifacts, and applicability to various measurement scenarios including clinical settings as well as the natural environment [5]. More importantly, fNIRS measurements have been proven to be similar to the blood oxygen level dependent (BOLD) response obtained by fMRI [20]. However, there are also several limitations of fNIRS techniques: the limited penetration depth, low signal-to-noise ratio, and low temporal resolution compared to EEG. EEG possess several advantages over fMRI for exploring dynamic brain activity: it is portable, inexpensive, and features a remarkably high temporal resolution (millisecond) compared to fNIRS and fMRI [21], though EEG is highly vulnerable to motion artifacts that would inhibit the EEG measurement in a natural settings [22].

To comprehensively explore the functional activity of the brain, multimodal approaches are needed. Integrated EEG–fNIRS approaches offer numerous benefits over single-modality methods by exploiting their individual strengths; EEG provides favorable temporal resolution while fNIRS offers better spatial resolution and is robust to noise [23,24]. Additionally, EEG and fNIRS signals are associated with the neuronal electrical activity and metabolic response, respectively, providing a built-in validation for identified activity. Measurements obtained from each of these two modalities thereby provide complementary information related to functional activity of the brain.

In addition to their complementary technical properties, the rationale behind the combination of EEG and fNIRS relies on a physiological phenomenon called neurovascular coupling within the brain [25]. Neural activity is inherently accompanied with the fluctuation of cerebral blood flow (CBF) that carries vital oxygen and nutrients to neurons. Specifically, when neurons are activated within a specific brain region, blood will flow to that brain region to meet the increased demand of glucose and oxygen, resulting in fluctuations of hemoglobin concentration (HbO and HbR) that can be detected by functional imaging techniques such as fNIRS and fMRI (Figure 2). The so-called neurovascular coupling forms the theoretical basis for integrated fNIRS–EEG imaging of brain activity. It has been shown in recent studies that impairment of neurovascular coupling could serve as a sign for several neurological diseases such as Alzheimer's disease and stroke [25–27], which might provide a new prospective for evaluation and diagnosis of neurological diseases as well as increase our understanding of mechanisms underlying neurovascular coupling.

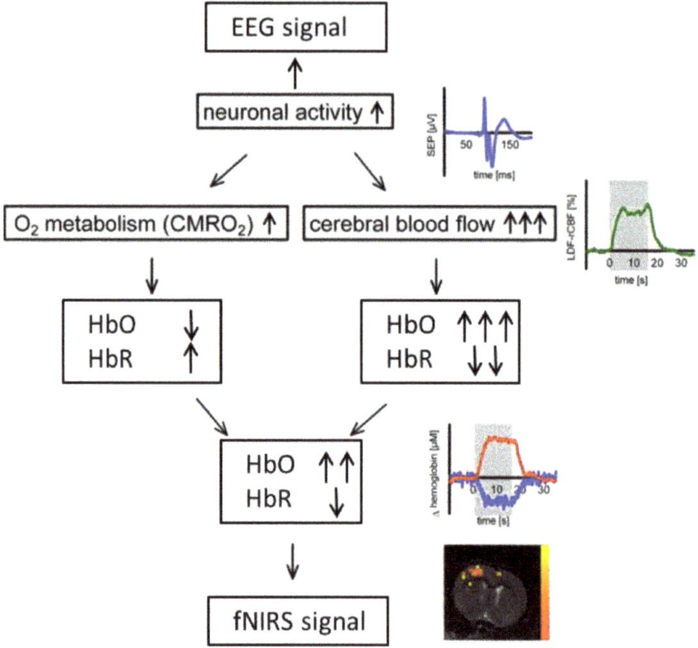

Figure 2. Demonstration of neurovascular coupling.

1.4. Motivation of the Present Review

The fact that integration of fNIRS and EEG provides complementary information about electrical and metabolic-hemodynamic activity of the brain activity has led to increasing investigations of the benefits of integrated EEG and fNIRS [27–29]. In the last decade, numerous studies utilizing integrated fNIRS–EEG systems have been reported on both nonclinical and clinical topics [30]. Data analysis of concurrent fNIRS–EEG recordings is a fundamental but essential step for fNIRS–EEG research studies. This step usually consists of several key processes, including raw data processing, feature extraction, and integrated/fused analysis of these two modalities. Although several recent reviews have been published to summarize the latest progress on applications of concurrent fNIRS–EEG recordings, such as brain–computer interface, development of wearable fNIRS–EEG devices, and neuromodulation, there is no comprehensive summary yet regarding the general analysis pipeline of simultaneously recorded fNIRS and EEG signals. To fill this knowledge gap, this review aims to systematically summarize the status of analyses methods used in concurrent fNIRS–EEG studies involving healthy individuals as well as patient populations. Specifically, we focus on multiple levels of integrated analyses of concurrent fNIRS–EEG recordings by critically evaluating the data processing methods, extracted features, and forms of integration of these two modalities. The present review differs from previous reviews in that this is the first systematic, methodology-focused review to describe which approaches were used in previous concurrent fNIRS–EEG studies and how these approaches were used, thus providing an up-to-date overview and technical guideline for future projects to conduct concurrent fNIRS-EEG studies.

This review is organized as follows: Section 1 is dedicated to the description of the origins, the main characteristics of fNIRS and EEG, and the rationale of combining fNIRS and EEG for multimodal brain imaging. Section 2 describes the strategy of our literature review and the criteria of identification and classification of published articles. Section 3 starts with a brief summary of the preprocessing of raw fNIRS and EEG data and then elaborates three main categories of analysis approaches in concurrent fNIRS–EEG studies. Finally, Section 4 is devoted to underlining the limitations, challenges, and future direction of data analysis of integrated fNIRS–EEG techniques.

2. Methodology

This review was conducted following the Preferred Reporting Items for Systematic Reviews and Meta-Analyses (PRISMA) protocol [31]. As shown in Figure 3, the flow diagram of PRISMA mainly includes three steps: (1) initial search: search related studies based on the defined keywords in selected databases; (2) prescreening: remove duplicated articles and select articles based on designed criteria; (3) qualifying: read through the full text of the selected articles to make sure they meet the eligibility and inclusion criteria.

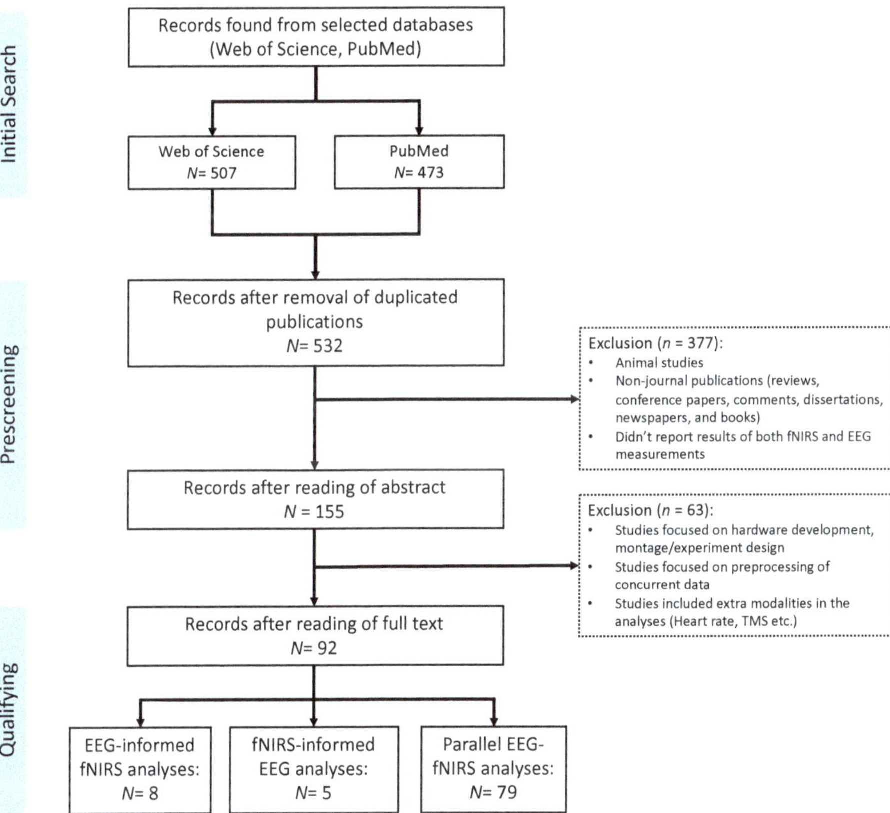

Figure 3. PRISMA flow diagram for the literature review and article selection.

2.1. Search Strategy

The search for relevant peer-reviewed articles describing the use of a concurrent fNIRS–EEG design was conducted on PubMed and Web of Science as literature sources. The following keyword combinations were used in the literature search: ("fNIRS" OR "NIRS" OR "functional near-infrared spectroscopy" OR "near-infrared spectroscopy")

AND ("EEG" OR "electroencephalography") AND ("Brain"). Only articles that were published in English through 31 August 2021 were included.

2.2. Prescreening and Qualifying Criteria

The prescreening criteria were based on the reading of titles and abstracts. First, duplicated articles under different titles were removed. Then, publications were excluded if they (1) were not in line with the topic, i.e., animal studies; (2) were non-journal publications, such as reviews, conference papers, comments, dissertations, newspapers, and books; and (3) did not report analysis results of both fNIRS and EEG measurements.

We then performed further screening and qualifying by reading through the full text of the articles. In this process, publications were excluded if they (1) focused on montage design, experimental design, or hardware development of concurrent fNIRS–EEG systems; (2) focused on preprocessing of fNIRS and/or EEG data; or (3) included extra modalities in the analyses, such as heart rate, electromyography, transcranial magnetic stimulation, etc. Furthermore, the following inclusion criteria for the review were considered: (1) articles focusing on brain function investigation using concurrent fNIRS–EEG were included; (2) articles with details of signal processing, feature extraction, and concurrent analysis of fNIRS–EEG were included.

3. Results

The search strategy resulted in a total of 980 records in the initial search from the selected databases (507 from Web of Science and 473 from PubMed, Figure 3). After the prescreening and qualifying stages, we obtained a total of 92 articles available for this review, including 5 studies focusing on fNIRS-informed EEG analyses, 8 studies focusing on EEG-informed fNIRS analyses, and 79 studies focusing on the parallel analyses of fNIRS–EEG (Figure 3). Figure 4A summarizes the number of concurrent fNIRS–EEG studies each year since 2012, and Figure 4B shows the percentage of each type of integrated analysis of fNIRS–EEG.

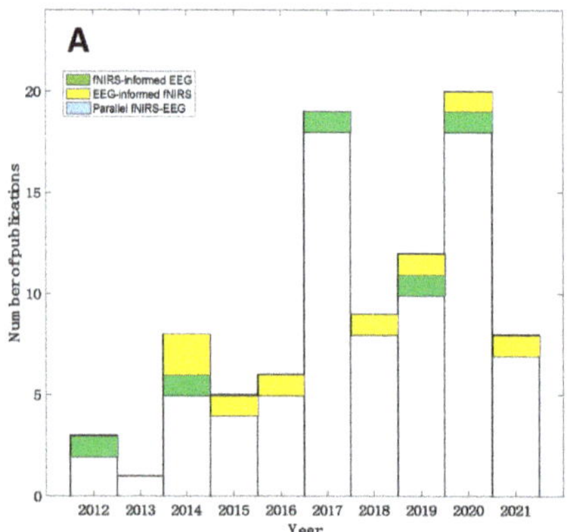

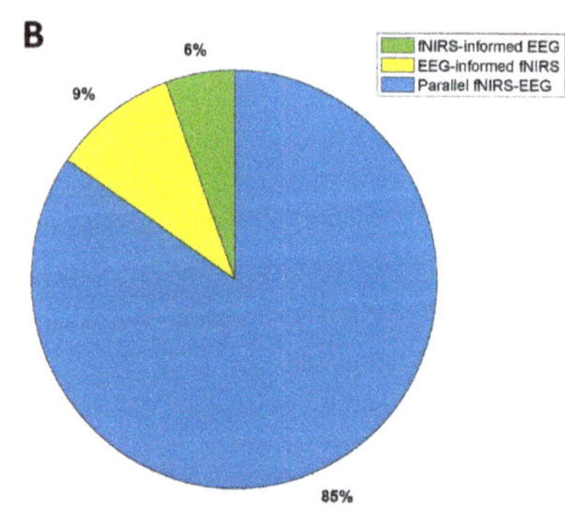

Figure 4. Literature summary of concurrent EEG–fNIRS studies: (**A**) Yearly publications from 2012 to 2021 and (**B**) distribution of each type of concurrent fNIRS-EEG studies.

3.1. Preprocessing of fNIRS and EEG Signal

Signal preprocessing is an essential step for any post-processing of integrated analysis of concurrent fNIRS–EEG data. Since the present review specifically focused on the integrated analysis of concurrent fNIRS–EEG data, here we only outline a general pipeline for the basic preprocessing of each modality.

3.1.1. Basic Preprocessing of fNIRS Signal

Basic preprocessing of fNIRS data is shown in Figure 5A. One particularly essential step in the preprocessing of fNIRS data is signal quality check and artifact correction. The quality of the fNIRS signal could be affected by several confounding noise sources, such as instrument noise (e.g., due to light source instability, electronic noise) [32], physiological interference (e.g., respiration, heartbeat) [33,34], or motion artifacts [35,36]. Instrument noise and physiological interference are mostly located within a constant frequency range. For instance, the instrument-degradation-induced noise is around 3~5 Hz, and respiration and heartbeat lie in 1~1.5 Hz and 0.2~0.5 Hz, respectively [5]. Thus, these noises can be easily removed by applying the band-pass filter/low pass filter. Motion artifact in the form of spikes or baseline shifts is a typical category of noise in raw fNIRS signal, especially in data collected from child populations or during experimental tasks that include motion (e.g., walking or speaking) [36,37]. Multiple algorithms have been developed to identify and correct motion artifacts in raw fNIRS signals, such as spline interpolation [38], wavelet-based methods [39,40], or principal component analysis [41]. We refer the readers to recently published articles for a more detailed overview of the preprocessing of fNIRS signal [42,43].

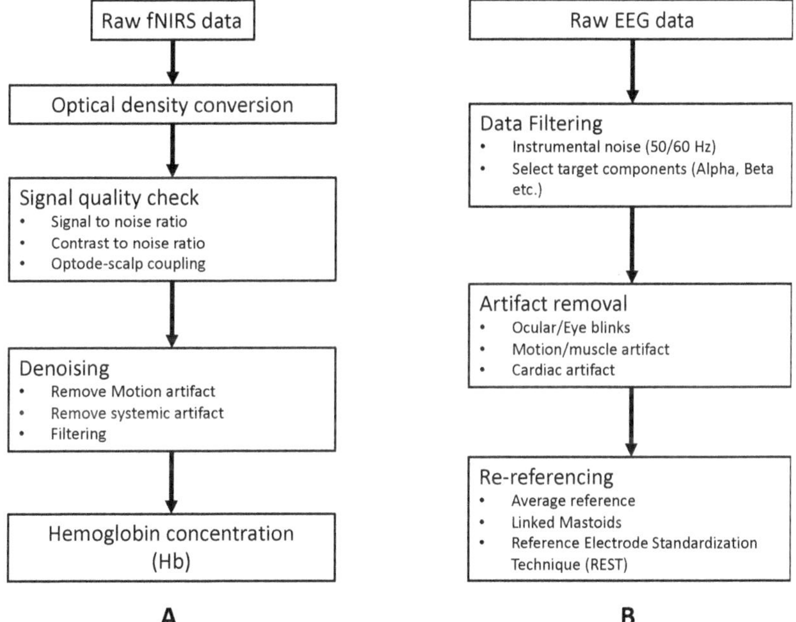

Figure 5. Basic preprocessing pipeline: (**A**) fNIRS raw data and (**B**) EEG raw data.

3.1.2. Basic Preprocessing of EEG Signal

We have outlined the basic preprocessing of EEG data in Figure 5B. Similar to fNIRS, EEG recordings are often contaminated by different artifacts that come from internal and external sources. Internal artifacts include physiological activities of the subject (e.g., ECG, muscle, and ocular artifacts) and movement [44,45]. External artifacts mainly include environmental/instrumental interference (50 Hz/60 Hz), electrode pop-up and cable movement. Elimination of internal artifacts relies on extra measurements (e.g., electrooculogram/electrocardiogram/accelerometer) or signal decomposition algorithms (e.g., ICA/PCA) [46,47]. External artifacts may be removed either by simple filters, signal decomposition algorithms (e.g., ICA), or artifactual segment rejection [48]. We refer the readers to [49,50] for a more detailed overview of the preprocessing of EEG signals.

3.2. EEG-Informed fNIRS Analyses

Neurovascular coupling demonstrates that regional neural activity is typically accompanied by the generation of electrical activity and the resulted metabolic variation, which is the fundamental principle of EEG and fNIRS measurements. Simultaneous fNIRS–EEG recording is therefore highly suited for neurovascular coupling investigation through various analysis approaches.

Among all the concurrent fNIRS–EEG studies, using EEG-derived characteristics to enhance fNIRS analyses, which is usually referred as EEG-informed fNIRS analyses, provides a particularly new and straightforward solution for investigating neurovascular coupling. Table 1 summarizes all studies that performed EEG-informed fNIRS analyses.

Table 1. Characteristics of studies that performed EEG-informed fNIRS analysis.

Authors	Tasks	Brain Regions	Features	Analysis Methods
Peng et al., 2014 [51]	Resting	fNIRS: Whole EEG: Whole	fNIRS: HbO/HbR/HbT concentration EEG: Amplitude	GLM
Pouliot et al., 2014 [52]	Resting	fNIRS: Whole EEG: Whole	fNIRS: HbO/HbR/HbT concentration EEG: Amplitude	GLM
Talukdar et al., 2015 [53]	Resting	fNIRS: Whole EEG: Whole	fNIRS: HbO concentration EEG: Power spectral envelopes	GLM
Peng et al., 2016 [54]	Simulation; Resting	fNIRS: Whole EEG: Whole	fNIRS: HbO/HbR/HbT concentration EEG: Amplitude	GLM
Khan et al., 2018 [55]	Motor	fNIRS: Left motor EEG: Left motor	fNIRS: HbO/HbR concentration EEG: Power spectrum	Vector-phase analysis
Zama et al., 2019 [56]	Motor	fNIRS: Motor EEG: Whole	fNIRS: HbO/HbR concentration EEG: ERD/ERS	GLM
Li et al., 2020 [57]	Motor	fNIRS: Motor EEG: Whole	fNIRS: HbO/HbR concentration EEG: Absolute Power (amplitude)	GLM
Sirpal et al., 2021 [58]	Resting	fNIRS: Whole EEG: Whole	fNIRS: HbO concentration EEG: Amplitude	Autoencoder

In typical fNIRS analyses (Figure 6), the fNIRS signal is commonly regressed via a general linear model (GLM) constructed by convolving the canonical hemodynamic response function (HRF) with a boxcar or impulse function representing the consistent temporal profile of the experimental paradigm to identify cortical regions activated by specific stimuli [59]. Briefly, for measured fNIRS signal Y in a channel, the GLM model is given by

$$Y = X\beta + \varepsilon \tag{1}$$

where X is the design matrix, β is the regression coefficients to be estimated, and ε is the error term. In the case of a block design experiment, X is commonly given by a convolution matrix of a chosen hemodynamic response function (HRF) and boxcar functions describing the latency and duration of the stimulus. Note that the HRF may use various type of shapes, such as canonical HRF, gamma-HRF, or Gaussian-HRF [35]. Columns of X are the regressors that represent conditions or tasks in the experiment, and additional nuisance terms or auxiliary measurements that usually account for the systemic physiology or motion artifacts.

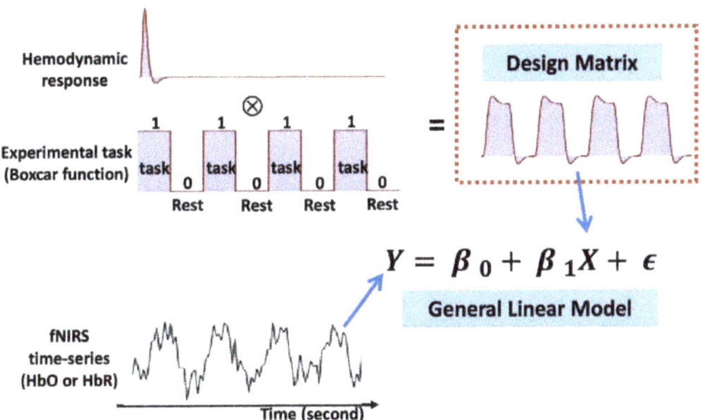

Figure 6. Basic principle of general linear model (GLM) in fNIRS analysis.

The estimated regression coefficient β and the error ϵ can be tested via a t-test to identify the channels that represent a significant contrast between different tasks. The t-test is calculated by

$$t = \frac{c^T \times \beta}{\sqrt{c^T cov(\beta) c}},\qquad(2)$$

where $cov(\beta)$ is the covariance matrix of β and c is the contrast vector, which determines the contrast between specific conditions.

The main limitation in common standalone fNIRS analysis is that neuronal response to repeated trials or stimuli is time-varying across the experiment in a realistic setting, which may be inconsistent with the boxcar function typically used in the construction of a GLM analysis design matrix. With this in mind, the core idea of EEG-informed fNIRS analysis is to replace or adjust the boxcar function in the fNIRS GLM analysis with temporal- or frequency-specific regressors of interest derived from EEG signals [57]. Based on the linear hypothesis of neurovascular coupling, the characteristics of the neural activity extracted in EEG may offer better estimation of the fNIRS response after convoluting the HRF, thus increasing the efficiency of identifying the related active region induced by experimental tasks.

Figure 7 summarizes a generalized analysis framework of EEG-informed fNIRS analysis. The selection of time-varying EEG features plays a crucial role in the construction of a fNIRS GLM analysis design matrix. Among all EEG-informed fNIRS analysis studies, amplitude information derived from EEG signals has been used as effective regressors of interest for improving the estimation of the active fNIRS response associated with different stimuli [56,57]. Li et al. collected concurrent EEG and fNIRS data from healthy participants during a repeated motor execution task and extracted the peak value and latency of the EEG signal within each trial to construct a series of frequency-specific design matrices [57]. Their results showed that amplitudes of frequency-specific EEG components, especially the alpha and beta band, could better capture the time-varying neural activity at single trial level and thus enhance the performance of fNIRS GLM analysis when compared with the classic boxcar function-based fNIRS method [57]. The potential value of EEG-informed fNIRS analysis in clinical applications was also explored, in particularly on the topic of epileptic activity, given the suitability of this technique for the localization of brain sites associated with epileptic discharges. A series of representative studies was performed by Pouliot and his colleagues, where the onsets and amplitudes of epileptic spikes were identified by EEG temporal traces and convolved with the HRF for fNIRS GLM estimation [51,52,54,60]. These studies demonstrated that an EEG-informed fNIRS approach revealed higher sensitivity and specificity than the classic GLM method in the detection of epileptic events such as seizures or interictal epileptiform discharges (IEDs). Their work provides evidence that EEG-informed fNIRS analysis could be a sensitive technique for monitoring epileptic activity.

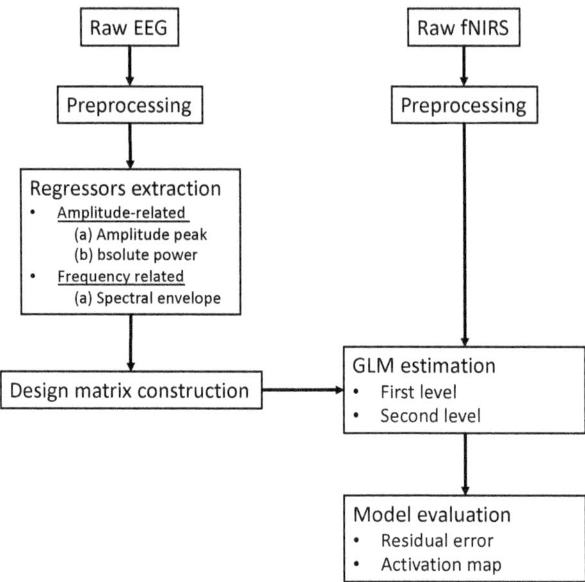

Figure 7. The conventional schematic of EEG-informed fNIRS GLM analysis framework.

In addition to the amplitude-specific information, frequency-related features were derived from EEG signals and used as regressors of interest for fNIRS GLM analysis. Talukdar et al. used gamma transfer functions to map EEG spectral envelopes that reflect time-varying power variations in neural rhythms to hemodynamics measured during median nerve stimulation [53]. The approach was evaluated through simulated EEG–fNIRS data and experimental EEG–NIRS data measured from three human subjects. Results indicated that fNIRS hemodynamics can be predicted by EEG spectral envelopes convoluted with multiple sets of gamma transfer functions, providing a new perspective for the modeling of neurovascular coupling.

3.3. FNIRS-Informed EEG Analyses

Studies using fNIRS to enhance the processing of EEG signals typically rely on the relatively robust spatial information of fNIRS compared to EEG. Within this context, fNIRS-informed EEG analyses, as summarized in Table 2, include two main levels of applications: fNIRS-informed EEG source localization and fNIRS-informed EEG channel selection. The former applies task-evoked information of fNIRS to enhance the mathematical estimation of active EEG source activity related to specific tasks [27], while the latter used fNIRS as a reliable reference for choosing the most representative task-related EEG channels for analysis [61].

3.3.1. FNIRS-Informed EEG Source Imaging Analysis

Due to its high temporal resolution and portability, EEG is by far the most widely used neuroimaging technique to measure rapid neuronal electrical activity. However, one limitation of scalp EEG is the volume conduction problem; a single electrode on the scalp picks up activity from multitude sources (cortical activity, subcortical activity, external noise, etc.), which results in difficulty accurately localizing the source activity [62]. Therefore, EEG source imaging (ESI) has been developed to overcome the limitation of scalp EEG in characterizing the spatial brain activity. Typically, ESI relies on the surface EEG signals and the anatomical structure and physiological properties of the brain to estimate sources within the brain. This allows for more accurate localization of the cortical regions contributing to EEG signals measured at the scalp. A common challenge for ESI is the ill-posed "inverse problem"; the number of sources that give rise to EEG signals vastly outnumbers the available measurements, making it impossible to localize the measured scalp EEG activity to the actual current-generating source within the brain with absolute certainty [63]. Given the good spatial resolution of fNIRS, the majority of fNIRS-informed EEG studies have focused on using fNIRS-based spatial priors to enhance the estimation of EEG source activity.

In summary of these studies, a traditional pipeline of fNIRS-informed EEG source imaging is shown in Figure 8. Briefly, this pipeline begins with the forward model of the ESI (Figure 8A):

$$Y = GJ + \varepsilon, \tag{3}$$

where $Y \in \mathbb{R}^{m \times d}$ is the scalp EEG signal consisting of m channels and d measurement samples, $J \in \mathbb{R}^{s \times d}$ is the unknown source activity of s dipole sources in the source space, $G \in \mathbb{R}^{m \times s}$ is the lead field matrix which describes the relationship between the source activity and the EEG electrodes, and ε represents the noise component in the sensor space. Using the EEG signals measured at the scalp, we can attempt to invert the forward model to determine which parts of the brain are active from their associated scalp potentials, which is the so-called inverse problem. A common solution of the inverse problem using classical minimum-norm estimate (MNE) is given as:

$$\hat{J} = RG^{T}\left(GRG^{T} + \lambda C\right)^{-1}Y, \tag{4}$$

where $\hat{J}$ is the estimated source activity, R is the source covariance matrix representing the prior knowledge about the distribution of source J, C is the noise covariance matrices, and λ is the regularization parameters representing the trade-off between model accuracy and complexity, which is traditionally determined using the L-curve method [64]. The source covariance matrix R and noise covariance matrix C are usually set to identity matrices when no prior information about the source space is available. With this in mind, spatial prior information provided by fNIRS, usually represented by t values of significant channels after GLM analysis, can be applied directly on the source covariance matrix R, changing the weight of each source according to whether or not it is within an fNIRS-active region. This results in improvement of EEG source activity estimation (Figure 8B). Note that the inverse problem can be solved by multiple approaches, such as MNE, weighted MNE, or probabilistic Bayesian methods, resulting in different forms of source covariance matrix R [65,66].

The analysis pipeline shown in Figure 8B has been adapted in all existing fNIRS-informed EEG analysis studies to investigate brain dynamics associated with typical brain function as well as brain disorders. The first fNIRS-informed ESI study was carried out by Aihara et al., in which the authors incorporated the fNIRS-based prior information in the current source estimation using a Variational Bayesian Multimodal EncephaloGraphy (VBMEG) method [67]. Using a simulation study and a finger tapping motor task, this study demonstrated that fNIRS-informed ESI can achieve results similar to fMRI-information ESI. Following a similar idea, Morioka et al. applied fNIRS-informed ESI to decode subjects' mental states in a spatial attention task and found that the fNIRS–EEG framework exhibited significant performance improvement over decoding methods based on EEG sensor signals alone [68]. Recently, Li et al. employed the fNIRS-informed ESI technique to explore the atypical brain dynamics associated with Alzheimer's disease and stroke, from which brain network alterations induced by these brain disorders were characterized in a high spatiotemporal manner [27,69].

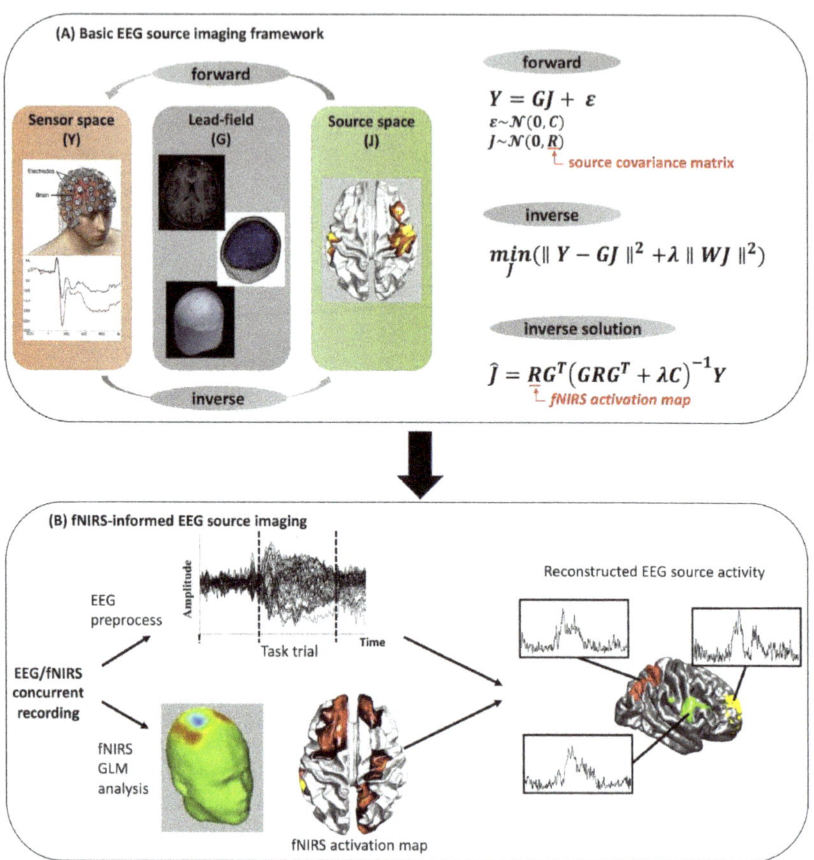

Figure 8. Basic concepts of EEG source imaging and traditional pipeline of fNIRS-informed EEG source imaging analysis (adapted with permission from Ref. [27]. 2019, Li et al.

3.3.2. FNIRS-Informed EEG Channel Selection for BCI Studies

FNIRS-informed EEG source imaging represents the deep fusion of fNIRS and EEG signals. In addition, one study published by Li et al. demonstrated that fNIRS-based spatial prior information can also be used to optimize processing of scalp EEG signal in BCI studies [61]. Briefly, a desirable BCI system should be portable, minimally invasive, and feature high classification accuracy and efficiency. However, the main challenge of hybrid EEG–fNIRS BCI systems is how to reduce the complexity of the system while achieving a satisfactory performance. To tackle this challenge, Li et al. proposed a fNIRS-based channel selection method to greatly reduce the number of fNIRS and EEG channels needed for BCI systems. In this fNIRS-based channel selection method, two fNIRS channels with strongest task-evoked response, as assessed by GLM analysis, were determined. Then only two EEG channels that were close to the selected fNIRS channels were selected for the performance assessment of the hybrid fNIRS–EEG BCI system. Results demonstrated that this approach could drastically minimized the burden (e.g., weight of cables, preparation time) on the user while achieving a good performance compared to BCI systems including large numbers of channels [61].

Overall, although limited studies focused on this topic were available or review, fNIRS-informed EEG source imaging analysis has potential for achieving a deep fusion of these two portable techniques. This multimodal approach holds promise for improving our understanding of the spatiotemporal dynamics of typical and atypical brain functions in various scenarios including naturalistic interaction and clinical settings.

Table 2. Characteristics of studies performed fNIRS-informed EEG analysis.

Authors	Tasks	Brain Regions	Features	Analysis Methods
Aihara et al., 2012 [67]	Motor (Simulation; Experiment)	fNIRS: Motor EEG: Whole	fNIRS: HbO peak EEG: Source current amplitude	EEG source imaging
Morioka et al., 2014 [68]	Mental	fNIRS: Parietal, occipital EEG: Whole	fNIRS: HbO t-statistic EEG: Source current amplitude	EEG source imaging
Li et al., 2017 [61]	Motor	fNIRS: Motor EEG: Whole	fNIRS: HBO/HbR concentrations and slope EEG: Wavelet transform coefficients	Binary classification
Li et al., 2019 [27]	Working memory	fNIRS: Frontal, central EEG: Whole	fNIRS: HbO t-statistic EEG: Functional connectivity	EEG source imaging, Brain network analysis
Li et al., 2020 [69]	Motor	fNIRS: Frontal, parietal EEG: Whole	fNIRS: HbO t-statistic EEG: Functional connectivity	EEG source imaging, Brain network analysis

3.4. Parallel Analysis of EEG-fNIRS

Sections 3.3 and 3.4 describe directional integration analyses of EEG and fNIRS. However, the majority of concurrent EEG-fNIRS studies available for review focused on parallel analysis/integration of the two complementary techniques (Figure 4). Such parallel analyses of concurrent fNIRS and EEG data usually seek to investigate the interaction between fNIRS and EEG signals through feature-based fusion analyses or correlational analyses without any directional interference from the two modalities.

3.4.1. Feature Fusion Based on fNIRS–EEG Signals for Classification

Hybrid fNIRS-EEG classification-based studies account for a significant portion of feature-based fusion analyses of concurrent fNIRS-EEG data. We roughly summarize these studies into two categories based on their study aims: (1) brain–computer interface (BCI) studies and, (2) characterization of typical and atypical brain functions.

The development of a BCI system allows users to control computers or external devices based directly on the modulation of brain activity. Active investigations of the benefits of hybrid EEG-fNIRS BCIs have been conducted and validated on healthy populations in a number of BCI studies [28,29,61,70]. Specifically, by fusing the features derived from two modalities, hybrid fNIRS-EEG studies have shown enhanced classification and decoding accuracy over a single modality in various tasks, such as motor imagery and execution [61,71].

On the other hand, the complementary properties of fNIRS and EEG have led to extensive investigations of the spatiotemporal hemodynamic and electrical patterns of brain activity associated with a variety of functions, such as mental workload [72–75], affective state [76], and intellectual function [77]. Similar analysis pipelines have also been adopted to identified atypical brain patterns associated with different brain disorders, from which multimodal features can be used to differentiate patients with Alzheimer's Disease [78] and Parkinson's Disease [79] from healthy controls.

Despite the different aims of studies within the above two categories, most studies tend to follow similar steps when processing concurrent fNIRS and EEG data, primarily consisting of feature extraction, feature fusion, and classification. Among the reviewed literature, widely used fNIRS features are commonly derived from the concentration changes of HbO and HbR, including the mean, slope, skewness, kurtosis, peak value, variance, and median of HbO/HbR [61,80–82]. Typical EEG features used in concurrent fNIRS-EEG analyses largely depend on the experimental tasks. In the case of a motor task, the power spectrum density and common spatial patterns are widely used [29,83–86], mainly due to the event-related desynchronization/event-related synchronization (ERD/ERS) observed in motor-evoked electrical potential [87]. Studies involving cognitive tasks usually adopt features related to band power of signals [72,82,88–91]. Additionally, the logarithmic band power features [84], time-frequency features [61,71], and amplitude-related properties [92–94] are often utilized in several studies involving motor and mental tasks. Definitions and calculations of these features are summarized and shown in Table 3. In terms of classification, most existing studies adopt traditional machine learning techniques such as decision tree [93,94], linear discriminant analysis (LDA) [28,29,81,82,86,89–91,95,96], support vector machine (SVM) [61,70,72,85], and k-nearest neighbors (KNN) [92–94]. Recent studies have demonstrated increasing interest in innovative deep learning techniques such as the convolutional neural network (CNN) [97] and recurrent neural networks (RNN) [98]. We refer the readers to [99,100] for a more detailed introduction of the state-of-the-art classification techniques.

Table 3. Definition and calculation of EEG and fNIRS features.

Features	Definitions
Mean (μ)	$\mu = \frac{1}{N} \sum\limits_{t=t_1}^{t_2} x(t)$
Slope (Sp)	$Sp = \frac{x(t_2)-x(t_1)}{t_2-t_1}$
Standard deviation (Sd)	$Sd = \sqrt{\frac{\sum(x(t)-\mu)^2}{N}}$
Skewness ($Skew$)	$Skew = \frac{1}{N} \frac{\sum_{t=t_1}^{t_2}(x(t)-\mu)^3}{Sd^3}$
Kurtosis ($Kurt$)	$Kurt = \frac{1}{N} \frac{\sum_{t=t_1}^{t_2}(x(t)-\mu)^4}{Sd^4}$
Median (Med)	$Med = \begin{cases} x\left(\frac{n}{2}\right) & \text{if } n \text{ is even} \\ \frac{x\left(\frac{n-1}{2}\right)+x\left(\frac{n+1}{2}\right)}{2} & \text{if } n \text{ is odd} \end{cases}$
Power spectral density (PSD)	$PSD_f^t = \frac{1}{N} \sum\limits_{t=1}^{N} \left| x(t)e^{-2\pi f t} \right|^2$
Logarithmic band power (PLB)	$PLB_f = log\left(PSD_f\right)$
Common spatial pattern (CSP)	$X_i = [C_i C_M] \begin{bmatrix} S_i \\ S_M \end{bmatrix}, (i=1,2)$
Phase locking value (PLV)	$PLV = \left| N^{-1} \sum\limits_{t=1}^{N} e^{i(\varnothing x(t)-\varnothing y(t))} \right|$
Pearson correlation coefficient (r)	$r = \frac{\sum(x(t)-\bar{x})\,(y(t)-\bar{y})}{\sqrt{\sum(x(t)-\bar{x})^2 \sum(y(t)-\bar{y})^2}}$

$x(t)$ is the input brain signals (i.e., EEG and fNIRS). N is the number of observations of the samples. $\varnothing x(t)$ and $\varnothing y(t)$ are instantaneous phase values at time point t. f refers to the f-th frequency band. X_i represent the measured signals of i-th tasks. S_i is the source signal related to the i-th task. S_M is the common source signal of both signals. C_i and C_M are the weight matrix of common spatial pattern. $x(t)$ and $y(t)$ present the signals from different channel. $\bar{x}$ and $\bar{y}$ refer to the mean value of the signals of $x(t)$ and $y(t)$, respectively.

3.4.2. Correlational Analysis of Concurrent fNIRS–EEG Data

The well-established phenomenon of neurovascular coupling (NVC) supports the premise that regional neural activity is accompanied by electrical activity generation and concurrent metabolic variation. Therefore, correlational analyses between concurrent fNIRS–EEG recordings have been extensively explored to investigate the spatiotemporal association between hemodynamic and electrical patterns of various brain functions. Among the eighteen articles reviewed here (Table 4), correlational analyses of concurrent fNIRS–EEG have mainly focused on correlation and coherence analyses. Pearson correlation, partial correlation, and simple linear regression are commonly used measures for assessing the relationship between the event-related potential pattern in EEG and hemodynamic changes in fNIRS [101–110]. Several studies assessed the relationship between EEG and fNIRS signal through cross-correlation analysis and canonical correlation analysis (CCA) [111–114]. Compared to the Pearson correlation method, cross-correlation can capture the delayed response of the hemodynamic compensation phenomenon after neural firing, while the CCA is a statistical method to identify a linear relationship between the two modality data sets by determining the inter-subject co-variances. Frequency and phase coupling were adopted in two studies to evaluate the interaction between electrical activation and hemodynamic response, in which spectral coherence and wavelet coherence were employed as metrics to assess the neurovascular coupling [115,116]. GLM-based analysis was also utilized to model the association of fNIRS and EEG in a recent study. Chaiarelli et al. proposed a novel general linear model-based algorithm to estimate the interaction of fNIRS and EEG signal in persons with Alzheimer's disease [117]. In the GLM, key components of the down-sampled EEG power spectrum (theta, alpha, and beta) were used as the independent variables. The fNIRS signal was treated as the dependent variable. Then the estimated β-weight was used to assess how well the frequency-specific neuronal electric activity correlated with the corresponding hemodynamic response. Similarly, Perpetuini et al. employed an entropy based GLM method to assess neurovascular coupling alternation for an Alzheimer's disease group relative to a healthy control group [118]. Due to the significant variation in the temporal scale of two signals, the EEG signal was first convolved with the canonical hemodynamic response and then downsampled. Compared with single EEG/fNIRS-based features, neurovascular coupling-based features achieved the highest classification accuracy for AD detection.

Table 4. Studies using parallel EEG–fNIRS analysis for neurovascular coupling investigation.

Authors	Task	Brain Regions	Features	Correlation Method
Chen et al., 2015 [101]	Visual and auditory	fNIRS: Temporal, occipital EEG: Whole	fNIRS: HbO/HbR concentrations EEG: ERP	Pearson correlation
Chen et al., 2020 [102]	Resting	Whole	fNIRS: HbO/HbR global amplitude EEG: Power Spectrum	Partial correlation
Balconi et al., 2016 [103]	Visual and auditory	fNIRS: Frontal EEG: Whole	fNIRS: HbO concentrations EEG: ERP	Pearson correlation
Zich et al., 2017 [104]	Motor execution	Central	fNIRS: HbO/HbR concentrations EEG: ERD	Pearson correlation
Borgheai et al., 2019 [105]	Mental arithmetic	fNIRS: Frontal EEG: Whole	fNIRS: HbO/HbR concentrations EEG: Power spectrum and ERP	Pearson correlation
Gentile et al., 2020 [106]	Finger tapping	fNIRS: Motor EEG: Whole	fNIRS: HbO/HbR concentrations EEG: ERP	Linear regression
Zhang et al., 2020 [107]	Resting	Whole	fNIRS: dynamic functional connectivity EEG: Microstate (amplitude)	Pearson correlation
Lin et al., 2020 [108]	Mental	Occipital and parietal	fNIRS: HbO concentration EEG: Power spectrum and ERD	Pearson correlation
Kaga et al., 2020 [109]	Working memory	fNIRS: Frontal EEG: Pz, Cz, Pz,	fNIRS: HbO concentration EEG: ERP	Pearson correlation
Suzuki et al., 2018 [110]	Working memory	fNIRS: Frontal EEG: Fz, O1, O2,	fNIRS: HbO concentration EEG: Power spectrum	Pearson correlation
Keles et al., 2016 [111]	Resting	Whole	fNIRS: HbO/HbR concentrations EEG: Power spectrum	Cross-correlation
Pinti et al., 2021 [112]	Visual stimulation	Occipital	fNIRS: HbO/HbR concentrations EEG: Power spectrum	Cross-correlation
Nair et al., 2021 [113]	Anesthesia	Frontal	fNIRS: HbO/HbR amplitude EEG: Amplitude	Cross-correlation and phase difference

Table 4. *Cont.*

Authors	Task	Brain Regions	Features	Correlation Method
Al-Shargie et al., 2017 [114]	Mental arithmetic	Frontal	fNIRS: HbO concentration EEG: Average power (amplitude)	Canonical correlation analysis
Govindan et al., 2016 [115]	Resting	Frontotemporal	fNIRS: difference between HbO and HbR EEG: Amplitude	Coherence and Phase Spectra
Chalak et al., 2017 [116]	Resting	Parietal	fNIRS: Cerebral tissue oxygen saturation EEG: Amplitude	Wavelet coherence
Chiarelli et al., 2021 [117]	Resting	Whole	fNIRS: HbO/HbR concentrations EEG: Power envelops	GLM-Standardized β-weight
Prepetuini et al., 2020 [118]	Working memory	fNIRS: Frontal EEG: Whole	fNIRS: HbO/HbR sample entropy EEG: Sample entropy	GLM-Standardized β-weight

4. Integrated Analysis of Concurrent fNIRS-EEG: Current Limitations and Future Directions

Both fNIRS and EEG are portable, non-invasive and cost-effective brain imaging techniques that enable researchers to study brain function in conditions not suited for other neuroimaging modalities such as fMRI and MEG. Accordingly, acquisition and analysis of concurrent, integrated fNIRS–EEG data can potentially reveal more comprehensive information associated with brain activity. The present review highlights what data processing and analysis approaches can be adopted to study brain functioning in healthy cohorts as well as those with brain disorders, thus serving as a foundation for future work. However, it should be acknowledged that further development of integrated analyses of the two modalities is required to fully benefit from the added value of each modality.

Neurovascular coupling in the brain is highly dynamic in nature, for both resting state and task-engaging states. While various fusion approaches of fNIRS and EEG signals allow for the imaging and investigation of brain activity with richer information, the majority of such integrated analyses still rely on a summary of signals extracted from fNIRS and EEG time series data. Neural activity is time-varying, thus requiring a more dynamic analytic approach to improve accuracy in modeling actual brain function. Therefore, it is important to explore the dynamic interaction of fNIRS and EEG signals with a more fine-grained temporal resolution. This is a challenge for fNIRS signals, which usually suffer reduced temporal resolution relative to EEG. Recently, effort has been made to tackle this challenge by growing interest in the temporal fluctuations of fNIRS-based functional connectivity across the brain, the so-called dynamic functional connectivity (dFC). Several studies have shown that resting-state and task-evoked hemodynamic responses can be characterized using dFC analysis to reflect a more dynamic and modular nature of neurovascular coupling during normal cognitive processing and atypical brain activity associated with Alzheimer's disease [119,120]. It is expected that fusion of the dynamic properties of fNIRS and EEG may open new lines of concurrent fNIRS–EEG analyses.

Despite the numerous approaches for integrated analysis of concurrent fNIRS–EEG, most studies have utilized feature-based fusion of these two modalities, such as hybrid BCI systems or correlation analyses between fNIRS-based (e.g., mean HbO) and EEG-based features (e.g., power spectrum). Such analyses only allow for a rough characterization of neurovascular coupling underlying brain activity. Questions remain as to how the findings obtained from the integrated analyses of fNIRS–EEG reflects the interaction between neuronal electric activity and the resulting hemodynamic response. Therefore, it is expected that more directionally integrated analyses of fNIRS and EEG data, such as the fNIRS-informed EEG analyses or the EEG-informed fNIRS analyses, can be explored in future work.

Combining fNIRS and EEG serves to bridge brain imaging techniques across laboratory settings to practical applications due to their high mobility, non-invasiveness, and low cost compared to MRI-based techniques. However, few studies, especially those focusing on hybrid fNIRS–EEG BCI systems, have validated the feasibility of using such multimodal approaches to address the needs of multiple practical scenarios, such as hybrid real-time BCI systems, bedside monitoring, or neuromodulation based on the so-called brain controllability analysis, to treat different neurological and psychiatric diseases [121–123]. Therefore, a prioritized goal of future research may focus on enhancing the ecological validity of experimental designs and analysis pipelines/algorithms that can be adopted in online or low time-delayed settings. In fact, as motivated by real-time BCI applications, progress has been made to increase the temporal response of fNIRS–BCI systems through single-trial analysis [124], early signal detection [61], and adaptive filtering [125]. We anticipate future solutions for real-time fNIRS signal processing may facilitate the development of real-time hybrid BCI systems that enable human–computer interaction with high spatial and temporal performance. Another typical experimental protocol of fNIRS is hyperscanning, where brain activities are recorded from two or more participants simultaneously, permitting a direct investigation of how multi-brains communicate to each other during social interaction [120,126]. Following this, we expect that the development of wearable fNIRS and EEG devices will likely drive the typical fNIRS-based hyperscanning studies toward multimodal fNIRS–EEG system-based hyperscanning research. This innovation will enable us to examine human interaction in a high spatiotemporal resolution perspective, thereby expanding our understanding of the neural mechanism underlying social interaction.

Apart from the perspective on methodological integration of fNIRS and EEG, we want to highlight challenges in instrument development that might affect study design and signal processing of concurrent fNIRS–EEG studies. In particular, conventional concurrent fNIRS–EEG studies usually connect separate fNIRS and EEG systems for data recording, which reduces the mobility of both systems and constrains the applications of concurrent fNIRS and EEG. Recent advances have been made toward fiberless and wearable integrated fNIRS–EEG systems that allows for broader research scenarios such as social interaction and outdoor activity [127,128]. However, further improvement of fNIRS and EEG instruments is necessary when applying these systems in clinical cohorts with psychological or psychiatric disorders. For example, patients with psychiatric disorders, such as ASD and ADHD, often display motor restlessness, anxiety, or hyperarousal symptoms that require specific considerations during development of integrated fNIRS–EEG instrumentation. Key factors to be considered may include (1) user-friendly materials for comfort contact between electrodes/optodes, (2) lightweight/highly integrated design for enhanced measurement experience, and (3) advanced signal processing algorithms for robust long-time real-world study. In addition, simultaneous multimodal data recording, including brain, physiological, and behavioral information, is important to the comprehensive understanding of disease-linked/function-specific brain activity. Physiological or auxiliary signals (e.g., blood pressure, respiration, and head movement) have been proven to greatly improve the filtering of physiological interference and motion artifacts during fNIRS signal processing [129–131]. In this context, one impactful direction of fNIRS–EEG instrument development should focus on the development of multimodal systems that are deeply integrated with these and other emerging modalities, such as eye tracking devices, physiology modules (e.g., heart rate, skin conductivity), and accelerometers as well as VR devices. From a clinical perspective, such multimodal systems could offer multi-dimensional brain–physiology–behavior biomarkers specifically linked to brain disorders at individual level. Together with powerful statistical/machine learning, we expect that future studies in the field will propose advanced algorithms to fuse such multimodal information for accurate monitoring of brain activity and facilitating personalized treatment protocols to obtain enhanced efficiency for each individual patient.

Author Contributions: Conceptualization, R.L.; literature search, R.L. and D.Y.; writing—original draft preparation, R.L., D.Y., F.F., K.-S.H., A.L.R. and Y.Z.; writing—review and editing, R.L. and D.Y.; supervision, Y.Z.; proofreading, R.L. and Y.Z. All authors have read and agreed to the published version of the manuscript.

Funding: This research received no external funding.

Institutional Review Board Statement: Not applicable.

Informed Consent Statement: Not applicable.

Data Availability Statement: Not applicable.

Conflicts of Interest: The authors declare no conflict of interest.

References

1. Herculano-Houzel, S. The human brain in numbers: A linearly scaled-up primate brain. *Front. Hum. Neurosci.* **2009**, *3*, 31. [CrossRef] [PubMed]
2. Pakkenberg, B.; Pelvig, D.; Marner, L.; Bundgaard, M.J.; Gundersen, H.J.; Nyengaard, J.R.; Regeur, L. Aging and the human neocortex. *Exp. Gerontol.* **2003**, *38*, 95–99. [CrossRef]
3. Jobsis, F.F. Noninvasive, infrared monitoring of cerebral and myocardial oxygen sufficiency and circulatory parameters. *Science* **1977**, *198*, 1264–1267. [CrossRef] [PubMed]
4. Ferrari, M.; Quaresima, V. A brief review on the history of human functional near-infrared spectroscopy (fNIRS) development and fields of application. *Neuroimage* **2012**, *63*, 921–935. [CrossRef]
5. Scholkmann, F.; Kleiser, S.; Metz, A.J.; Zimmermann, R.; Pavia, J.M.; Wolf, U.; Wolf, M. A review on continuous wave functional near-infrared spectroscopy and imaging instrumentation and methodology. *Neuroimage* **2014**, *85*, 6–27. [CrossRef] [PubMed]
6. Berger, H. Über das elektroenkephalogramm des menschen. *Arch. Für Psychiatr. Und Nervenkrankh.* **1929**, *87*, 527–570. [CrossRef]
7. Buzsaki, G.; Anastassiou, C.A.; Koch, C. The origin of extracellular fields and currents—EEG, ECoG, LFP and spikes. *Nat. Rev. Neurosci.* **2012**, *13*, 407–420. [CrossRef]
8. Schomer, D.L.; Da Silva, F.L. *Niedermeyer's Electroencephalography: Basic Principles, Clinical Applications, and Related Fields*; Lippincott Williams & Wilkins: Philadelphia, PA, USA, 2012.
9. Pizzagalli, D.A. Electroencephalography and high-density electrophysiological source localization. *Handb. Psychophysiol.* **2007**, *3*, 56–84.
10. Wu, J.; Quinlan, E.B.; Dodakian, L.; McKenzie, A.; Kathuria, N.; Zhou, R.J.; Augsburger, R.; See, J.; Le, V.H.; Srinivasan, R.; et al. Connectivity measures are robust biomarkers of cortical function and plasticity after stroke. *Brain* **2015**, *138*, 2359–2369. [CrossRef]
11. Zhang, Q.; Hu, Y.; Potter, T.; Li, R.; Quach, M.; Zhang, Y. Establishing functional brain networks using a nonlinear partial directed coherence method to predict epileptic seizures. *J. Neurosci. Methods* **2020**, *329*, 108447. [CrossRef]
12. Gao, Y.; Gao, B.; Chen, Q.; Liu, J.; Zhang, Y. Deep convolutional neural network-based epileptic electroencephalogram (EEG) signal classification. *Front. Neurol.* **2020**, *11*, 375. [CrossRef] [PubMed]
13. Gao, Y.; Wang, X.; Potter, T.; Zhang, J.; Zhang, Y. Single-trial EEG Emotion recognition using granger causality/transfer entropy analysis. *J. Neurosci. Methods* **2020**, *346*, 108904. [CrossRef]
14. Ma, Y.; Chen, B.; Li, R.; Wang, C.; Wang, J.; She, Q.; Luo, Z.; Zhang, Y. Driving fatigue detection from EEG using a modified PCANet method. *Comput. Intell. Neurosci.* **2019**, *2019*, 4721863. [CrossRef] [PubMed]
15. Ren, Z.; Li, R.; Chen, B.; Zhang, H.; Ma, Y.; Wang, C.; Lin, Y.; Zhang, Y. EEG-based driving fatigue detection using a two-level learning hierarchy radial basis function. *Front. Neurorobotics* **2021**, *15*, 618408. [CrossRef]
16. Li, R.; Rui, G.; Chen, W.; Li, S.; Schulz, P.E.; Zhang, Y. Early detection of Alzheimer's disease using non-invasive near-infrared spectroscopy. *Front. Aging Neurosci.* **2018**, *10*, 366. [CrossRef] [PubMed]
17. Nguyen, T.; Zhou, T.; Potter, T.; Zou, L.; Zhang, Y. The cortical network of emotion regulation: Insights from advanced EEG-fMRI integration analysis. *IEEE Trans. Med. Imaging* **2019**, *38*, 2423–2433. [CrossRef]
18. Fang, F.; Potter, T.; Nguyen, T.; Zhang, Y. Dynamic reorganization of the cortical functional brain network in affective processing and cognitive reappraisal. *Int. J. Neural Syst.* **2020**, *30*, 2050051. [CrossRef]
19. Li, R.; Rui, G.; Zhao, C.; Wang, C.; Fang, F.; Zhang, Y. Functional network alterations in patients with amnestic mild cognitive impairment characterized using functional near-infrared spectroscopy. *IEEE Trans. Neural Syst. Rehabil. Eng.* **2019**, *28*, 123–132. [CrossRef]
20. Cui, X.; Bray, S.; Bryant, D.M.; Glover, G.H.; Reiss, A.L. A quantitative comparison of NIRS and fMRI across multiple cognitive tasks. *Neuroimage* **2011**, *54*, 2808–2821. [CrossRef]
21. Pfurtscheller, G.; Neuper, C. Motor imagery and direct brain-computer communication. *Proc. IEEE* **2001**, *89*, 1123–1134. [CrossRef]
22. He, B.; Yang, L.; Wilke, C.; Yuan, H. Electrophysiological imaging of brain activity and connectivity-challenges and opportunities. *IEEE Trans. BioMed. Eng.* **2011**, *58*, 1918–1931. [CrossRef] [PubMed]
23. Nicolas-Alonso, L.F.; Gomez-Gil, J. Brain computer interfaces, a review. *Sensors* **2012**, *12*, 1211–1279. [CrossRef] [PubMed]
24. Waldert, S.; Tushaus, L.; Kaller, C.P.; Aertsen, A.; Mehring, C. fNIRS exhibits weak tuning to hand movement direction. *PLoS ONE* **2012**, *7*, e49266. [CrossRef] [PubMed]
25. Girouard, H.; Iadecola, C. Neurovascular coupling in the normal brain and in hypertension, stroke, and Alzheimer disease. *J. Appl. Physiol.* **2006**, *100*, 328–335. [CrossRef] [PubMed]
26. D'Esposito, M.; Deouell, L.Y.; Gazzaley, A. Alterations in the BOLD fMRI signal with ageing and disease: A challenge for neuroimaging. *Nat. Rev. Neurosci.* **2003**, *4*, 863–872. [CrossRef] [PubMed]

27. Li, R.; Nguyen, T.; Potter, T.; Zhang, Y. Dynamic cortical connectivity alterations associated with Alzheimer's disease: An EEG and fNIRS integration study. *Neuroimage Clin.* **2019**, *21*, 101622. [CrossRef]
28. Fazli, S.; Mehnert, J.; Steinbrink, J.; Curio, G.; Villringer, A.; Muller, K.R.; Blankertz, B. Enhanced performance by a hybrid NIRS-EEG brain computer interface. *Neuroimage* **2012**, *59*, 519–529. [CrossRef]
29. Buccino, A.P.; Keles, H.O.; Omurtag, A. Hybrid EEG-fNIRS Asynchronous Brain-Computer Interface for Multiple Motor Tasks. *PLoS ONE* **2016**, *11*, e0146610. [CrossRef]
30. Chiarelli, A.M.; Zappasodi, F.; Di Pompeo, F.; Merla, A. Simultaneous functional near-infrared spectroscopy and electroencephalography for monitoring of human brain activity and oxygenation: A review. *Neurophotonics* **2017**, *4*, 041411. [CrossRef]
31. Moher, D.; Liberati, A.; Tetzlaff, J.; Altman, D.G.; Group, P. Preferred reporting items for systematic reviews and meta-analyses: The PRISMA statement. *PLoS Med.* **2009**, *6*, e1000097. [CrossRef]
32. Hernandez, S.M.; Pollonini, L. NIRSplot: A tool for quality assessment of fNIRS scans. In *Optics and the Brain*; Optical Society of America: Washington, DC, USA, 2020.
33. Obrig, H.; Neufang, M.; Wenzel, R.; Kohl, M.; Steinbrink, J.; Einhaupl, K.; Villringer, A. Spontaneous low frequency oscillations of cerebral hemodynamics and metabolism in human adults. *Neuroimage* **2000**, *12*, 623–639. [CrossRef] [PubMed]
34. Yucel, M.A.; Selb, J.; Aasted, C.M.; Lin, P.Y.; Borsook, D.; Becerra, L.; Boas, D.A. Mayer waves reduce the accuracy of estimated hemodynamic response functions in functional near-infrared spectroscopy. *Biomed. Opt. Express* **2016**, *7*, 3078–3088. [CrossRef] [PubMed]
35. Barker, J.W.; Aarabi, A.; Huppert, T.J. Autoregressive model based algorithm for correcting motion and serially correlated errors in fNIRS. *Biomed. Opt. Express* **2013**, *4*, 1366–1379. [CrossRef] [PubMed]
36. Brigadoi, S.; Ceccherini, L.; Cutini, S.; Scarpa, F.; Scatturin, P.; Selb, J.; Gagnon, L.; Boas, D.A.; Cooper, R.J. Motion artifacts in functional near-infrared spectroscopy: A comparison of motion correction techniques applied to real cognitive data. *Neuroimage* **2014**, *85 Pt 1*, 181–191. [CrossRef] [PubMed]
37. Novi, S.L.; Roberts, E.; Spagnuolo, D.; Spilsbury, B.M.; Price, D.C.; Imbalzano, C.A.; Forero, E.; Yodh, A.G.; Tellis, G.M.; Tellis, C.M.; et al. Functional near-infrared spectroscopy for speech protocols: Characterization of motion artifacts and guidelines for improving data analysis. *Neurophotonics* **2020**, *7*, 015001. [CrossRef] [PubMed]
38. Scholkmann, F.; Spichtig, S.; Muehlemann, T.; Wolf, M. How to detect and reduce movement artifacts in near-infrared imaging using moving standard deviation and spline interpolation. *Physiol. Meas.* **2010**, *31*, 649–662. [CrossRef]
39. Molavi, B.; Dumont, G.A. Wavelet-based motion artifact removal for functional near-infrared spectroscopy. *Physiol. Meas.* **2012**, *33*, 259–270. [CrossRef]
40. Chiarelli, A.M.; Maclin, E.L.; Fabiani, M.; Gratton, G. A kurtosis-based wavelet algorithm for motion artifact correction of fNIRS data. *Neuroimage* **2015**, *112*, 128–137. [CrossRef]
41. Zhang, X.; Noah, J.A.; Hirsch, J. Separation of the global and local components in functional near-infrared spectroscopy signals using principal component spatial filtering. *Neurophotonics* **2016**, *3*, 015004. [CrossRef]
42. Pinti, P.; Scholkmann, F.; Hamilton, A.; Burgess, P.; Tachtsidis, I. Current Status and Issues Regarding Pre-processing of fNIRS Neuroimaging Data: An Investigation of Diverse Signal Filtering Methods Within a General Linear Model Framework. *Front. Hum. Neurosci.* **2018**, *12*, 505. [CrossRef]
43. Yucel, M.A.; Luhmann, A.V.; Scholkmann, F.; Gervain, J.; Dan, I.; Ayaz, H.; Boas, D.; Cooper, R.J.; Culver, J.; Elwell, C.E.; et al. Best practices for fNIRS publications. *Neurophotonics* **2021**, *8*, 012101. [CrossRef] [PubMed]
44. Chan, H.L.; Tsai, Y.T.; Meng, L.F.; Wu, T. The removal of ocular artifacts from EEG signals using adaptive filters based on ocular source components. *Ann. Biomed. Eng.* **2010**, *38*, 3489–3499. [CrossRef] [PubMed]
45. Gao, J.F.; Yang, Y.; Lin, P.; Wang, P.; Zheng, C.X. Automatic removal of eye-movement and blink artifacts from EEG signals. *Brain Topogr.* **2010**, *23*, 105–114. [CrossRef] [PubMed]
46. Wang, G.; Teng, C.; Li, K.; Zhang, Z.; Yan, X. The Removal of EOG Artifacts From EEG Signals Using Independent Component Analysis and Multivariate Empirical Mode Decomposition. *IEEE J. Biomed. Health Inform.* **2016**, *20*, 1301–1308. [CrossRef]
47. Frolich, L.; Dowding, I. Removal of muscular artifacts in EEG signals: A comparison of linear decomposition methods. *Brain Inform.* **2018**, *5*, 13–22. [CrossRef]
48. Uriguen, J.A.; Garcia-Zapirain, B. EEG artifact removal-state-of-the-art and guidelines. *J. Neural Eng.* **2015**, *12*, 031001. [CrossRef]
49. Jiang, X.; Bian, G.B.; Tian, Z. Removal of Artifacts from EEG Signals: A Review. *Sensors* **2019**, *19*, 987. [CrossRef]
50. Gao, J.; Yang, Y.; Sun, J.; Yu, G. Automatic removal of various artifacts from EEG signals using combined methods. *J. Clin. Neurophysiol.* **2010**, *27*, 312–320. [CrossRef]
51. Peng, K.; Nguyen, D.K.; Tayah, T.; Vannasing, P.; Tremblay, J.; Sawan, M.; Lassonde, M.; Lesage, F.; Pouliot, P. fNIRS-EEG study of focal interictal epileptiform discharges. *Epilepsy Res.* **2014**, *108*, 491–505. [CrossRef]
52. Pouliot, P.; Tran, T.P.Y.; Birca, V.; Vannasing, P.; Tremblay, J.; Lassonde, M.; Nguyen, D.K. Hemodynamic changes during posterior epilepsies: An EEG-fNIRS study. *Epilepsy Res.* **2014**, *108*, 883–890. [CrossRef]
53. Talukdar, M.T.; Frost, H.R.; Diamond, S.G. Modeling Neurovascular Coupling from Clustered Parameter Sets for Multimodal EEG-NIRS. *Comput. Math. Methods Med.* **2015**, *2015*, 830849. [CrossRef] [PubMed]
54. Peng, K.; Nguyen, D.K.; Vannasing, P.; Tremblay, J.; Lesage, F.; Pouliot, P. Using patient-specific hemodynamic response function in epileptic spike analysis of human epilepsy: A study based on EEG-fNIRS. *Neuroimage* **2016**, *126*, 239–255. [CrossRef] [PubMed]

55. Khan, M.J.; Ghafoor, U.; Hong, K.S. Early Detection of Hemodynamic Responses Using EEG: A Hybrid EEG-fNIRS Study. *Front. Hum. Neurosci.* **2018**, *12*, 479. [CrossRef]

56. Zama, T.; Takahashi, Y.; Shimada, S. Simultaneous EEG-NIRS Measurement of the Inferior Parietal Lobule During a Reaching Task With Delayed Visual Feedback. *Front. Hum. Neurosci.* **2019**, *13*, 301. [CrossRef] [PubMed]

57. Li, R.; Zhao, C.; Wang, C.; Wang, J.; Zhang, Y. Enhancing fNIRS analysis using EEG rhythmic signatures: An EEG-informed fNIRS analysis study. *IEEE Trans. BioMed. Eng.* **2020**, *67*, 2789–2797. [CrossRef] [PubMed]

58. Sirpal, P.; Damseh, R.; Peng, K.; Nguyen, D.K.; Lesage, F. Multimodal Autoencoder Predicts fNIRS Resting State From EEG Signals. *Neuroinformatics*, 2021; *online ahead of print*. [CrossRef]

59. Tak, S.; Ye, J.C. Statistical analysis of fNIRS data: A comprehensive review. *Neuroimage* **2014**, *85*, 72–91. [CrossRef]

60. Peng, K.; Pouliot, P.; Lesage, F.; Nguyen, D.K. Multichannel continuous electroencephalography-functional near-infrared spectroscopy recording of focal seizures and interictal epileptiform discharges in human epilepsy: A review. *Neurophotonics* **2016**, *3*, 031402. [CrossRef]

61. Li, R.; Potter, T.; Huang, W.; Zhang, Y. Enhancing Performance of a Hybrid EEG-fNIRS System Using Channel Selection and Early Temporal Features. *Front. Hum. Neurosci.* **2017**, *11*, 462. [CrossRef]

62. Borich, M.R.; Brown, K.E.; Lakhani, B.; Boyd, L.A. Applications of Electroencephalography to characterize brain activity: Perspectives in stroke. *J. Neurol. Phys.* **2015**, *39*, 43–51. [CrossRef]

63. He, B.; Sohrabpour, A.; Brown, E.; Liu, Z.M. Electrophysiological source imaging: A noninvasive window to brain dynamics. *Annu. Rev. Biomed. Eng.* **2018**, *20*, 171–196. [CrossRef]

64. Hansen, P.C. Analysis of discrete ill-posed problems by means of the L-Curve. *Siam. Rev.* **1992**, *34*, 561–580. [CrossRef]

65. Michel, C.M.; Brunet, D. EEG source imaging: A practical review of the analysis steps. *Front. Neurol.* **2019**, *10*, 325. [CrossRef]

66. Grech, R.; Cassar, T.; Muscat, J.; Camilleri, K.P.; Fabri, S.G.; Zervakis, M.; Xanthopoulos, P.; Sakkalis, V.; Vanrumste, B. Review on solving the inverse problem in EEG source analysis. *J. Neuroeng. Rehabil.* **2008**, *5*, 25. [CrossRef]

67. Aihara, T.; Takeda, Y.; Takeda, K.; Yasuda, W.; Sato, T.; Otaka, Y.; Hanakawa, T.; Honda, M.; Liu, M.G.; Kawato, M.; et al. Cortical current source estimation from electroencephalography in combination with near-infrared spectroscopy as a hierarchical prior. *Neuroimage* **2012**, *59*, 4006–4021. [CrossRef]

68. Morioka, H.; Kanemura, A.; Morimoto, S.; Yoshioka, T.; Oba, S.; Kawanabe, M.; Ishii, S. Decoding spatial attention by using cortical currents estimated from electroencephalography with near-infrared spectroscopy prior information. *Neuroimage* **2014**, *90*, 128–139. [CrossRef] [PubMed]

69. Li, R.; Li, S.; Roh, J.; Wang, C.; Zhang, Y. Multimodal Neuroimaging Using Concurrent EEG/fNIRS for Poststroke Recovery Assessment: An Exploratory Study. *Neurorehab. Neural Repair* **2020**, *34*, 1099–1110. [CrossRef] [PubMed]

70. Putze, F.; Hesslinger, S.; Tse, C.; Huang, Y.; Herff, C.; Guan, C.; Schultz, T. Hybrid fNIRS-EEG based classification of auditory and visual perception processes. *Front. Neurosci.* **2014**, *8*, 373. [CrossRef]

71. Yin, X.X.; Xu, B.L.; Jiang, C.H.; Fu, Y.F.; Wang, Z.D.; Li, H.Y.; Shi, G. A hybrid BCI based on EEG and fNIRS signals improves the performance of decoding motor imagery of both force and speed of hand clenching. *J. Neural Eng.* **2015**, *12*, 036004. [CrossRef]

72. Aghajani, H.; Garbey, M.; Omurtag, A. Measuring mental workload with EEG+fNIRS. *Front. Hum. Neurosci.* **2017**, *11*, 359. [CrossRef]

73. Al-Shargie, F.; Tang, T.B.; Kiguchi, M. Stress assessment based on decision fusion of EEG and fNIRS signals. *IEEE Access* **2017**, *5*, 19889–19896. [CrossRef]

74. Al-Shargie, F.; Kiguchi, M.; Badruddin, N.; Dass, S.C.; Hani, A.F.; Tang, T.B. Mental stress assessment using simultaneous measurement of EEG and fNIRS. *Biomed. Opt. Express* **2016**, *7*, 3882–3898. [CrossRef] [PubMed]

75. Omurtag, A.; Aghajani, H.; Keles, H.O. Decoding human mental states by whole-head EEG+fNIRS during category fluency task performance. *J. Neural Eng.* **2017**, *14*, 066003. [CrossRef] [PubMed]

76. Sun, Y.; Ayaz, H.; Akansu, A.N. Multimodal Affective State Assessment Using fNIRS + EEG and Spontaneous Facial Expression. *Brain Sci.* **2020**, *10*, 85. [CrossRef] [PubMed]

77. Firooz, S.; Setarehdan, S.K. IQ estimation by means of EEG-fNIRS recordings during a logical-mathematical intelligence test. *Comput. Biol. Med.* **2019**, *110*, 218–226. [CrossRef] [PubMed]

78. Cicalese, P.A.; Li, R.; Ahmadi, M.B.; Wang, C.; Francis, J.T.; Selvaraj, S.; Schulz, P.E.; Zhang, Y. An EEG-fNIRS hybridization technique in the four-class classification of alzheimer's disease. *J. Neurosci. Methods* **2020**, *336*, 108618. [CrossRef]

79. Abtahi, M.; Bahram Borgheai, S.; Jafari, R.; Constant, N.; Diouf, R.; Shahriari, Y.; Mankodiya, K. Merging fNIRS-EEG brain monitoring and body motion capture to distinguish Parkinson's disease. *IEEE Trans. Neural Syst. Rehabil. Eng.* **2020**, *28*, 1246–1253. [CrossRef]

80. Hong, K.S.; Khan, M.J.; Hong, M.J. Feature Extraction and Classification Methods for Hybrid fNIRS-EEG Brain-Computer Interfaces. *Front. Hum. Neurosci.* **2018**, *12*, 246. [CrossRef]

81. Khan, M.J.; Hong, M.J.Y.; Hong, K.S. Decoding of four movement directions using hybrid NIRS-EEG brain-computer interface. *Front. Hum. Neurosci.* **2014**, *8*, 244. [CrossRef]

82. Khan, M.J.; Hong, K.S. Hybrid EEG-fNIRS-based eight-command decoding for bci: Application to quadcopter control. *Front. Neurorobotics* **2017**, *11*, 6. [CrossRef]

83. Koo, B.; Lee, H.G.; Nam, Y.; Kang, H.; Koh, C.S.; Shin, H.C.; Choi, S. A hybrid NIRS-EEG system for self-paced brain computer interface with online motor imagery. *J. Neurosci. Methods* **2015**, *244*, 26–32. [CrossRef]

84. Lee, M.-H.; Fazli, S.; Mehnert, J.; Lee, S.-W. Subject-dependent classification for robust idle state detection using multi-modal neuroimaging and data-fusion techniques in BCI. *Pattern. Recogn.* **2015**, *48*, 2725–2737. [CrossRef]
85. Ge, S.; Ding, M.-Y.; Zhang, Z.; Lin, P.; Gao, J.-F.; Wang, R.-M.; Sun, G.-P.; Iramina, K.; Deng, H.-H.; Yang, Y.-K. Temporal-spatial features of intention understanding based on EEG-fNIRS bimodal measurement. *IEEE Access* **2017**, *5*, 14245–14258. [CrossRef]
86. Kwon, J.; Shin, J.; Im, C.H. Toward a compact hybrid brain-computer interface (BCI): Performance evaluation of multi-class hybrid EEG-fNIRS BCIs with limited number of channels. *PLoS ONE* **2020**, *15*, e0230491. [CrossRef] [PubMed]
87. Padfield, N.; Zabalza, J.; Zhao, H.; Masero, V.; Ren, J. EEG-based brain-computer interfaces using motor-imagery: Techniques and challenges. *Sensors* **2019**, *19*, 1423. [CrossRef] [PubMed]
88. Rezazadeh Sereshkeh, A.; Yousefi, R.; Wong, A.T.; Rudzicz, F.; Chau, T. Development of a ternary hybrid fNIRS-EEG brain–computer interface based on imagined speech. *Brain Comput. Interfaces* **2019**, *6*, 128–140. [CrossRef]
89. Ahn, S.; Nguyen, T.; Jang, H.; Kim, J.G.; Jun, S.C. Exploring Neuro-Physiological Correlates of Drivers' Mental Fatigue Caused by Sleep Deprivation Using Simultaneous EEG, ECG, and fNIRS Data. *Front. Hum. Neurosci.* **2016**, *10*, 219. [CrossRef]
90. Liu, Y.; Ayaz, H.; Shewokis, P.A. Multisubject "Learning" for Mental Workload Classification Using Concurrent EEG, fNIRS, and Physiological Measures. *Front. Hum. Neurosci.* **2017**, *11*, 389. [CrossRef]
91. Liu, Y.; Ayaz, H.; Shewokis, P.A. Mental workload classification with concurrent electroencephalography and functional near-infrared spectroscopy. *Brain Comput. Interfaces* **2017**, *4*, 175–185. [CrossRef]
92. Alhudhaif, A. An effective classification framework for brain-computer interface system design based on combining of fNIRS and EEG signals. *PEER J. Comput. Sci.* **2021**, *7*, e537. [CrossRef]
93. Hasan, M.A.H.; Khan, M.U.; Mishra, D. A Computationally Efficient Method for Hybrid EEG-fNIRS BCI Based on the Pearson Correlation. *Biomed. Res. Int.* **2020**, *2020*, 1838140. [CrossRef]
94. Khan, M.U.; Hasan, M.A.H. Hybrid EEG-fNIRS BCI Fusion Using Multi-Resolution Singular Value Decomposition (MSVD). *Front. Hum. Neurosci.* **2020**, *14*, 599802. [CrossRef] [PubMed]
95. Shin, J.; Kim, D.W.; Muller, K.R.; Hwang, H.J. Improvement of Information Transfer Rates Using a Hybrid EEG-NIRS Brain-Computer Interface with a Short Trial Length: Offline and Pseudo-Online Analyses. *Sensors* **2018**, *18*, 1827. [CrossRef] [PubMed]
96. Shin, J.; Kwon, J.; Im, C.H. A Ternary Hybrid EEG-NIRS Brain-Computer Interface for the Classification of Brain Activation Patterns during Mental Arithmetic, Motor Imagery, and Idle State. *Front. Neuroinform.* **2018**, *12*, 5. [CrossRef] [PubMed]
97. Rahman, M.A.; Uddin, M.S.; Ahmad, M. Modeling and classification of voluntary and imagery movements for brain–computer interface from fNIR and EEG signals through convolutional neural network. *Health Inf. Sci. Syst.* **2019**, *7*, 1–22. [CrossRef]
98. Sirpal, P.; Kassab, A.; Pouliot, P.; Nguyen, D.K.; Lesage, F. fNIRS improves seizure detection in multimodal EEG-fNIRS recordings. *J. Biomed. Opt.* **2019**, *24*, 1–9. [CrossRef]
99. Litjens, G.; Kooi, T.; Bejnordi, B.E.; Setio, A.A.A.; Ciompi, F.; Ghafoorian, M.; van der Laak, J.; van Ginneken, B.; Sanchez, C.I. A survey on deep learning in medical image analysis. *Med. Image Anal.* **2017**, *42*, 60–88. [CrossRef]
100. Khan, H.; Naseer, N.; Yazidi, A.; Eide, P.K.; Hassan, H.W.; Mirtaheri, P. Analysis of Human Gait using Hybrid EEG-fNIRS-based BCI System: A review. *Front. Hum. Neurosci.* **2020**, *14*, 605. [CrossRef]
101. Chen, L.C.; Sandmann, P.; Thorne, J.D.; Herrmann, C.S.; Debener, S. Association of Concurrent fNIRS and EEG Signatures in Response to Auditory and Visual Stimuli. *Brain Topogr.* **2015**, *28*, 710–725. [CrossRef]
102. Chen, Y.; Tang, J.; Chen, Y.; Farrand, J.; Craft, M.A.; Carlson, B.W.; Yuan, H. Amplitude of fNIRS Resting-State Global Signal Is Related to EEG Vigilance Measures: A Simultaneous fNIRS and EEG Study. *Front. Neurosci.* **2020**, *14*, 560878. [CrossRef]
103. Balconi, M.; Vanutelli, M.E. Hemodynamic (fNIRS) and EEG (N200) correlates of emotional inter-species interactions modulated by visual and auditory stimulation. *Sci. Rep.* **2016**, *6*, 23083. [CrossRef]
104. Zich, C.; Debener, S.; Thoene, A.K.; Chen, L.C.; Kranczioch, C. Simultaneous EEG-fNIRS reveals how age and feedback affect motor imagery signatures. *Neurobiol. Aging* **2017**, *49*, 183–197. [CrossRef] [PubMed]
105. Borgheai, S.B.; Deligani, R.J.; McLinden, J.; Zisk, A.; Hosni, S.I.; Abtahi, M.; Mankodiya, K.; Shahriari, Y. Multimodal exploration of non-motor neural functions in ALS patients using simultaneous EEG-fNIRS recording. *J. Neural Eng.* **2019**, *16*, 066036. [CrossRef] [PubMed]
106. Gentile, E.; Brunetti, A.; Ricci, K.; Delussi, M.; Bevilacqua, V.; de Tommaso, M. Mutual interaction between motor cortex activation and pain in fibromyalgia: EEG-fNIRS study. *PLoS ONE* **2020**, *15*, e0228158. [CrossRef]
107. Zhang, Y.; Zhu, C. Assessing brain networks by resting-state dynamic functional connectivity: An fNIRS-EEG Study. *Front. Neurosci.* **2019**, *13*, 1430. [CrossRef] [PubMed]
108. Lin, C.T.; King, J.T.; Chuang, C.H.; Ding, W.; Chuang, W.Y.; Liao, L.D.; Wang, Y.K. Exploring the brain responses to driving fatigue through simultaneous EEG and fNIRS measurements. *Int. J. Neural Syst.* **2020**, *30*, 1950018. [CrossRef] [PubMed]
109. Kaga, Y.; Ueda, R.; Tanaka, M.; Kita, Y.; Suzuki, K.; Okumura, Y.; Egashira, Y.; Shirakawa, Y.; Mitsuhashi, S.; Kitamura, Y.; et al. Executive dysfunction in medication-naive children with ADHD: A multi-modal fNIRS and EEG study. *Brain Dev.* **2020**, *42*, 555–563. [CrossRef]
110. Suzuki, K.; Okumura, Y.; Kita, Y.; Oi, Y.; Shinoda, H.; Inagaki, M. The relationship between the superior frontal cortex and alpha oscillation in a flanker task: Simultaneous recording of electroencephalogram (EEG) and near infrared spectroscopy (NIRS). *Neurosci. Res.* **2018**, *131*, 30–35. [CrossRef]
111. Keles, H.O.; Barbour, R.L.; Omurtag, A. Hemodynamic correlates of spontaneous neural activity measured by human whole-head resting state EEG plus fNIRS. *Neuroimage* **2016**, *138*, 76–87. [CrossRef]

112. Pinti, P.; Siddiqui, M.F.; Levy, A.D.; Jones, E.J.H.; Tachtsidis, I. An analysis framework for the integration of broadband NIRS and EEG to assess neurovascular and neurometabolic coupling. *Sci. Rep.* **2021**, *11*, 3977. [CrossRef]

113. Vijayakrishnan Nair, V.; Kish, B.R.; Yang, H.S.; Yu, Z.; Guo, H.; Tong, Y.; Liang, Z. Monitoring anesthesia using simultaneous functional Near Infrared Spectroscopy and Electroencephalography. *Clin. Neurophysiol.* **2021**, *132*, 1636–1646. [CrossRef]

114. Al-Shargie, F.; Tang, T.B.; Kiguchi, M. Assessment of mental stress effects on prefrontal cortical activities using canonical correlation analysis: An fNIRS-EEG study. *Biomed. Opt. Express* **2017**, *8*, 2583–2598. [CrossRef] [PubMed]

115. Govindan, R.B.; Massaro, A.; Chang, T.; Vezina, G.; du Plessis, A. A novel technique for quantitative bedside monitoring of neurovascular coupling. *J. Neurosci. Methods* **2016**, *259*, 135–142. [CrossRef] [PubMed]

116. Chalak, L.F.; Tian, F.; Adams-Huet, B.; Vasil, D.; Laptook, A.; Tarumi, T.; Zhang, R. Novel Wavelet Real Time Analysis of Neurovascular Coupling in Neonatal Encephalopathy. *Sci. Rep.* **2017**, *7*, 45958. [CrossRef] [PubMed]

117. Chiarelli, A.M.; Perpetuini, D.; Croce, P.; Filippini, C.; Cardone, D.; Rotunno, L.; Anzoletti, N.; Zito, M.; Zappasodi, F.; Merla, A. Evidence of Neurovascular Un-Coupling in Mild Alzheimer's Disease through Multimodal EEG-fNIRS and Multivariate Analysis of Resting-State Data. *Biomedicines* **2021**, *9*, 337. [CrossRef]

118. Perpetuini, D.; Chiarelli, A.M.; Filippini, C.; Cardone, D.; Croce, P.; Rotunno, L.; Anzoletti, N.; Zito, M.; Zappasodi, F.; Merla, A. Working Memory Decline in Alzheimer's Disease Is Detected by Complexity Analysis of Multimodal EEG-fNIRS. *Entropy* **2020**, *22*, 1380. [CrossRef]

119. Niu, H.; Zhu, Z.; Wang, M.; Li, X.; Yuan, Z.; Sun, Y.; Han, Y. Abnormal dynamic functional connectivity and brain states in Alzheimer's diseases: Functional near-infrared spectroscopy study. *Neurophotonics* **2019**, *6*, 025010. [CrossRef]

120. Li, R.; Mayseless, N.; Balters, S.; Reiss, A.L. Dynamic inter-brain synchrony in real-life inter-personal cooperation: A functional near-infrared spectroscopy hyperscanning study. *Neuroimage* **2021**, *238*, 118263. [CrossRef]

121. Gu, S.; Pasqualetti, F.; Cieslak, M.; Telesford, Q.K.; Alfred, B.Y.; Kahn, A.E.; Medaglia, J.D.; Vettel, J.M.; Miller, M.B.; Grafton, S.T. Controllability of structural brain networks. *Nat. Commun.* **2015**, *6*, 1–10. [CrossRef]

122. Fang, F.; Gao, Y.; Schulz, P.E.; Selvaraj, S.; Zhang, Y. Brain controllability distinctiveness between depression and cognitive impairment. *J. Affect. Disord.* **2021**, *294*, 847–856. [CrossRef]

123. Scheid, B.H.; Ashourvan, A.; Stiso, J.; Davis, K.A.; Mikhail, F.; Pasqualetti, F.; Litt, B.; Bassett, D.S. Time-evolving controllability of effective connectivity networks during seizure progression. *Proc. Natl. Acad. Sci. USA* **2021**, *118*, e2006436118. [CrossRef]

124. von Luhmann, A.; Ortega-Martinez, A.; Boas, D.A.; Yucel, M.A. Using the General Linear Model to Improve Performance in fNIRS Single Trial Analysis and Classification: A Perspective. *Front. Hum. Neurosci.* **2020**, *14*, 30. [CrossRef] [PubMed]

125. Ortega-Martinez, A.; Von Luhmann, A.; Farzam, P.; Rogers, D.; Mugler, E.M.; Boas, D.A.; Yucel, M.A. Multivariate Kalman filter regression of confounding physiological signals for real-time classification of fNIRS data. *Neurophotonics* **2022**, *9*, 025003. [CrossRef] [PubMed]

126. Cui, X.; Bryant, D.M.; Reiss, A.L. NIRS-based hyperscanning reveals increased interpersonal coherence in superior frontal cortex during cooperation. *Neuroimage* **2012**, *59*, 2430–2437. [CrossRef]

127. Ban, H.Y.; Barrett, G.M.; Borisevich, A.; Chaturvedi, A.; Dahle, J.L.; Dehghani, H.; Dubois, J.; Field, R.M.; Gopalakrishnan, V.; Gundran, A.; et al. Kernel Flow: A high channel count scalable time-domain functional near-infrared spectroscopy system. *J Biomed. Opt.* **2022**, *27*, 074710. [CrossRef] [PubMed]

128. von Luhmann, A.; Wabnitz, H.; Sander, T.; Muller, K.R. M3BA: A Mobile, Modular, Multimodal Biosignal Acquisition Architecture for Miniaturized EEG-NIRS-Based Hybrid BCI and Monitoring. *IEEE Trans. Biomed. Eng.* **2017**, *64*, 1199–1210. [CrossRef] [PubMed]

129. Gagnon, L.; Cooper, R.J.; Yucel, M.A.; Perdue, K.L.; Greve, D.N.; Boas, D.A. Short separation channel location impacts the performance of short channel regression in NIRS. *Neuroimage* **2012**, *59*, 2518–2528. [CrossRef]

130. Caldwell, M.; Scholkmann, F.; Wolf, U.; Wolf, M.; Elwell, C.; Tachtsidis, I. Modelling confounding effects from extracerebral contamination and systemic factors on functional near-infrared spectroscopy. *Neuroimage* **2016**, *143*, 91–105. [CrossRef]

131. von Luhmann, A.; Boukouvalas, Z.; Muller, K.R.; Adali, T. A new blind source separation framework for signal analysis and artifact rejection in functional Near-Infrared Spectroscopy. *Neuroimage* **2019**, *200*, 72–88. [CrossRef]

Article

Four-Class Classification of Neuropsychiatric Disorders by Use of Functional Near-Infrared Spectroscopy Derived Biomarkers

Sinem Burcu Erdoğan * and Gülnaz Yükselen

Department of Biomedical Engineering, Acıbadem Mehmet Ali Aydınlar University, Istanbul 34684, Turkey; gulnaz.yukselen@acibadem.edu.tr
* Correspondence: sinem.erdogan@acibadem.edu.tr

Abstract: Diagnosis of most neuropsychiatric disorders relies on subjective measures, which makes the reliability of final clinical decisions questionable. The aim of this study was to propose a machine learning-based classification approach for objective diagnosis of three disorders of neuropsychiatric or neurological origin with functional near-infrared spectroscopy (fNIRS) derived biomarkers. Thirteen healthy adolescents and sixty-seven patients who were clinically diagnosed with migraine, obsessive compulsive disorder, or schizophrenia performed a Stroop task, while prefrontal cortex hemodynamics were monitored with fNIRS. Hemodynamic and cognitive features were extracted for training three supervised learning algorithms (naïve bayes (NB), linear discriminant analysis (LDA), and support vector machines (SVM)). The performance of each algorithm in correctly predicting the class of each participant across the four classes was tested with ten runs of a ten-fold cross-validation procedure. All algorithms achieved four-class classification performances with accuracies above 81% and specificities above 94%. SVM had the highest performance in terms of accuracy ($85.1 \pm 1.77\%$), sensitivity ($84 \pm 1.7\%$), specificity ($95 \pm 0.5\%$), precision ($86 \pm 1.6\%$), and F1-score ($85 \pm 1.7\%$). fNIRS-derived features have no subjective report bias when used for automated classification purposes. The presented methodology might have significant potential for assisting in the objective diagnosis of neuropsychiatric disorders associated with frontal lobe dysfunction.

Keywords: fNIRS; BCI; classification; schizophrenia; obsessive compulsive disorder; migraine; Stroop test

Citation: Erdoğan, S.B.; Yükselen, G. Four-Class Classification of Neuropsychiatric Disorders by Use of Functional Near-Infrared Spectroscopy Derived Biomarkers. *Sensors* **2022**, *22*, 5407. https://doi.org/10.3390/s22145407

Academic Editor: Rebecca Re

Received: 23 May 2022
Accepted: 22 June 2022
Published: 20 July 2022

1. Introduction

In clinical practice, the majority of neuropsychiatric disorders are diagnosed with a clinician-dependent interpretation of patient information, which is obtained through a variety of subjectively biased sources such as clinical interviews, self-reports, observational data, and behavioral measures [1–5]. The potential of introducing subjectivity during both interpretation and acquisition of these diagnostic measures may have a prominent impact on the final clinical decision and highlights the critical need for developing accurate, objective, and reliable clinical decision support systems. Such decision support systems should ideally analyze objective and quantitative measures of the distinct characteristics of the neurobiological changes that are gradually induced by each neuropsychiatric disorder [6,7].

Within this context, various functional brain imaging modalities, such as functional magnetic resonance imaging (fMRI), positron emission tomography (PET), electroencephalography (EEG), and functional near-infrared spectroscopy (fNIRS), have been utilized for characterizing the neurobiological underpinnings of a variety of neuropsychiatric disorders [8–10]. Among these modalities, fNIRS systems have stepped forward for extracting informative, neuronally induced hemodynamic markers of cognition during altered brain states in naturalistic settings [11,12]. Consequently, fNIRS systems have also received increasing interest in the field of psychiatry for assisting diagnosis, prognosis, and follow-up of treatment procedures thanks to their: (1) portability, (2) non-invasive nature, (3) modest equipment size, (4) robustness to electrogenic or motion artifacts, (5) low operating cost, (6) quick set-up time and calibration, (7) ability to collect biological information at any desired frequency and duration, and (8) ease of application in ecologically valid settings to a broad range of patient populations involving children and elderly adults [13].

Indeed, recent studies have presented compelling evidence that a wide variety of neuropsychiatric disorders can be characterized by functional alterations in the hemodynamic activity of the prefrontal cortex (PFC), which can be detected with fNIRS [13]. For instance, hypoactivation in frontal lobe regions has been detected in patients with schizophrenia (SCZ) and major depressive disorder (MDD) during verbal fluency tasks when compared to their healthy counterparts [5,14]. Similarly, hyper- and hypo-connectivity between different brain regions during resting state have been identified in patients with schizophrenia (SCZ) [15–19] and major depressive disorder (MDD) [8,9,14], while decreased cerebral blood flow in bilateral symmetric regions of the inferior PFC has been detected in patients with obsessive compulsive disorder (OCD) when compared to their healthy counterparts [20,21]. PFC dysfunction in the form of hypo- or hyper-connectivity during resting state or hypoactivation during various cognitive tests (e.g., Stroop and verbal fluency test) has been extensively observed and reported in patient groups diagnosed with a variety of major neuropsychiatric disorders, which include SCZ, MDD, bipolar disorder (BD), post-traumatic stress disorder (PTSD), and attention deficiency and hyperactivity disorder (ADHD). Results from meta-analysis studies indicated that the topographical distributions of functional abnormalities observed in these patient groups are likely to have disorder-specific patterns [14]. Overall, these studies have highlighted the potential of exploring PFC-based neurofunctional features as objective and distinctive biomarkers of various major neuropsychiatric disorder states. They also showed that information from practical and preferably field-deployable cerebral physiology monitoring tools such as fNIRS systems can quantify and parameterize abnormalities in frontal lobe function and may have a great potential for assisting in the objective diagnosis and classification of major psychiatric disorders which, in most cases, have overlapping behavioral symptoms across each other and are difficult to distinguish when decisions are based solely on observation, self-report, interview, and/or rating scales.

Considering the critical demand to integrate more objective measures of neurophysiological alterations into diagnostic clinical decision processes, the presented study aimed to assess the feasibility and applicability of an fNIRS-based automated classification approach for accurate prediction and objective identification of the presence of three distinct neuropsychiatric or neurological disorder states which are known to induce alterations in frontal lobe function. In our recent work, we demonstrated the feasibility and applicability of an fNIRS-assisted automated classification approach for accurate prediction of the presence of impulsivity in adolescents [22]. More specifically, our results suggested that training computationally efficient supervised learning algorithms with informative features obtained from clinical, behavioral, and fNIRS-derived hemodynamic measures could serve as a decision support system for recognizing the presence of impulsivity in individuals. However, the clinical features included in the feature sets still had the potential to present subjective bias when used for algorithm training purposes because there always existed some probability that the subjects could provide false reports in the clinical interviews.

Based on the promising performance of integrating fNIRS-derived features and clinical features for recognizing the presence of impulsivity in our recent work, the objective of this study was to introduce a more reliable, machine learning-based classification approach for correct identification of the presence of three distinct neuropsychiatric and/or neurological disorder states. The proposed machine learning-based classification approach involved training three supervised learning algorithms with (i) fNIRS-derived informative biomarkers only and (ii) a combination of fNIRS-derived biomarkers and performance measures obtained during a cognitive test, named the Stroop task. We tested the feasibility of the proposed approach with three distinct supervised learning algorithms and by extending our classification problem to include four classes of subjects. The ultimate goal was to demonstrate the feasibility of an fNIRS-based automated classification methodology for predicting the presence of a neuropsychiatric disease, where input features are of pure biological origin and can be derived non-invasively in naturalistic settings by use of ergonomic fNIRS probes. For this purpose, hemodynamic information obtained from concurrent fNIRS recordings during a Stroop task was processed to extract global efficiency metrics which are indicative of the strength of functional connectivity among different PFC regions. The efficacy of training three distinct supervised machine learning algorithms, namely naive Bayes (NB), linear discriminant analysis (LDA), and support vector machines (SVM), with (a) fNIRS-derived neuronally induced biomarkers and (b) a combination of fNIRS-derived biomarkers and cognitive performance measures obtained during the Stroop task, was evaluated. The performance metrics of possible combinations of each classification algorithm and feature set combination were assessed by whether each subject was correctly labeled among the four classes, which included healthy controls (HC), patients diagnosed with migraine without aura (MIG), schizophrenia (SCZ), and obsessive compulsive disorder (OCD).

Our study presents the following novelties with respect to the current literature: To date, there have been no studies that attempted to identify the presence of a neuropsychiatric or neurological disorder by use of a four-class automated classification scheme based on a combination of fNIRS-derived neuronally induced metrics obtained during a neuropsychological test and supervised machine learning methods. The efficacy of an automated classification approach which aims to correctly label a neuropsychiatric disorder into one of four categories has not been evaluated before with structural or functional neuroimaging measures. The efficacy of combining fNIRS-derived global efficiency metrics of the PFC as sole informative features of a neuropsychiatric and/or neurological disorder with supervised learning methods has also not been evaluated before.

2. Materials and Methods

2.1. Subjects

In this study, 13 healthy control (HC) subjects (6 female (F), mean age 26), 20 migraine (MIG) patients without aura (12 F, mean age 27), 26 patients with obsessive compulsive disorder (OCD) (11 F, mean age 29), and 21 schizophrenia (SCZ) patients (10 F, mean age 28) participated. Each subject provided informed consent before participating in the experiment. The study protocol was approved by the Ethics Committee of Pamukkale University, Denizli, Turkey. All experiments were conducted according to the latest Declaration of Helsinki. Parts of these datasets have been utilized in previous works performed by our group and coworkers [23–29].

2.2. Experimental Protocol

During the experiments, subjects sat on a comfortable chair in front of a computer screen which was placed approximately 1 m away from their eyes. All experiments were carried out in a dimly illuminated, silent room. The experimental protocol was briefly explained to each subject prior to the onset of each experiment. They were requested to sit relaxed and refrain from moving their head during the fNIRS recordings. During the experiment, their task was to carefully complete a color–word Stroop task which was adapted to Turkish from a pioneer protocol proposed by Zysset et al. [30]. Each experiment began with 30 s of a baseline recording followed by presentation of alternating blocks of 3 stimulus conditions which consisted of neutral (N), congruent (C), and incongruent (IC) stimuli (Figures 1 and 2). There was a total of 5 stimulus blocks for each condition (i.e., N, C, IC) and all task blocks were presented in a randomized order that changed for every experimental session. Each stimulus block consisted of 6 different trials of the same condition. Within a block, each trial appeared on the screen for 2.5 s followed by a 4 s blank screen. Task blocks were separated with 20 s periods of rest (Figure 2).

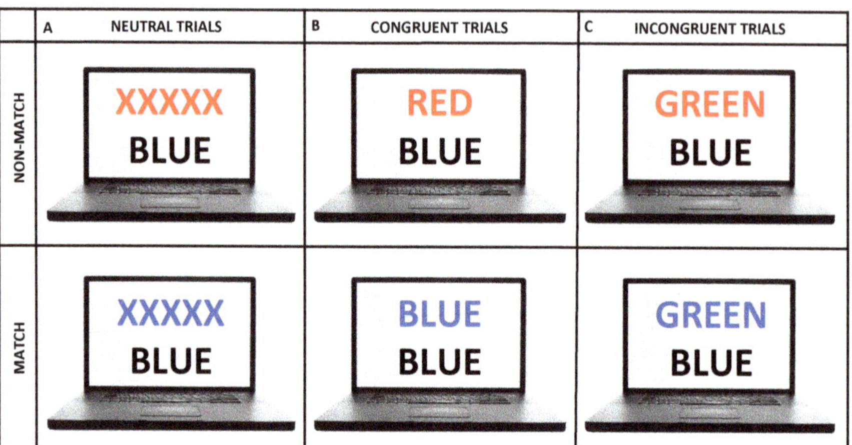

Figure 1. Types of stimuli that were presented within the color–word Stroop experiment. Samples of trial presentations are schematically represented for match (top row) and non-match conditions (bottom row) of (**A**) neutral, (**B**) congruent, and (**C**) incongruent stimuli.

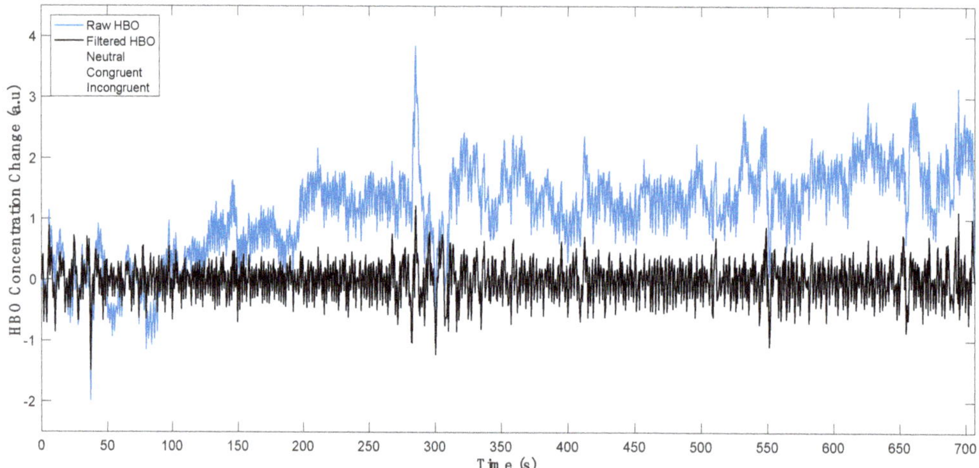

Figure 2. Experimental protocol depicting stimuli timing for neutral, congruent, and incongruent stimuli blocks in a sample session. HBO time series from a representative channel of a subject are plotted before (black) and after an 8th order Butterworth high-pass filter is applied with a cut-off frequency of 0.009 Hz (blue). a.u: arbitrary units.

During each stimulus presentation, two rows of letters were displayed on the screen. The task was to evaluate whether the color of the letters displayed at the top row matched with the meaning of the word displayed at the bottom row. Subjects were asked to press the left button of the mouse if the color of the upper row letters matched with the meaning of the bottom row word. These cases were called match cases (Figure 1, top panel). They were asked to press the right button if the color of the upper row letters did not match with the meaning of the bottom row word for non-match cases (Figure 1, bottom panel). The letters in all trials were printed in one of four basic colors, which were yellow, red, blue, or green. In N trials, top row letters were written in yellow, red, blue, or green but did not form a meaningful word, and a color name was typed in black on the bottom row. For C trials, a word with the meaning of a color was typed in the same congruent color in the top row. For the IC trials, a word with the meaning of a color was typed in another color (i.e., incongruent) in the top row. Hence, subjects had to suppress processing the color information and evaluate the meaning information of the top row letters while making a comparison with the meaning information of the bottom row letters to provide a correct answer for the IC trials. Such an interference between two competing cognitive inputs induced a Stroop effect [23–27,30]. The number of match and non-match cases for each trial type was balanced during the experiment. The average reaction time and error rate were calculated for N, IC, and C trials separately.

2.3. fNIRS Data Acquisition

Hemodynamic signals were collected from the prefrontal cortex region with a wireless ARGES-CEREBRO system (Hemosoft Information Technology and Training Services Inc., Ankara, Turkey) [24–26,31–33] which has a flexible forehead probe equipped with 4 light emitting diodes (LEDs) and 10 photodetectors (Figure 3A). The LED–photodetector pairs with 2.5 cm distance were accepted as channels and a total of 16 equidistant channels were formed which covered parts medial, orbitofrontal, and dorsolateral cortices (Figure 3B). Each LED emits near-infrared light at 750 and 850 nm in continuous wave mode and the sampling rate of the system is 1.77 Hz. The ability of this probe design to allow light penetration through the cortical tissue and collect hemodynamic information from the anterior part of the PFC has been discussed extensively in previous work by our group [23,33]. Wavelength-specific light intensity changes were detected at each detector separately, and this information was converted to optical density (OD) changes of each wavelength for each channel. Channels whose raw light intensity signals presented coefficient of variability (C.V) above 7.5% (C.V = 100 × standard deviation(signal)/mean(signal)) were not included in the analyses [34]. Time series of OD changes were provided as inputs to the modified Beer–Lambert law to compute channel-specific changes in localized HBO and HBR concentrations [10,11,35,36]. The partial pathlength factor was taken as 6 for both wavelengths [37–39]. HBO signals were visually inspected to exclude trial blocks which had motion artifacts within a time window spanning 5 s pre- and post-stimulus duration. Changes in HBO concentration have been reported to be a better indicator of alterations in neuronal metabolism induced by cognitive tasks [40–44], while having a higher signal-to-noise ratio when compared to HBR signals [40,41,44]. Hence, the efficacy of only HBO-derived hemodynamic features was tested for classification purposes.

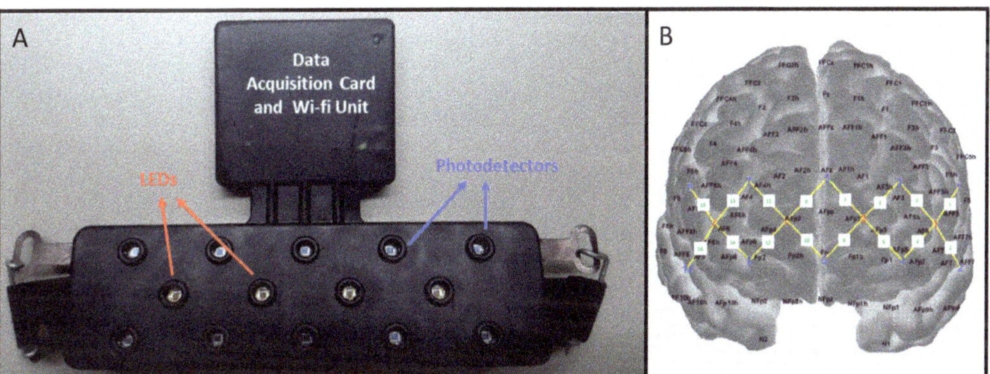

Figure 3. (**A**) Configuration of the forehead probe and (**B**) approximate location of the LEDs and photodetectors with respect to the international 10–20 system for electrode placement. LEDs are demonstrated with red dots and detectors are represented with blue dots. Yellow lines are drawn between the LED–photodetector pairs, which form a total of 16 channels.

2.4. Data Analysis

2.4.1. Processing of fNIRS Signals

The fNIRS-HBO signals are composed of neuronally and systemically induced hemodynamic components which are intermixed with each other over a broad range of frequencies. The neuronally induced hemodynamic variations in the HBO signal are caused by both spontaneous and task-related neuronal activity, while the systemic physiological activity-related hemodynamic components have multiple origins, which include variations in heartbeat, respiration, blood pressure, and vascular tone. Hence, prior to obtaining correlation-based functional connectivity metrics between HBO signals of different channel pairs, the impact of common, global systemic effects of non-neuronal origin inherent in both channel data had to be reduced. Such a procedure is necessary to isolate the extent of correlation caused by only neuronally induced hemodynamic effects, since common physiological effects of non-neuronal origin could inflate the correlation between signals of channel pairs. A partial correlation approach was adapted from the works of Akin [28] and Akin [29] to reduce the impact of common systemic interference to Pearson's correlation coefficients calculated between HBO signals of each channel pair. Similar to these works, HBO signals of all channels were initially high-pass-filtered with a cut-off frequency of 0.009 Hz using an 8th order Butterworth filter. The high-pass-filtered HBO signals were then averaged to have a single global signal regressor, which was utilized as the partial regressor for modeling and removing the impact of common systemic noise from the correlations between each channel pair in the subsequent step of the analysis [22,24–29].

Time traces used for computing the correlation between each channel pair were obtained as follows. For each channel, HBO signals corresponding to each stimulus block were truncated from the onset to the end of that block. These time segments were then concatenated in time to obtain a single task-related HBO signal for each channel of each subject. Similarly, the partial correlation regressor was obtained by truncating and concatenating the time segments belonging to all task blocks in the global signal regressor. Then, 16-by-16 partial correlation (PC)-corrected functional connectivity (FC) matrices for each subject were generated after removing the impact of this partial correlation regressor [22,29].

2.4.2. Computation of Cognitive Quotient and Global Efficiency Features

Two groups of features were extracted from the behavioral and hemodynamic data obtained during the Stroop task. Similar to our previous work [22], the behavioral performance was quantified with a feature named the cognitive quotient (CQ), which could be considered as a generalized cognitive performance indicator of each subject during the Stroop task. The accuracy and reaction time metrics obtained from all trials of the Stroop experiment were fused in this single metric by dividing the overall accuracy performance (i.e., percentage of correct answers over all trials) with the average reaction time for all trials.

Regarding the hemodynamic features, a relatively novel functional connectivity metric called global efficiency (GE) was obtained from the 16-by-16 partial correlation-corrected FC matrices obtained for each subject. The GE metric was obtained from a graph theoretical network analysis approach, and its efficacy in demonstrating the degree of connectedness and information transfer between cortical regions during various cognitive tasks has been shown in previous studies [22,27–29].

After the partial correlation-corrected FC matrices were obtained for each subject, these matrices were decomposed into two matrices, which represent the degree of connectedness of the default mode (DM) and the cognitive mode (CM) networks of the brain. This decomposition was established by applying principal component analysis to the FC matrix, the details of which are extensively explained in the recent work of Akin [29]. Briefly, principal component (PC) decomposition was applied to the 16 by 16 FC matrices and the weights of the PCs were thresholded using an optimization procedure described in [29]. The DM and CM components of the FC matrices were reconstructed by weighting and summing the PC regressors that had weights below and above the threshold, separately.

GE values of the DM and the CM components were computed separately for each subject by using the formula of Latora and Marchiori [45]. The GE feature for the DM network was named GE_{dm}, and similarly, the GE feature for the CM was named GE_{cm} (Figure 4).

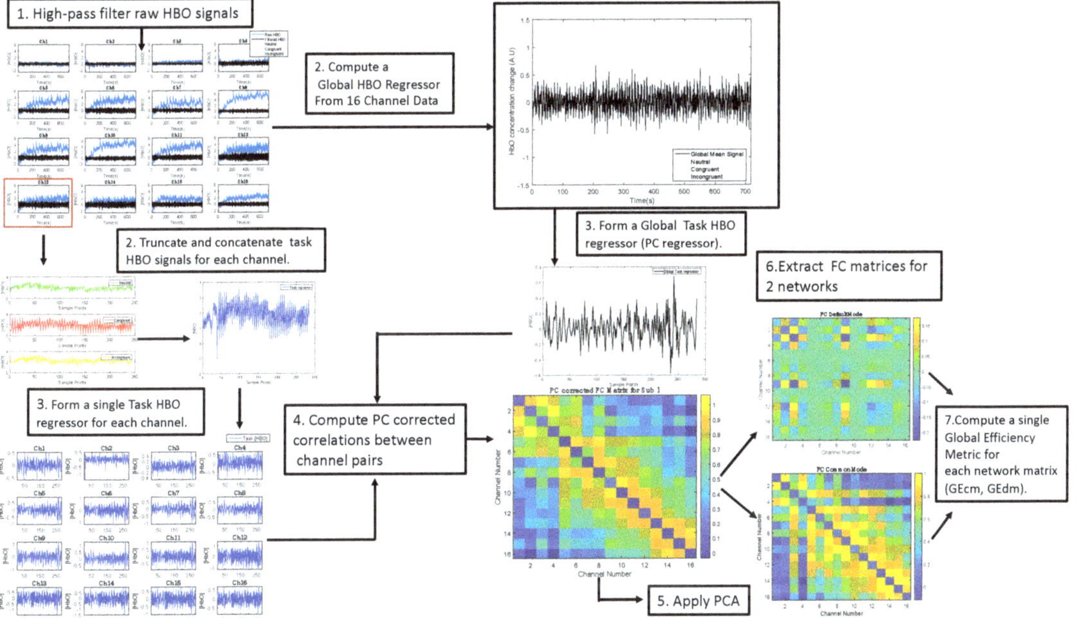

Figure 4. Pipeline for extracting fNIRS derived features.

2.4.3. Classification Methods

The feasibility of training fNIRS-derived GE features with machine learning classifiers for correct identification of the presence of a disorder in each subject was evaluated and compared for three distinct algorithms. These algorithms were naive Bayes (NB), linear discriminant analysis (LDA), and support vector machines (SVM). These algorithms were selected for several reasons: (1) They have been shown to perform well with small sample sizes (n < 200) [5,15,22,46–57]. (2) Their computational cost is low. (3) They have performed successfully in a variety of classification problems where fNIRS features extracted during cognitive and motor tasks were utilized [25,26,28–31]. (4) Their good performance for classification at the single subject level has been reported for previous neuropsychiatry studies with similar sample sizes, but a lower number of classes [1,5,15,58–62]. The mathematical architecture of these algorithms has been extensively explained in previous work performed by our group [22,46] and others [52–56].

Each classification algorithm was constructed by using the libraries of the WEKA platform (version 3.8.5) [63]. The sequential minimal optimization (SMO) algorithm was utilized for training the SVM classifier [64]. SMO was run with the Pearson VII universal kernel [65], also known as the PUK kernel. To avoid overfitting, the regularization parameter (C) of SMO and PUK kernel parameters (i.e., omega (ω) and sigma (σ)) was optimized by maximizing the accuracy with a grid-search procedure. Assigning C = 10 and $\omega = \sigma = 1$ yielded the best results. LDA and NB classifiers were constructed with the default parameters implemented in the WEKA software. A brief flowchart of the processing pipeline is demonstrated in Figure 5.

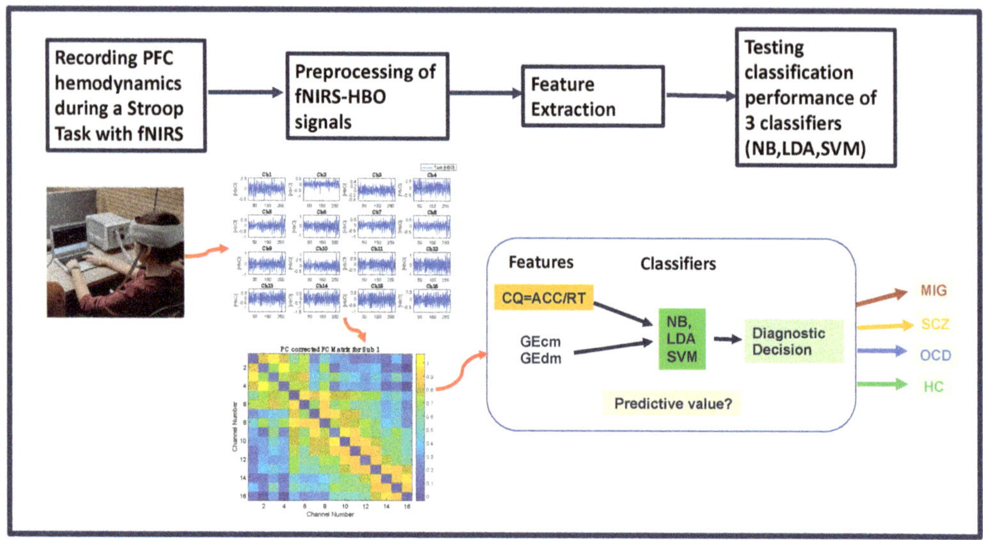

Figure 5. Flowchart of the system design.

2.4.4. Performance Evaluation

To evaluate and compare the classification performances of each algorithm, accuracy, precision, sensitivity, specificity, and F1-score were calculated through a comparison of the actual and predicted labels of test data [61]. For each algorithm, performance metrics were obtained after 10 runs of a 10-fold cross-validation (C.V) procedure, where in each run, 1/10th of the subject data were separated for testing the algorithm and the remainder of subject data were used for training, and this procedure was repeated 10 times. For each performance metric, the mean scores across all runs and their standard deviation were computed (Tables 1 and 2). This procedure was conducted for cases when each algorithm was trained with (i) fNIRS only features (i.e., GE_{cm} and GE_{dm}) and (ii) a combination of fNIRS-derived features (i.e., GE_{cm}, GE_{dm}) and a behavioral feature (i.e., CQ). All features were computed for each subject separately.

Table 1. Four-class classification performances of NB, LDA, and SVM algorithms when trained with fNIRS only features (i.e., GE_{cm} and GE_{dm}). Each performance metric is represented in percentages (%) as the mean value across all runs $\pm$ standard deviation of the mean.

Method	Accuracy	Precision	Recall	Specificity	F1-Score
NB	81.77 ± 1.06	82.1 ± 1	81 ± 0.01	94 ± 0.004	81 ± 1
LDA	83.8 ± 1	85 ± 0.01	83 ± 0.01	95 ± 0.01	84 ± 0.01
SVM	81 ± 0.84	80 ± 0.01	79 ± 0.01	94 ± 0.003	80 ± 0.008

Pairwise comparisons between the performance metrics obtained from each possible algorithm (i.e., NB, LDA, SVM) and feature set (i.e., $GE_{cm} + GE_{dm}$ or $GE_{cm} + GE_{dm} + CQ$) combination were performed with two-tailed, two-sample t-tests. Comparisons of each performance metric (i.e., accuracy, precision, recall, specificity, and F1-score) among different combinations of algorithm and feature set choices aimed to assess: (i) whether training each algorithm with only fNIRS features resulted in a statistically significantly different classification performance when compared to training the same algorithm with a combination of fNIRS and behavioral features, and (ii) whether there exists an algorithm and feature set combination with a statistically significantly higher performance when compared to all other options.

Table 2. Four-class classification performances of NB, LDA, and SVM when trained with fNIRS and behavioral features (i.e., GE_{cm}, GE_{dm}, and CQ). Each performance metric is represented in percentages (%) as the mean value across all runs ± standard deviation of the mean. Bold-typed results denote significantly higher performance of the corresponding algorithm with respect to the results when the algorithm is fed with fNIRS only features.

Method	Accuracy	Precision	Recall	Specificity	F1-Score
NB	**84.68 ± 1.3**	**85 ± 0.01**	**83 ± 0.01**	**95 ± 0.01**	**84 ± 0.01**
LDA	83.8 ± 1.6	84 ± 1.1	83 ± 1.4	94 ± 0.04	84 ± 1.2
SVM	**85 ± 1.77**	**86 ± 1.6**	**84 ± 1.7**	**95 ± 0.5**	**85 ± 1.7**

3. Results

Table 1 presents the four-class classification performances of NB, LDA, and SVM classifiers when they were trained with two fNIRS-derived features (i.e., GE_{cm}, GE_{dm}). All algorithms achieved accuracy, precision, recall, and F1-score performances above 81%, while the specificity scores were all above 94%. It should be noted that LDA performed significantly higher than both SVM and NB (Figure 6) in terms of accuracy ($83.8 \pm 1\%$, $p < 0.05$), precision ($85 \pm 0.01\%$, $p < 0.05$), recall ($83 \pm 0.01\%$, $p < 0.05$), specificity ($95 \pm 0.01\%$, $p < 0.05$), and F1-score ($84 \pm 0.01\%$, $p < 0.05$). The performances of NB and SVM were not statistically significantly different in terms of the reported metrics.

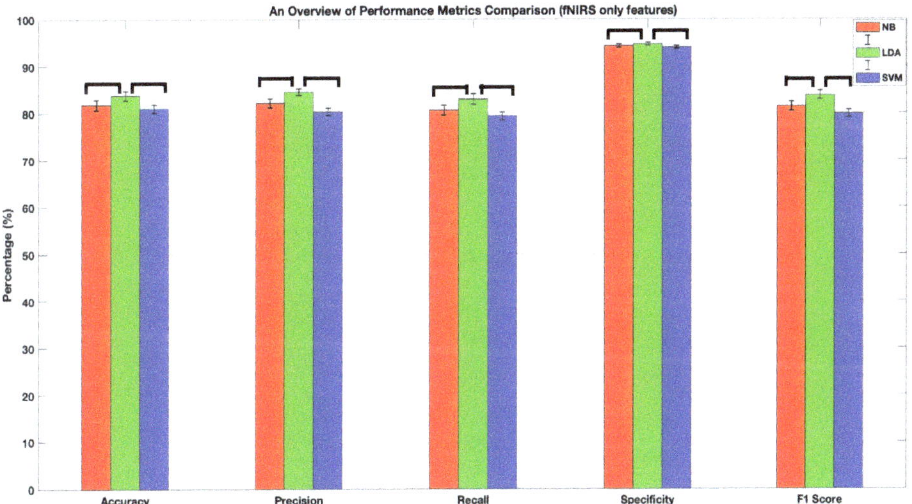

Figure 6. Classification performances of NB, LDA, and SVM algorithms after being trained with fNIRS-derived features (i.e., GE_{cm} and GE_{dm}). Horizontal lines depict statistically significant differences between performances of different algorithm pairs. All algorithms achieved accuracy, precision, recall, and F1-score performances above 80%, while the specificity scores were above 94%. LDA performed significantly higher than both SVM and NB in terms of accuracy, precision, recall, specificity, and F1-score. The error bars represent standard error of the mean performance after 10 runs of a 10-fold C.V.

Table 2 presents the four-class classification performances of NB, LDA, and SVM classifiers when they were trained with fNIRS and behavioral features (i.e., GE_{cm}, GE_{dm}, and CQ). Comparisons between the performance of each tabulated algorithm with respect to the corresponding performance obtained with fNIRS only features (Table 1) were performed with paired *t*-tests, and bold-typed results (Table 2) denote significantly higher performance of the corresponding algorithm when compared to the results when the same algorithm is fed with fNIRS only features. All algorithms achieved accuracy, precision, recall, and F1-score performances above 83%, while the specificity scores were all above 94%. Feeding NB and SVM with a combination of fNIRS and behavioral features resulted in a statistically significantly higher performance in each metric when compared to the performance obtained by training the same algorithm with fNIRS only features. However, LDA achieved a similar performance in each metric regardless of the type of feature set combination utilized for training. There were no statistically significant differences in accuracy, recall, specificity, and F1-scores among the three algorithms. Nonetheless, the precision score obtained with SVM was statistically significantly higher than both LDA and NB ($86 \pm 1.6\%$, $p < 0.05$, Table 2 and Figure 7).

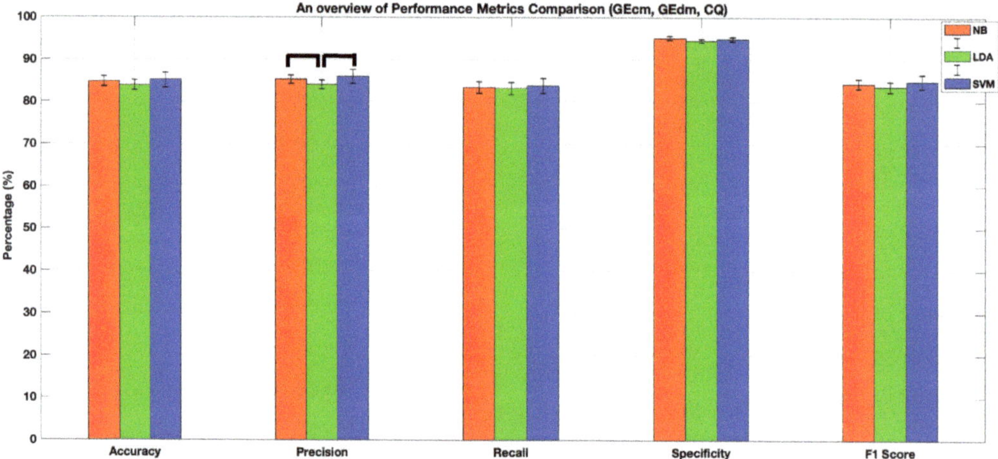

Figure 7. Classification performances of NB, LDA, and SVM algorithms after being trained with a combination of fNIRS and behavioral features (i.e., GE_{cm}, GE_{dm}, and CQ). Horizontal lines depict statistically significant differences between performances of different algorithm pairs. All algorithms achieved accuracy, sensitivity, precision, and recall performances above 83%, while the specificity scores were all above 94%. There was no statistically significant difference among accuracy, recall, specificity, and F1-score performances of the three algorithms. The error bars represent standard error of the mean performance after 10 runs of a 10-fold C.V.

Training LDA with fNIRS only features resulted in a comparable performance with the performance metrics obtained when the same algorithm was trained with a combination of fNIRS and behavioral features. A statistical comparison of the performance of the best-performing algorithm (LDA) and fNIRS only feature set combination of Table 1 with the performance metrics of NB and SVM of Table 2 demonstrated that no significant difference existed between any algorithm pair for accuracy, recall, specificity, and F1-scores.

To sum up, we conclude that training LDA with fNIRS only features results in a comparable performance with training the three supervised algorithms with a combination of fNIRS and behavioral features. Regarding the best performance, although there were no statistically significant differences among the three algorithms for accuracy, recall, specificity, and F1-scores (Figure 6), we should still note that SVM had the best performance in all metrics when trained with a combination of fNIRS and behavioral features obtained during the Stroop task (Tables 1 and 2 and Figure 7).

Figure 8 presents the confusion matrices for each algorithm, which demonstrate the true-positive and false-negative predictions attributed to each class. All algorithms achieved classification accuracies above 70% for each class. All algorithms demonstrated the highest true-positive prediction rate for SCZ patients, which was followed by OCD, HC, and MIG. SCZ and OCD subjects were not misclassified as HCs for any of the algorithms.

This result is significant as these two patient groups are expected to have the most distinct alterations in cognitive performance and cerebral hemodynamic activity during the Stroop task when compared to the HC group [13,17,66–71]. The fact that HC subjects were not misclassified as OCD or SCZ for any of the algorithms suggests the distinctive and physiology-related informative power of the selected features. However, HC subjects could be falsely attributed to the MIG class (SVM: 1.05%, NB: 4.74%, LDA: 4.21%) regardless of the algorithm type. This result is not surprising as MIG subjects were tested during the interictal period while they were exempt from attacks, hence their cognitive performance and the relevant spatial and topographic distribution of functional activation might have been similar to HCs during the interictal period. The consistencies in the classification performance patterns of the three algorithms as well as the consistency of performance results with physiology-related information highlight the distinctive power and biologically informative nature of the fNIRS-derived features utilized in the study. It can be concluded that training NB, LDA, and SVM with fNIRS-derived metrics demonstrates a differential diagnosis potential, regardless of the mathematical architecture of the algorithm.

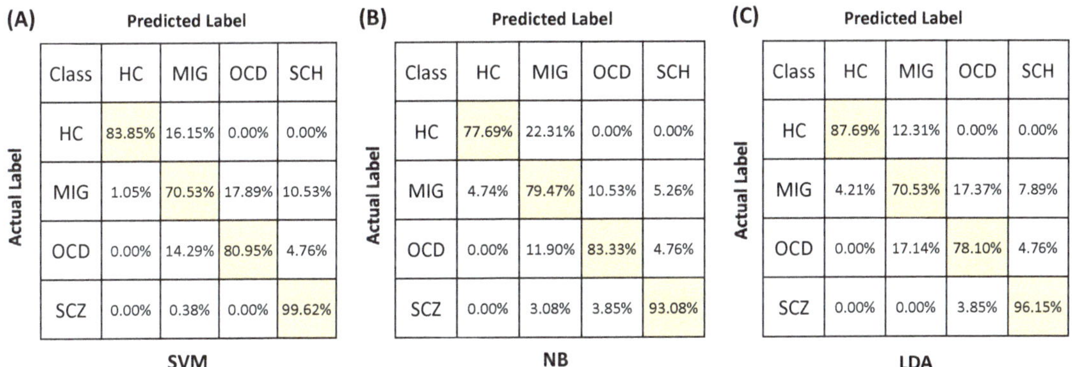

Figure 8. Confusion matrices depicting the true-positive (shaded in yellow) and false-negative predictions of (**A**) SVM, (**B**) NB, (**C**) LDA when they are trained with a combination of fNIRS and behavioral features.

4. Discussion

The current diagnostic model for a majority of neuropsychiatric disorders relies on evaluation of measures which include clinical, observational, and/or behavioral scales that are obtained through interviews, questionnaires, observations, self-reports, and/or neuropsychiatric test batteries [3–5]. However, subjectivity introduced during both collection and clinical interpretation of these multi-domain measures brings forth the demand for more objective diagnostic markers. The high variability in clinical decisions for similar cases observed across different clinicians, cultures, and countries highlights the critical need for developing more objective decision support systems for diagnosis, which should ideally be based on quantitative measures of the neurophysiological alterations underlying each disorder.

Taking this critical demand into consideration, the presented study aimed to assess the feasibility and applicability of an fNIRS-based automated classification approach for accurate prediction and objective identification of the presence of three distinct neuropsychiatric or neurological disorder states which are known to induce alterations in frontal lobe function. The proposed machine learning-based classification approach involved training various supervised learning algorithms with (i) novel fNIRS-derived informative biomarkers and (ii) a combination of fNIRS-derived biomarkers and performance measures obtained during a neuro-cognitive test, named the Stroop task. We tested and compared the efficacy of training three commonly employed and computationally efficient supervised learning algorithms with these neuronally induced biomarkers, and their comparably high performances were demonstrated with accuracy, precision, recall, specificity, and F1-scores. The performance of each algorithm in the correct identification of the presence of a disorder in each subject was evaluated by whether the subject was correctly labeled among the four classes, which included HCs, MIG, SCZ, and OCD. Hence, four-class brain–computer interface system designs were formulated which simply included the collection of hemodynamic signals with an fNIRS system while the subject was engaged in a Stroop task. Two global efficiency features were obtained from the PFC HBO signals, and accuracy and reaction rate performance obtained during the Stroop task were fused in a single behavioral feature, named the cognitive quotient (CQ). The comparably high performance scores obtained with the three classification algorithms, which have distinct mathematical architectures, highlighted the informative nature of these neuronally induced features. They also demonstrated the promising nature of integrating fNIRS-derived features together with cognitive performance scores from neuropsychiatric test measures and multivariate pattern analysis (MVPA) approaches for accurate recognition of neuropsychiatric disorder states. Our methodological approach resulted in increased classification accuracy when compared to the brain–computer interface (BCI) study designs conducted with fNIRS for other classification purposes, such as decoding mental thought processes or motor imagery signals [71,72].

In the following sections, we first evaluate the efficacy of NB, LDA, and SVM in correct identification of the presence of a disorder at the single subject level and we discuss the differential diagnostic potential of the proposed approach. We then highlight the importance of our findings, discuss the limitations of our study, and propose recommendations for future work.

4.1. Comparison of the Classification Performances of LDA, NB, and SVM

Training NB, LDA, and SVM with two fNIRS-derived functional connectivity metrics resulted in accuracy, precision, recall, and F1-score performances above 81%, while the specificity scores were all above 94%. While the performance metrics obtained with each algorithm had a very close range, it should be noted that LDA performed significantly higher than both SVM and NB in terms of accuracy ($83.8 \pm 1\%$, $p < 0.05$), precision ($85 \pm 0.01\%$, $p < 0.05$), recall ($83 \pm 0.01\%$, $p < 0.05$), specificity ($95 \pm 0.01\%$, $p < 0.05$), and F1-score ($84 \pm 0.01\%$, $p < 0.05$) when trained with fNIRS only features. A statistical comparison of the performance of the best-performing algorithm (LDA) and fNIRS only feature set combination of Table 1 with the performance metrics obtained by training each algorithm with a combination of fNIRS and behavioral features demonstrated that no significant difference existed between the performances of any algorithm pair for accuracy, recall, specificity, and F1-scores. Hence, we conclude that training LDA with fNIRS only features results in a comparable performance with training the three supervised algorithms with a combination of fNIRS and behavioral metrics. Regarding the best performance, we should note that SVM had the best performance in all metrics when trained with a combination of fNIRS and behavioral features obtained during a Stroop task (Tables 1 and 2 and Figure 6). However, we should also note that SVM did not have a statistically significantly higher performance than the rest of the algorithms for the majority of the performance metrics (i.e., accuracy, recall, specificity, and F1-scores reported in Figure 6). Hence, we can conclude that the utilized features are distinctive in nature as they performed well with all three classifiers regardless of the mathematical architecture of the algorithm. Obtaining a high classification performance with all classifiers highlights the feasibility and applicability of feeding machine learning-based methods with fNIRS-derived neuro-cognitive biomarkers for classification of disorder states associated with alterations in frontal lobe function.

With the recent advances in the computational power of daily used computers, MVPA methods have received increasing interest for automated identification and objective recognition of neurological and neuropsychiatric disorder states by use of structural and functional neuroimaging features. The majority of these studies examined the diagnostic potential of utilizing multivariate features for: (i) correct identification of the presence of a disease state, (ii) rating the severity of a clinical state, or (iii) differentiating subgroups of patients. Arabshirani et al. provided an excellent review of previous neuroimaging studies that aimed at single-subject prediction of neurological, neurodegenerative, or neuropsychiatric disorders by use of structural and functional imaging features [61], while Orru et al. presented an extensive summary of the previous studies that utilized SVM for differentiating a neuropsychiatric disease state from a healthy state [73]. Regarding automated recognition of SCZ, Steardo et al. provided a review of classification studies that utilized a combination of SVM and neuroimaging markers [58]. The majority of these studies reported binary classification performances for differentiating a disorder state from a healthy state and the reported accuracies ranged from 67% to 100%. Regarding differentiation of OCD from a healthy state, the highest performance metrics were reported by Sen et al., who proposed the utility of resting state functional connectivity-derived network features with SVM [74]. They achieved 80% accuracy, 81% sensitivity, and 77% specificity with a relatively small sample size (n = 16 for OCD and n = 13 for HC). Similarly, three studies utilized MVPA methods and MRI-based neuroimaging markers for accurate prediction of the presence of migraine by use of two-class classification schemes, and the reported accuracies ranged between 80% and 96% [75–77].

Among three-class classification studies, Yu et al. reported a study where they used several MVPA methods to discriminate healthy controls (n = 38), schizophrenic patients (n = 32), and patients diagnosed with major depression disorder (n = 19). They achieved a correct classification rate of 81% using functional connectivity features from resting state fMRI scans [59]. Their sample size was also similar to our study. Kawazaki et al. built a binary classification model for differentiating SCZ from HC utilizing voxel-based morphometry features from MRI with a small dataset (n = 30 per class). Their classification accuracy performance was 80% [78]. Yassin et al. performed a three-class classification study where they trained several machine learning algorithms for accurate identification of autism spectrum disorder, healthy controls, and SCZ patients. The best results were achieved with MRI-derived cortical thickness parameters using a logistic regression (LR) classifier. Their overall maximum classification accuracy was reported as 69%. The maximum binary classification accuracies between different class pairs were less than 80% when tested with several classifiers, including SVM [79].

We should note that an objective comparison of our performance results with the performances reported in previous studies is complicated since the study designs differed in terms of sample size, number of classes, type and number of features, disorder types, C.V procedure, and the selected classifiers (Table 3). Nonetheless, we can still conclude that the performance metrics achieved with our four-class classification methodology fall in the high-performance spectrum among the performance metrics reported in previous studies, which targeted classification of various neuropsychiatric populations from healthy counterparts by use of structural and functional neuroimaging measures.

Table 3. Comparison of the classification performances of the discussed studies.

Author/s (Year)	Sample Size	Classifier(s)	Number of Classes	Features	Mean Accuracy (%)
Sen et al. (2017) [74]	16 OCD, 13 HC	SVM	2	Resting state network features derived from fMRI data	80
Chong et al. (2016) [75]	58 MIG, 50 HC	Quadratic Discriminate Analysis	2	Resting state network features derived from fMRI data	86
Yang et al. (2018) [76]	21 MIG without aura, 15 MIG with aura, 28 HC	Convolutional Neural Networks	2 and 3	Resting state network features derived from fMRI data	85–99 (2 class), 87(3 class)
Hernandez et al. (2014) [77]	15 HC, 20 MIG, 19 Medication Abuse	SVM	2	Graph theoretical features derived from fMRI data	87

Table 3. *Cont.*

Author/s (Year)	Sample Size	Classifier(s)	Number of Classes	Features	Mean Accuracy (%)
Yu et al.(2013) [59]	32 SCZ, 19 MDD, 38 HC	SVM	3	Resting state network features derived from fMRI data	81
Kawasaki et al.(2007) [78]	30 SCZ, 30 HC	Multivariate Linear Model	2	Voxel based morphometry features extracted from MRI data	80
Yassin et al. (2020) [79]	64 SCZ, 36 ASD, 106 HC	Logistic Regression	3	Cortical thickness and subcortical volume features derived from MRI data	69
Pardo et al. (2006) [80]	10 SCZ, 10 BP, 8 HC	LDA	3	Neuropsychiatric test scores and structural metrics derived from MRI data	96
Present work	20 MIG, 26 OCD, 21 SCZ, 13 HC	LDA, SVM, NB	4	Cognitive quotient and Global Efficiency metrics derived from fNIRS data	84.7 (LDA), 83.8(NB), 85 (SVM)

4.2. Potential of the Proposed Methodology for Differential Diagnosis

Comorbidities often exist among major neuropsychiatric disorders in the form of overlapping behavioral symptoms and similar neurobiological alterations. Hence, one of the major challenges for a precise diagnostic decision is to be able to differentially diagnose neuropsychiatric disorders which have overlapping symptoms, such as SCZ, MDD, and BD [14,57]. While differential diagnosis of the patient groups presented in this study would be easy to decipher at the clinical stage, we should emphasize the fact that our work serves as a proof-of-concept study to demonstrate the utility of combining fNIRS-derived functional connectivity metrics obtained during a cognitive test with machine learning-based classification methods for assisting accurate classification and objective identification of neuropsychiatric disorder states associated with frontal lobe functional abnormalities.

Recent studies have presented compelling evidence that a wide variety of neuropsychiatric disorders are characterized with alterations in the neural activity of the PFC [13]. However, whether there exists a distinct topographical distribution of functional abnormalities specific to each neuropsychiatric disorder and whether each neuropsychiatric disorder can be associated with a distinct abnormality in cerebral activation that can be recognized by fNIRS during a cognitive test remains unclear. In our study, all algorithms achieved classification accuracies above 70% for each class. All algorithms demonstrated the highest true-positive prediction rate for SCZ patients, which was followed by OCD, HC, and MIG. HC subjects were not misclassified as OCD or SCZ for any of the algorithms. These two patient groups are expected to have the most distinct alterations in cognitive performance and cerebral hemodynamic activity during the Stroop tasks when compared to the HC group [13,17,66–70]. Hence, the fact that HC subjects were not misclassified as OCD or SCZ for any of the algorithms suggests the distinctive and physiology-related informative power of the selected features. HC subjects could be falsely attributed to the MIG class. This result is not surprising as MIG subjects were tested during the interictal period which might be cognitively similar to a healthy state, and hence the spatial and topographic distribution of their functional activation might not be significantly different from HCs during the Stroop task. OCD and SCZ subjects were not misclassified as HCs for any of the algorithms. The consistencies in the classification performance patterns of the three algorithms as well as the consistency of performance results with physiology-related information highlight the distinctive power and biologically informative nature of the fNIRS-derived features utilized in the study.

To sum up, our results suggest that training NB, SVM, or LDA with the fNIRS-derived global efficiency metrics obtained during a Stroop task demonstrates a differential diagnosis potential, regardless of the mathematical architecture of the algorithm. Our findings also support the notion that some novel neuro-biological features obtained with fNIRS methodology during cognitive tasks might serve as distinct signatures of the spatiotemporal characteristics of different neuropsychiatric disorder states which are associated with frontal lobe function abnormalities. Exploration of such informative and biologically derived features and combining them with machine learning-based classification approaches may have significant potential for differential diagnoses of psychopathologies which have comorbidities and overlapping symptoms.

4.3. Limitations of the Study and Recommendations for Future Work

We should note that the sample sizes of our subject groups were still small, although they exceeded the sample sizes reported in many of the previously reported classification studies in neuropsychiatry literature [58,61,73–80]. As a continuation of this study, we will test the performance of our methodology on a larger subject cohort. Our classification problems will include a higher number of disorder types and we will test the efficacy of identifying patients with comorbidities. We will also test the informative power of extracting hemodynamic and cognitive features from concurrent fNIRS recordings taken during a variety of neuropsychological tests which target different aspects of cognition.

Deep learning (DL) techniques have a great potential to improve the performance of fNIRS-based BCI systems if sufficiently large training sets are available [81,82]. The major advantages of these techniques rely on their ability to capture the complexity of neural information embedded in the HBO signal patterns through optimization of the network structures [81]. Indeed, there exists some successfully implemented DL classifiers with fNIRS and EEG signals [82–86]. However, we avoided testing the utility of DL algorithms in the presented work because of the limited cohort size of each group. Models constructed with DL algorithms have a tendency to overfit data when they are trained with small sample sizes (i.e., n < 5000) [81]. Future work will involve testing the efficacy of DL algorithms for addressing the presented classification problem in a larger cohort size and by utilizing data augmentation procedures.

In the presented study, clinical diagnosis of each participant was performed by experienced psychiatrists after careful follow-up procedures, and their final clinical decision was considered the golden standard. Hence, we could test and report the performance of each algorithm by whether it could correctly predict the final clinical decision of an experienced psychiatrist whose decision is considered as ground truth. Although the participants included in the study were reported to have strong and distinct symptoms and the clinicians had good clinical expertise for making a correct diagnosis, there still exists a possibility that some of the patients might have been given a different diagnosis by a different group of clinicians and might be incorrectly labeled. Hence, we can only report the value and high performance of combining fNIRS only markers and supervised learning algorithms in correctly predicting the clinical decision of an experienced clinician. Nonetheless, such a decision support system still might assist young clinicians who have not gained enough expertise with patients.

While the differential diagnosis of the patient classes reported in this study might not be a difficult problem in the clinics, we should note that this is a proof-of-concept study for demonstrating the potential of predicting a clinical decision through analysis of informative hemodynamic features obtained noninvasively in a clinical setting with a wearable and ergonomic fNIRS system design. Hemodynamic information can be collected with similar system designs during similar cognitive tests or vasomechanical challenges and can be processed to extract biomarkers which can be used for differential diagnosis of neurological or neuropsychiatric disorders that are known to induce abnormalities in PFC function.

5. Conclusions

The overarching goal of this study was to test the feasibility of an fNIRS-based BCI system design for accurate and objective identification of the presence of neuropsychiatric or neurological disorders. Our results demonstrate the potential of training supervised learning algorithms with fNIRS-derived hemodynamic and cognitive features for precise recognition of the presence of a neurological or neuropsychiatric disorder at the single-subject level. They also highlight the promise of exploring PFC-based neurofunctional features as distinctive and objective biomarkers of neuropsychiatric or neurological disorders which are associated with alterations in frontal lobe function. Neuronally induced biomarkers can be easily obtained in clinical settings with portable, wearable fNIRS systems. Such system designs might also have great potential for objective classification and differential diagnosis of major neuropsychiatric disorders which, in most cases, have overlapping behavioral symptoms across each other and are hard to distinguish when decisions are based solely on observation, self-report, interview, and/or rating scales.

Author Contributions: Conceptualization, S.B.E. and G.Y.; methodology, S.B.E. and G.Y.; validation, S.B.E. and G.Y.; formal analysis, S.B.E. and G.Y.; investigation, S.B.E. and G.Y.; data curation, S.B.E. and G.Y.; writing—original draft preparation, S.B.E.; writing—review and editing, S.B.E. and G.Y.; visualization, S.B.E. and G.Y.; supervision, S.B.E. All authors have read and agreed to the published version of the manuscript.

Funding: This research received no external funding.

Institutional Review Board Statement: The study was conducted in accordance with the Declaration of Helsinki, and approved by the Institutional Review Board of Pamukkale University (protocol code B.30.2.PAU.0.01.00.00-200/130-0146, 15 January 2007).

Informed Consent Statement: Informed consent was obtained from all subjects involved in the study.

Data Availability Statement: Data is available upon request.

Acknowledgments: The authors would like to thank Ata Akın for his generosity in sharing the data set and the fruitful discussions.

Conflicts of Interest: The authors declare no conflict of interest.

References

1. Klöppel, S.; Abdulkadir, A.; Jack, C.R.; Koutsouleris, N.; Mourão-Miranda, J.; Vemuri, P. Diagnostic neuroimaging across diseases. *Neuroimage* **2012**, *61*, 457–463. [CrossRef] [PubMed]
2. Singh, I.; Rose, N. Biomarkers in psychiatry. *Nature* **2009**, *460*, 202–207. [CrossRef] [PubMed]
3. World Health Organization. *International Statistical Classification of Diseases*, 10th ed.; World Health Organization: Geneva, Switzerland, 1992.
4. American Psychiatric Association. *Diagnostic and Statistical Manual of Mental Disorders*, 4th ed.; American Psychiatric Press: Washington, DC, USA, 1994.
5. Azechi, M.; Iwase, M.; Ikezawa, K.; Takahashi, H.; Canuet, L.; Kurimoto, R.; Nakahachi, T.; Ishii, R.; Fukumoto, M.; Ohi, K.; et al. Discriminant analysis in schizophrenia and healthy subjects using prefrontal activation during frontal lobe tasks: A near-infrared spectroscopy. *Schizophr. Res.* **2010**, *117*, 52–60. [CrossRef] [PubMed]
6. Aboraya, A.; Rankin, E.; France, C.; El-Missiry, A.; John, C. The Reliability of Psychiatric Diagnosis Revisited: The Clinician's Guide to Improve the Reliability of Psychiatric Diagnosis. *Psychiatry* **2006**, *3*, 41–50.
7. Ward, C.H.; Beck, A.T.; Mendelson, M.; Mock, J.E.; Erbaugh, J.K. The psychiatric nomenclature. Reasons for diagnostic disagreement. *Arch. Gen. Psychiatry* **1962**, *7*, 198–205. [CrossRef]
8. Drevets, W.C. Neuroimaging studies of mood disorders. *Biol. Psychiatry* **2000**, *48*, 813–829. [CrossRef]
9. Okada, G.; Okamoto, Y.; Yamashita, H.; Ueda, K.; Takami, H.; Yamawaki, S. Attenuated prefrontal activation during a verbal fluency task in remitted major depression. *Psychiatry Clin. Neurosci.* **2009**, *63*, 423–425. [CrossRef]
10. Weyandt, L.; Swentosky, A.; Gudmundsdottir, B.G. Neuroimaging and ADHD: fMRI, PET, DTI Findings, and Methodological Limitations. *Dev. Neuropsychol.* **2013**, *38*, 211–225. [CrossRef]
11. Ferrari, M.; Quaresima, V. A brief review on the history of human functional near-infrared spectroscopy (fNIRS) development and fields of application. *Neuroimage* **2012**, *63*, 921–935. [CrossRef]
12. Scholkmann, F.; Kleiser, S.; Metz, A.J.; Zimmermann, R.; Pavia, J.M.; Wolf, U.; Wolf, M. A review on continuous wave functional near-infrared spectroscopy and imaging instrumentation and methodology. *Neuroimage* **2014**, *85*, 6–27. [CrossRef]
13. Ehlis, A.-C.; Schneider, S.; Dresler, T.; Fallgatter, A.J. Application of functional near-infrared spectroscopy in psychiatry. *Neuroimage* **2014**, *85*, 478–488. [CrossRef] [PubMed]
14. Yeung, M.K.; Lin, J. Probing depression, schizophrenia, and other psychiatric disorders using fNIRS and the verbal fluency test: A systematic review and meta-analysis. *J. Psychiatr. Res.* **2021**, *140*, 416–435. [CrossRef] [PubMed]
15. Song, H.; Chen, L.; Gao, R.; Bogdan, I.I.M.; Yang, J.; Wang, S.; Dong, W.; Quan, W.; Dang, W.; Yu, X. Automatic schizophrenic discrimination on fNIRS by using complex brain network analysis and SVM. *BMC Med. Inform. Decis. Mak.* **2017**, *17* (Suppl. S3), 166. [CrossRef] [PubMed]

16. Garrity, A.G.; Pearlson, G.D.; McKiernan, K.; Lloyd, D.; Kiehl, K.A.; Calhoun, V.D. Aberrant "Default Mode" Functional Connectivity in Schizophrenia. *Am. J. Psychiatry* **2007**, *164*, 450–457. [CrossRef] [PubMed]
17. Woodward, T.S.; Leong, K.; Sanford, N.; Tipper, C.M.; Lavigne, K.M. Altered balance of functional brain networks in Schizophrenia. *Psychiatry Res. Neuroimaging* **2016**, *248*, 94–104. [CrossRef]
18. Sharma, A.; Weisbrod, M.; Kaiser, S.; Markela-Lerenc, J.; Bender, S. Deficits in fronto-posterior interactions point to inefficient resource allocation in schizophrenia. *Acta Psychiatr. Scand.* **2010**, *123*, 125–135. [CrossRef]
19. Venkataraman, A.; Whitford, T.J.; Westin, C.-F.; Golland, P.; Kubicki, M. Whole brain resting state functional connectivity abnormalities in schizophrenia. *Schizophr. Res.* **2012**, *139*, 7–12. [CrossRef]
20. Narayanaswamy, J.; Hazari, N.; Venkatasubramanian, G. Neuroimaging findings in obsessive–compulsive disorder: A narrative review to elucidate neurobiological underpinnings. *Indian J. Psychiatry* **2019**, *61* (Suppl. S1), S9–S29. [CrossRef]
21. Fajnerova, I.; Gregus, D.; Francova, A.; Noskova, E.; Koprivova, J.; Stopkova, P.; Hlinka, J.; Horacek, J. Functional Connectivity Changes in Obsessive–Compulsive Disorder Correspond to Interference Control and Obsessions Severity. *Front. Neurol.* **2020**, *11*, 568. [CrossRef]
22. Erdoğan, S.B.; Yükselen, G.; Yegül, M.M.; Usanmaz, R.; Kıran, E.; Derman, O.; Akın, A. Identification of impulsive adolescents with a functional near infrared spectroscopy (fNIRS) based decision support system. *J. Neural Eng.* **2021**, *18*, 056043. [CrossRef]
23. Çiftçi, K.; Sankur, B.; Kahya, Y.P.; Akin, A. Multilevel Statistical Inference from Functional Near-Infrared Spectroscopy Data during Stroop Interference. *IEEE Trans. Biomed. Eng.* **2008**, *55*, 2212–2220. [CrossRef]
24. Aydore, S.; Mihcak, M.K.; Ciftci, K.; Akin, A.; Mihak, M.; Cifti, K. On Temporal Connectivity of PFC via Gauss–Markov Modeling of fNIRS Signals. *IEEE Trans. Biomed. Eng.* **2009**, *57*, 761–768. [CrossRef] [PubMed]
25. Dadgostar, M.; Setarehdan, S.K.; Shahzadi, S.; Akin, A. Functional connectivity of the PFC via partial correlation. *Optik* **2016**, *127*, 4748–4754. [CrossRef]
26. Einalou, Z.; Maghooli, K.; Setarehdan, S.K.; Akin, A. Effective channels in classification and functional connectivity pattern of prefrontal cortex by functional near infrared spectroscopy signals. *Optik* **2016**, *127*, 3271–3275. [CrossRef]
27. Einalou, Z.; Maghooli, K.; Setarehdan, S.K.; Akin, A. Graph theoretical approach to functional connectivity in prefrontal cortex via fNIRS. *Neurophotonics* **2017**, *4*, 041407. [CrossRef] [PubMed]
28. Akin, A. Partial correlation-based functional connectivity analysis for functional near-infrared spectroscopy signals. *J. Biomed. Opt.* **2017**, *22*, 126003. [CrossRef]
29. Akın, A. fNIRS-derived neurocognitive ratio as a biomarker for neuropsychiatric diseases. *Neurophotonics* **2021**, *8*, 035008. [CrossRef]
30. Zysset, S.; Müller, K.; Lohmann, G.; von Cramon, D. Color-Word Matching Stroop Task: Separating Interference and Response Conflict. *Neuroimage* **2001**, *13*, 29–36. [CrossRef]
31. Erdogan, S.B.; Özsarfati, E.; Dilek, B.; Sogukkanlı Kadak, K.; Hanoglu, L.; Akin, A. Classification of motor imagery and execution signals with population-level feature sets: Implications for probe design in fNIRS based BCI. *J. Neural Eng.* **2019**, *16*, 026029. [CrossRef]
32. Akgül, C.B.; Akin, A.; Sankur, B. Extraction of cognitive activity-related waveforms from functional near-infrared spectroscopy signals. *Med. Biol. Eng. Comput.* **2006**, *44*, 945–958. [CrossRef]
33. Erdoğan, S.B.; Yucel, M.; Akin, A. Analysis of task-evoked systemic interference in fNIRS measurements: Insights from fMRI. *Neuroimage* **2014**, *87*, 490–504. [CrossRef] [PubMed]
34. Pollonini, L.; Bortfeld, H.; Oghalai, J.S. PHOEBE: A method for real time mapping of optodes-scalp coupling in functional near-infrared spectroscopy. *Biomed. Opt. Express* **2016**, *7*, 5104–5119. [CrossRef] [PubMed]
35. Jobsis, F.F. Noninvasive, infrared monitoring of cerebral and myocardial oxygen sufficiency and circulatory parameters. *Science* **1977**, *198*, 1264–1267. [CrossRef]
36. Firbank, M.; Eiji, O.; Delpy, D.T. A theoretical study of the signal contribution of regions of the adult head to near-infrared spectroscopy studies of visual evoked responses. *Neuroimage* **1998**, *8*, 69–78. [CrossRef] [PubMed]
37. Cope, M.; Delpy, D.T. System for long-term measurement of cerebral blood and tissue oxygenation on newborn infants by near infra-red transillumination. *Med. Biol. Eng. Comput.* **1988**, *26*, 289–294. [CrossRef]
38. Delpy, D.T.; Cope, M.; Zee, P.V.D.; Arridge, S.; Wray Sand Wyatt, J. Estimation of optical pathlength through tissue from direct time of flight measurement. *Phys. Med. Biol.* **1988**, *33*, 1433–1442. [CrossRef]
39. Boas, D.A.; Dale, A.M.; Franceschini, M.A. Diffuse optical imaging of brain activation: Approaches to optimizing image sensitivity, resolution, and accuracy. *Neuroimage* **2004**, *23*, S275–S288. [CrossRef]
40. Sutoko, S.; Monden, Y.; Tokuda, T.; Ikeda, T.; Nagashima, M.; Funane, T.; Atsumori, H.; Kiguchi, M.; Maki, A.; Yamagata, T.; et al. Atypical dynamic-connectivity recruitment in attention-deficit/hyperactivity disorder children: An insight into task-based dynamic connectivity through an fNIRS study. *Front. Hum. Neurosci.* **2020**, *14*, 3. [CrossRef]
41. Al-Shargie, F.; Kiguchi, M.; Badruddin, N.; Dass, S.C.; Hani AF, M.; Tang, T.B. Mental stress assessment using simultaneous measurement of EEG and fNIRS. *Biomed. Opt. Express* **2016**, *7*, 3882–3898. [CrossRef]
42. Al-Shargie, F.; Tang, T.B.; Kiguchi, M. Assessment of mental stress effects on prefrontal cortical activities using canonical correlation analysis: An fNIRS-EEG study. *Biomed. Opt. Express* **2017**, *8*, 2583–2598. [CrossRef]
43. Hoshi, Y. Functional near-infrared optical imaging: Utility and limitations in human brain mapping. *Psychophysiology* **2003**, *40*, 511–520. [CrossRef] [PubMed]

44. Hoshi, Y.; Kobayashi, N.; Tamura, M. Interpretation of near-infrared spectroscopy signals: A study with a newly developed perfused rat brain model. *J. Appl. Physiol.* **2001**, *90*, 1657–1662. [CrossRef] [PubMed]
45. Latora, V.; Marchiori, M. Efficient behavior of small-world networks. *Phys. Rev. Lett.* **2001**, *87*, 198701. [CrossRef]
46. Cui, X.; Bray, S.; Reiss, A.L. Speeded near infrared spectroscopy (NIRS) response detection. *PLoS ONE* **2010**, *5*, e15474. [CrossRef] [PubMed]
47. Hawkins, D.M.; Basak, S.C.; Mills, D. Assessing model fit by cross-validation. *J. Chem. Inf. Comput. Sci.* **2003**, *43*, 579–586. [CrossRef] [PubMed]
48. Tai, K.; Chau, T. Single-trial classification of NIRS signals during emotional induction tasks: Towards a corporeal machine interface. *J. Neuroeng. Rehabil.* **2009**, *6*, 39. [CrossRef]
49. Sitaram, R.; Zhang, H.; Guan, C.; Thulasidas, M.; Hoshi, Y.; Ishikawa, A.; Shimizu, K.; Birbaumer, N. Temporal classification of multichannel near-infrared spectroscopy signals of motor imagery for developing a brain–computer interface. *Neuroimage* **2007**, *34*, 1416–1427. [CrossRef]
50. Tanaka, H.; Katura, T. Classification of change detection and change blindness from near-infrared spectroscopy signals. *J. Biomed. Opt.* **2011**, *16*, 087001. [CrossRef]
51. Misawa, T.; Takano, S.; Shimokawa, T.; Hirobayashi, S. A brain-computer interface for motor assist by the prefrontal cortex. *Electron. Commun. Jpn.* **2012**, *95*, 1–8. [CrossRef]
52. Naseer, N.; Hong, K.-S. fNIRS-based brain-computer interfaces: A review. *Front. Hum. Neurosci.* **2015**, *9*, 3. [CrossRef]
53. Naseer, N.; Qureshi, N.K.; Noori, F.M.; Hong, K.-S. Analysis of Different Classification Techniques for Two-Class Functional Near-Infrared Spectroscopy-Based Brain-Computer Interface. *Comput. Intell. Neurosci.* **2016**, *2016*, 5480760. [CrossRef] [PubMed]
54. Khan, R.A.; Naseer, N.; Qureshi, N.K.; Noori, F.M.; Nazeer, H.; Khan, M.U. fNIRS-based Neurorobotic Interface for gait rehabilitation. *J. Neuroeng. Rehabil.* **2018**, *15*, 7. [CrossRef] [PubMed]
55. Shin, J.; Jeong, J. Multiclass classification of hemodynamic responses for performance improvement of functional near-infrared spectroscopy-based brain-computer interface. *J. Biomed. Opt.* **2014**, *19*, 67009. [CrossRef] [PubMed]
56. Shoushtarian, M.; Alizadehsani, R.; Khosravi, A.; Acevedo, N.; McKay, C.M.; Nahavandi, S.; Fallon, J.B. Objective measurement of tinnitus using functional near-infrared spectroscopy and machine learning. *PLoS ONE* **2020**, *15*, e0241695. [CrossRef]
57. Cearns, M.; Hahn, T.; Baune, B.T. Recommendations and future directions for supervised machine learning in psychiatry. *Transl. Psychiatry* **2019**, *9*, 271. [CrossRef]
58. Steardo, L.; Carbone, E.A.; De Filippis, R.; Pisanu, C.; Segura-Garcia, C.; Squassina, A.; De Fazio, P. Application of Support Vector Machine on fMRI Data as Biomarkers in Schizophrenia Diagnosis: A Systematic Review. *Front. Psychiatry* **2020**, *11*, 588. [CrossRef]
59. Yu, Y.; Shen, H.; Zeng, L.-L.; Ma, Q.; Hu, D. Convergent and Divergent Functional Connectivity Patterns in Schizophrenia and Depression. *PLoS ONE* **2013**, *8*, e68250. [CrossRef]
60. Sakai, K.; Yamada, K. Machine learning studies on major brain diseases: 5-year trends of 2014–2018. *Jpn. J. Radiol.* **2018**, *37*, 34–72. [CrossRef]
61. Arbabshirani, M.R.; Plis, S.; Sui, J.; Calhoun, V.D. Single subject prediction of brain disorders in neuroimaging: Promises and pitfalls. *Neuroimage* **2016**, *145*, 137–165. [CrossRef]
62. Stephan, K.E.; Schlagenhauf, F.; Huys, Q.; Raman, S.; Aponte, E.; Brodersen, K.; Rigoux, L.; Moran, R.; Daunizeau, J.; Dolan, R.; et al. Computational neuroimaging strategies for single patient predictions. *Neuroimage* **2016**, *145*, 180–199. [CrossRef]
63. Hall, M.; Frank, E.; Holmes, G.; Pfahringer, B.; Reutemann, P.; Witten, I.H. The WEKA data mining software: An update SIGKDD. *Explorations* **2009**, *11*, 10–18. [CrossRef]
64. Keerthi, S.S.; Shevade, S.K.; Bhattacharyya, C.; Murthy, K.R.K. Improvements to Platt's SMO algorithm for SVM classifier design. *Neural Comput.* **2001**, *13*, 637–649. [CrossRef]
65. Üstün, B.; Melssen, W.J.; Buydens, L.M. Facilitating the application of support vector regression by using a universal Pearson VII function based kernel. *Chemom. Intell. Lab. Syst.* **2006**, *81*, 29–40. [CrossRef]
66. Zhang, S.; Yang, G.; Ou, Y.; Guo, W.; Peng, Y.; Hao, K.; Zhao, J.; Yang, Y.; Li, W.; Zhang, Y.; et al. Abnormal default-mode network homogeneity and its correlations with neurocognitive deficits in drug-naive first-episode adolescent-onset schizophrenia. *Schizophr. Res.* **2019**, *215*, 140–147. [CrossRef] [PubMed]
67. Yamamuro, K.; Kimoto, S.; Iida, J.; Kishimoto, N.; Tanaka, S.; Toritsuka, M.; Ikawa, D.; Yamashita, Y.; Ota, T.; Makinodan, M.; et al. Distinct patterns of blood oxygenation in the prefrontal cortex in clinical phenotypes of schizophrenia and bipolar disorder. *J. Affect. Disord.* **2018**, *234*, 45–53. [CrossRef] [PubMed]
68. Zhang, Z.; Wang, Y.; Zhang, Q.; Zhao, W.; Chen, X.; Zhai, J.; Chen, M.; Du, B.; Deng, X.; Ji, F.; et al. The effects of CACNA1C gene polymorphism on prefrontal cortex in both schizophrenia patients and healthy controls. *Schizophr. Res.* **2019**, *204*, 193–200. [CrossRef] [PubMed]
69. Okada, K.; Ota, T.; Iida, J.; Kishimoto, N.; Kishimoto, T. Lower prefrontal activity in adults with obsessive–compulsive disorder as measured by near-infrared spectroscopy. *Prog. Neuro-Psychopharmacol. Biol. Psychiatry* **2013**, *43*, 7–13. [CrossRef]
70. Ota, T.; Iida, J.; Sawada, M.; Suehiro, Y.; Yamamuro, K.; Matsuura, H.; Tanaka, S.; Kishimoto, N.; Negoro, H.; Kishimoto, T. Reduced Prefrontal Hemodynamic Response in Pediatric Obsessive–Compulsive Disorder as Measured by Near-Infrared Spectroscopy. *Child Psychiatry Hum. Dev.* **2012**, *44*, 265–277. [CrossRef]
71. Ahn, S.; Jun, S.C. Multi-Modal Integration of EEG-fNIRS for Brain-Computer Interfaces—Current Limitations and Future Directions. *Front. Hum. Neurosci.* **2017**, *11*, 503. [CrossRef]

72. Batula, A.M.; Kim, Y.E.; Ayaz, H. Virtual and Actual Humanoid Robot Control with Four-Class Motor-Imagery-Based Optical Brain-Computer Interface. *BioMed Res. Int.* **2017**, *2017*, 1463512. [CrossRef]
73. Orru, G.; Pettersson-Yeo, W.; Marquand, A.F.; Sartori, G.; Mechelli, A. Using support vector machine to identify imaging biomarkers of neurological and psychiatric disease: A critical review. *Neurosci. Biobehav. Rev.* **2012**, *36*, 1140–1152. [CrossRef] [PubMed]
74. Sen, B.; Bernstein, G.A.; Xu, T.; Mueller, B.A.; Schreiner, M.W.; Cullen, K.R.; Parhi, K.K. Classification of obsessive-compulsive disorder from resting-state fMRI. In Proceedings of the 2016 38th Annual International Conference of the IEEE Engineering in Medicine and Biology Society (EMBC), Orlando, FL, USA, 16–20 August 2016; pp. 3606–3609. [CrossRef]
75. Chong, C.D.; Gaw, N.; Fu, Y.; Li, J.; Wu, T.; Schwedt, T.J. Migraine classification using magnetic resonance imaging resting-state functional connectivity data. *Cephalalgia* **2016**, *37*, 828–844. [CrossRef] [PubMed]
76. Yang, H.; Zhang, J.; Liu, Q.; Wang, Y. Multimodal MRI-based classification of migraine: Using deep learning convolutional neural network. *Biomed. Eng. Online* **2018**, *17*, 138. [CrossRef] [PubMed]
77. Jorge-Hernandez, F.; Chimeno, Y.G.; Garcia-Zapirain, B.; Zubizarreta, A.C.; Beldarrain, M.A.G.; Fernandez-Ruanova, B. Graph theory for feature extraction and classification: A migraine pathology case study. *Bio-Med. Mater. Eng.* **2014**, *24*, 2979–2986. [CrossRef]
78. Kawasaki, Y.; Suzuki, M.; Kherif, F.; Takahashi, T.; Zhou, S.-Y.; Nakamura, K.; Matsui, M.; Sumiyoshi, T.; Seto, H.; Kurachi, M. Multivariate voxel-based morphometry successfully differentiates schizophrenia patients from healthy controls. *Neuroimage* **2007**, *34*, 235–242. [CrossRef]
79. Yassin, W.; Nakatani, H.; Zhu, Y.; Kojima, M.; Owada, K.; Kuwabara, H.; Gonoi, W.; Aoki, Y.; Takao, H.; Natsubori, T.; et al. Machine-learning classification using neuroimaging data in schizophrenia, autism, ultra-high risk and first-episode psychosis. *Transl. Psychiatry* **2020**, *10*, 278. [CrossRef]
80. Pardo, P.J.; Georgopoulos, A.P.; Kenny, J.T.; Stuve, T.A.; Findling, R.L.; Schulz, S.C. Classification of adolescent psychotic disorders using linear discriminant analysis. *Schizophr. Res.* **2006**, *87*, 297–306. [CrossRef]
81. D'souza, R.N.; Huang, P.Y.; Yeh, F.C. Structural Analysis and Optimization of Convolutional Neural Networks with a Small Sample Size. *Sci. Rep.* **2020**, *10*, 834. [CrossRef]
82. Wickramaratne, S.D.; Mahmud, M.S. Conditional-GAN Based Data Augmentation for Deep Learning Task Classi-fier Improvement Using fNIRS Data. *Front. Big Data* **2021**, *4*, 659146. [CrossRef]
83. Khalil, K.; Asgher, U.; Ayaz, Y. Novel fNIRS study on homogeneous symmetric feature-based transfer learning for brain-computer interface. *Sci. Rep.* **2022**, *12*, 3198. [CrossRef]
84. Lyu, B.; Pham, T.; Blaney, G.; Haga, Z.; Sassaroli, A.; Fantini, S.; Aeron, S. Domain adaptation for robust workload level alignment between sessions and subjects using fNIRS. *J. Biomed. Opt.* **2021**, *26*, 022908. [CrossRef] [PubMed]
85. Hennrich, J.; Herff, C.; Heger, D.; Schultz, T. Investigating Deep Learning for fNIRS Based BCI. In Proceedings of the 2015 37th Annual International Conference of the IEEE Engineering in Medicine and Biology Society (EMBC), Milan, Italy, 25–29 August 2015; pp. 2844–2847. [CrossRef]
86. Chiarelli, A.M.; Croce, P.; Merla, A.; Zappasodi, F. Deep Learning for Hybrid EEG-fNIRS Brain-Computer In-terface: Application to Motor Imagery Classification. *J. Neural Eng.* **2018**, *15*, 036028. [CrossRef] [PubMed]

Article

Unmanned Aerial Vehicle for Laser Based Biomedical Sensor Development and Examination of Device Trajectory

Usman Masud [1,2,*,†], Tareq Saeed [3,†], Faraz Akram [4,‡], Hunida Malaikah [5,‡] and Altaf Akbar [6,*]

[1] Faculty of Electrical and Electronics Engineering, University of Engineering and Technology, Taxila 47050, Pakistan

[2] Department of Electrical Communication Engineering, University of Kassel, 34127 Kassel, Germany

[3] Nonlinear Analysis and Applied Mathematics (NAAM)-Research Group, Department of Mathematics, Faculty of Science, King Abdulaziz University, P.O. Box 80203, Jeddah 21589, Saudi Arabia; tsalmalki@kau.edu.sa

[4] Faculty of Engineering and Applied Sciences, Riphah International University, Islamabad 44000, Pakistan; faraz.akram@riphah.edu.pk

[5] Department of Mathematics, Faculty of Science, King Abdulaziz University, P.O. Box 80203, Jeddah 21589, Saudi Arabia; hmalaikah@kau.edu.sa

[6] Department of Economics, Management, Industrial Engineering and Tourism (DEGEIT), University of Aveiro, 3800-000 Aveiro, Portugal

* Correspondence: usmanmasud123@hotmail.com (U.M.); altafakbar@ua.pt (A.A.)

† These authors contributed equally to this work.

‡ These authors contributed equally to this work.

Citation: Masud, U.; Saeed, T.; Akram, F.; Malaikah, H.; Akbar, A. Unmanned Aerial Vehicle for Laser Based Biomedical Sensor Development and Examination of Device Trajectory. *Sensors* **2022**, *22*, 3413. https://doi.org/10.3390/s22093413

Academic Editors: Noman Naseer, Imran Khan Niazi and Hendrik Santosa

Received: 16 March 2022
Accepted: 25 April 2022
Published: 29 April 2022

Publisher's Note: MDPI stays neutral with regard to jurisdictional claims in published maps and institutional affiliations.

Abstract: Controller design and signal processing for the control of air-vehicles have gained extreme importance while interacting with humans to form a brain–computer interface. This is because fewer commands need to be mapped into multiple controls. For our anticipated biomedical sensor for breath analysis, it is mandatory to provide medication to the patients on an urgent basis. To address this increasingly tense situation in terms of emergencies, we plan to design an unmanned vehicle that can aid spontaneously to monitor the person's health, and help the physician spontaneously during the rescue mission. Simultaneously, that must be done in such a computationally efficient algorithm that the minimum amount of energy resources are consumed. For this purpose, we resort to an unmanned logistic air-vehicle which flies from the medical centre to the affected person. After obtaining restricted permission from the regional administration, numerous challenges are identified for this design. The device is able to lift a weight of 2 kg successfully which is required for most emergency medications, while choosing the smallest distance to the destination with the GPS. By recording the movement of the vehicle in numerous directions, the results deviate to a maximum of 2% from theoretical investigations. In this way, our biomedical sensor provides critical information to the physician, who is able to provide medication to the patient urgently. On account of reasonable supply of medicines to the destination in terms of weight and time, this experimentation has been rendered satisfactory by the relevant physicians in the vicinity.

Keywords: unmanned aerial vehicle; spectroscopy; brain–computer interface application; mathematical modelling; semiconductor laser

1. Introduction

Over the past decades, computer and communication technologies have developed rapidly in the form of brain–computer interface (BCI). As evident from its underlying nomenclature, BCI enables a user to control a computer or any other device with signals of the brain. In the past couple of decades, researchers have developed several applications of the BCI including, but not limited to, character spelling [1,2] and word typing [3,4], wheelchair control [5,6], prosthetics control [7,8], neurological rehabilitation [9], home control [10], virtual reality control [11], gaming [12,13] and quadcopter control [14,15].

Taking these into account, this paper aims at developing an Unmanned Aerial Vehicle (UAV) for the BCI application for better operating of air-vehicles.

Therefore, these techniques have now become an indispensable tool for patients' daily life. The purpose of BCI is to make patients' life more appropriate and natural in a daily living environment [16]. The fundamental aim of BCI is to assist patients (particularly, in locked-in state) to interact with the living environment using only brain signals [17,18]. After the obtained command from the brain, another challenge is to properly control the applications that manipulate wheelchairs, robotic arms or drones [19–25] as in this study. Many signal processing techniques have previously been created, however, in this study, a framework is devised that can be helpful in finer operation of air-vehicles. Several techniques can be implemented for this purpose, as in [26,27].

Biomedical logistic sensors are gaining interest in the aerospace industry, extending their applications from solar-powered drones to origami-style space-based solar power stations due to their flexibility, light weight and transparency. The basic element is an autonomous air-vehicle which is a quad-rotor without the presence of a human pilot onboard [28–30]. Its navigation or movement is controlled through a control system on board or it can also be navigated manually by remote control from the ground [31–34]. A quad-rotor has three translational and rotational movements through which it can achieve six degrees of freedom. For this purpose, the rotational and translational motion have to couple with the help of rotors [35,36]. The quad-copter has four arms and each arm contains an independent rotor. Generally, quad-rotors use two pairs of propellers with identical parameters (i.e., two clockwise and two anti-clockwise). The variation in speed of each rotor makes it possible to achieve the manoeuvring of the quad-rotor.

The resulting dynamics of the model are highly nonlinear, especially after accounting for the complicated aerodynamics effects and unlike ground vehicles which have much friction during their motion, the quad-rotors may have little friction to gain their movement (as per their system model or design), and they must provide their own damping effects to eliminate all these nonlinear factors. For this purpose, these vehicles use an electronic control system to maintain the stability of quad-rotors using electronic board and sensors, i.e., an accelerometer [37–43]. The quad-rotors were among the first vehicles to take off and land vertically. The agile and revolutionary design of quad-rotors not only makes them capable of exploring an unknown locus, but they can also move with precision and much faster pace than any other vehicle in a dense environment [44,45]. This paper is a demonstration of one of the versatile applications of quad-rotors to transfer product autonomously to the required destination.

With this background, the main goal of our work is to develop a microsensor which is based on the fundamentals of two modes (wavelengths), utilising the principles of intracavity absorption spectroscopy [38,44]. This is a laser-based setup that consists of two wavelengths (specified by the term *modes*), as shown in Figure 1. The scheme works in a way in which the light from the Semiconductor Optical Amplifier (SOA) is reflected by two Fibre Bragg Gratings (FBGs), each specifying a wavelength (mode) that creates a competition between both modes. The difference between the intensities of both modes (SOA and human breath) aids in the detection of diseases of elderly patients, monitored continuously by the respective physician. The Variable Couplers (VCs) divide the light intensity, and the isolators ensure its flow in one direction (for instance, to the Optical Spectrum Analyzer (OSA)) (complete experimental details can be found in [46,47], with latest developments regarding expansion to wireless channel in [48]).

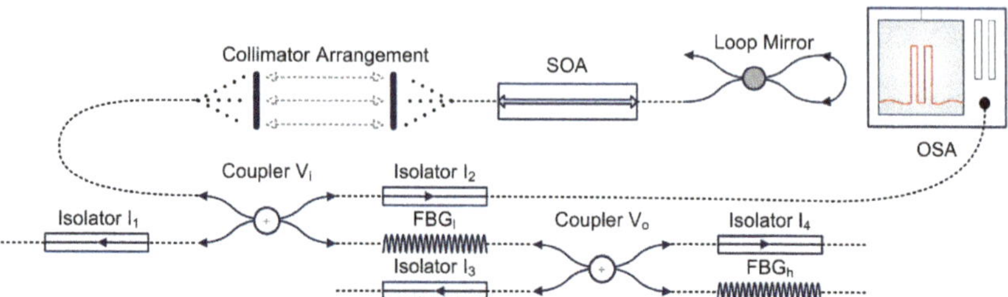

Figure 1. Design of the laser based two mode experimental scheme [38].

In case of any emergency situation, it is obvious from the description provided above that this demands spontaneous action by the physician in the form of transportation of medication to the said patient [39,44,45,49]. In light of this fact, we plan to make a product that is capable of transferring products, in particular, medical items from one place to another autonomously, with a limited time span. The physicians dealing with emergency situations were consulted for this purpose, and the foremost requirements were discussed. According to them, the said device must have reliable accuracy and efficiency that is mandatory in transportation procedure. As illustrated above, in case of an emergency, the patient must be immediately provided with aid in such a way that the life threatening situation can be avoided. This can be some suitable medicine or some medical instruments that can help as a life saver. This, in turn, means that the hardware and software parts must both be developed in a strictly technical way that helps in the operation of the device afterwards [40,42,43,50].

The proposed sensor will be used for the detection of Volatile Organic Compunds (VOCs) in the human breath which is exhaled from the lungs. This can provide useful information about specific diseases, hence the motivation behind this work. Certain sensors exist for the checking of specific health characteristics in human beings [51,52]. For instance, wearable sensors are being specifically developed that help to check the health of human beings on a continuous basis [51]. Molecularly imprinted materials are being widely used to sense human attributes [53]. Recent advancements include, but are not limited to, behaviour verification, movement monitoring, surveillance, characterization and other applications for human beings and animals [54,55]. On account of these notions, there exists a wide range of work for future in this specific area of development, as predicted in [56]. Hence, this is the first attempt that targets human breath and its diagnosis [38,55,57]. The list of abbreviations is outlined in Table 1.

Motivation and objectives

- BCIs have been successfully incorporated into numerous areas of the world, with successful outcomes;
- Biomedical sensors have been evolving fast for the past few decades and BCIs are proving to be a very important tool for that;
- During the state of an emergency for elderly patients, it is mandatory to attend and provide medication at the earliest to the relevant person, for which UAVs can be utilized effectively;
- Considering the development of biomedical sensor for the targeted patients, a UAV is planned to be designed that can aid in the remote monitoring and first aid to the said individuals;
- The quadcopter designed in this work has presented a highly stable operation, which is mandatory for the supply of medical equipment from the hospital to the patient.

Table 1. List of symbols.

VC_i/VC1	Variable Coupler corresponding to the inner cavity
VC_o/VC2	Variable Coupler corresponding to the outer cavity
SOA	Semiconductor Optical Amplifier
OSA	Optical Spectrum Analyzer
M_i/M1	Mode corresponding to the inner cavity
M_o/M2	Mode corresponding to the outer cavity
UAV	Unmanned Air Vehicle
ESC	Electronic Speed Controller
BLDC	Brushless Direct Current
BEC	Battery Elimination Circuit
Li-Po	Lithium Polymer
ϕ	Roll
θ	Pitch
Ψ	Yaw
V_b	Back Electromotive Force
T_e	Electromagnetic Torque
$T_{\omega\prime}$	Torque due to rotational acceleration of motor
T_ω	Torque generated due to velocity of the motor
T_L	Torque due to mechanical load across motor
K_t	Torque constant
J	Inertia of constant
K_p	Coefficient for Proportional term
K_i	Coefficient for Integral term
K_d	Coefficient for Derivative term

2. Related Work

In the last few years, the relevance and applications of unmanned air-vehicles have multiplied significantly in different areas of life ([30,32,36,45,58,59]), based on various operational mechanisms. There are many advantages to these devices, including, but not limited to, the cheap potential use of these vehicles, which gives opportunities to perform tasks which are very difficult and sometimes impossible otherwise, especially for emergency operations [44,60]. Most of these vehicles are not only able to fly in complex and busy places, but also can reach them in much less time than people themselves, which is desirable in the modern era. Thanks to these merits, many research organizations in collaboration with vendors have started delivering products through multi-rotors after recognizing the capabilities of these flying vehicles [61–63]. However, being a recent area of investigation, these devices lack stability and appropriate designs for specific applications, which are the foremost disadvantages of using them. Another problem is the usage of expensive equipment that cannot be purchased for normal air-vehicles. For these reasons, the current focus is on the design and modelling of quad-rotors in order to make them stable, reliable and cheap. Frame design and modelling are the first steps toward the journey of building any air-vehicle, as the overall flight dynamics and parameters are based on its frame. It is therefore important to mention that the design of a quad-copter's dynamics is classified with respect to two reference systems, i.e., body frame and inertial frame.

After discussions with the physicians working in the emergency, it is very important to realize that the device is designed for the transportation of medical items. For this purpose, there are several factors that must be considered beforehand. These include, but should not be limited to, efficient utilization of resources, supplying driving energy to the device in the form of its voltage, spectrum allocation and obtaining permission from the telecommunication authorities. Another prominent factor would be that the medical aspects have to be checked and examined to the maximum possible level of emergency items. To the best of our knowledge, this is the first approach in our vicinity in this connection, with limited financial resources and overall high efficiency, as discussed in the forthcoming sections in detail. The cases have been discussed in appropriate details in [64–67]. The trajectory of the quad-copter must be controlled in a reasonable way to do this, which involves its propulsion [68,69]. Propulsion means to push forward or to drive an object. In terms of an air-vehicle, propulsion is the force through which propellers push the air down and gain thrust. Brushless motors are mostly used to achieve propulsion in a quad-copter. These motors not only have very high power-to-weight ratio, but can also spin with thousands of revolutions per minute (rpm) [63]. The motor's speed is controlled through an electronic device called an electronic speed controller (ESC) [70,71]. The ESC can switch motors from on to off state and vice versa, thereby maintaining the desired speed and thrust of quad-copter. Therefore, controlling the air-vehicle's flight dynamics is a complex and interesting problem. The core unit of an air-vehicle is its control mechanism, which works like a brain in humans, and therefore has been the focus of interest in the last few years by the scientists via different approaches. Some model the control system directly by calculating their required parameters from the system and some use different theories to achieve these tasks (i.e., through classical control theory phenomenon) [36,37,58,62,64].

3. Hardware Considerations

Since the device is aimed for a biomedical application, there are numerous aspects that have to be brought into attention beforehand.

3.1. FRAME

The standout amongst the most vital piece of any quad-copter is its frame. It ought to be lightweight and flexible with the goal that it can hold the weight of different segments [72,73]. This is essential because the medical equipment can be of various weights. Moreover, the object to be moved might be sensitive in the sense that a slight damage during the operation could cause serious consequences [64]. Therefore, the option of constructing an edge at home, with the assistance of aluminium or balsa sheet, is eradicated. We resort to using a pre-fabricated casing whose parts are moderately low-cost and simple to supplant. In case of a crash, the arms ought to be somewhat resistant in a way that they should be the ones which undergo damage, thereby averting any harm to engines or costly gadgets on the edge. On the other hand, we really need them to be somewhat fragile, to accommodate the object on quad-copter.

Likewise, the arms assume an essential part in the battle against vibrations, which can cause various diverse issues. Flight controllers, with their touchy accelerometers and spinners, do not normally respond well to unremitting shaking. Any vibrations during the flight in a non-friendly environment may cause damage [74].

If we use an arm which has too much flexibility, it can reverberate and create harmonics that are transferred across the multi-rotor. On the contrary, arms that are too stiff would pass on vibrations with no hosing, bringing similar issues. There is a fine line between these two matters which needs to be drawn, and we attempt to use carbon fibre which is one of the most widely recognized materials for multi-rotor outlines. A large number of its physical properties suit our application [44,64].

3.2. Selection of Brush-Less DC (BLDC) Motor

BLDC (brush-less DC) motors do not utilise brushes for compensation. They are electronically commutated and have the following properties [72,75]:

- High effectiveness with noiseless tasks;

- Better speed versus torque attributes; and
- Quite high speed and longer life.

To control the speed of the motor, an ESC is required. The life of BLDC motor is quite long as there are no brushes which might be a source of damage. There is no starting and considerably less electrical commotion. In addition, considering the device for a biomedical application, the advantage of a brushless motor is its better preliminary cost ([44,65]).

We require four numbers of BLDC Motors for the copter. For the major part of the device, brushless motors are stated in kVs, such as, 810 kV, 1100 kV, 1400 kV, and 1800 kV. At this point, it must be recalled that the kV rating indicates the number of revolutions per minute (rpm) of the motor. For instance, a BLDC motor which has a kV rating of 1000 kV will turn at 1000 rpm when a voltage of 1 volt is applied at its input. Similarly, on applying 12 volts, the motor will turn at 12,000 rpm, and so on [63,75].

3.3. Electronic Speed Controller (ESC)

It is important to control the mechanism of the BLDC in a reliable way, for which an ESC is used. An ESC in a quad-copter performs two essential functions ([69,70]). First, it works as a Battery Elimination Circuit (BEC), a device that enables both the motors and the recipient to be powered by a solitary battery. The second function is to take the receiver's feedback, in the form of the flight controller's signal, and provide the required amount of current to the motors.

Each motor requires an ESC that regulates the amount of power to the engine, as indicated by the input throttle level. It additionally gives +5 V power to the flight hardware. ESC is based on a 32-bit microcontroller (ARM/AVR) and it has a variety of MOSFETs to drive the BLDC engine [63]. The firmware of ESC can be customized during the manufacturing phase.

In this way, the ESCs perform and ideal job of controlling BLDC. In simple words, an ESC is just a brushless motor controller board with battery input and a three-phase yield for the motor. ESCs can be found in a wide range of variations, where the input current is the most critical factor. This enables an appropriate control of the cut-off voltage, timing, acceleration and braking mechanism of the system [63,71,75].

3.4. Power Input

Lithium Polymer (Li-Po) battery is a kind of rechargeable battery that has overwhelmed the electric world, particularly for quad-copters. They are the primary reason why electric flight is presently an exceptionally feasible choice over fuel-controlled models.

Li-Po batteries are light in weight and hold enormous power in a little bundle. However, they are costly and have a lifetime of just 400 to 500 charge cycles. This means that a lot of care has to be to be taken while using this device. Otherwise, due to the unpredictable electrolyte utilized as a part of Li-Po batteries, they can blast or burst into flames easily when misused. Every cell of the Li-Po battery is rated at 3.7 V. The Li-Po battery, with four cells each of 3.7 V, used for our project is rated at 14.8 V. The amount of power the battery pack can hold is called the limit, and it is shown in milliamp hours (mAh). The battery we use for our project is of 3000 mAh. This means that 3000 mAh would be totally released in one hour when a 3000 mA stack is placed on it.

3.5. Propellers

It is imperative to rotate the motor, and hence the device. Therefore, on each of the brushless motors, a propeller is mounted. The four propellers are really not indistinguishable. When one takes a photo from below, one sees that the front and the back propellers are tilted to one side, while the left and right propellers are tilted to one side.

4. Design Methodology

Here, the procedure we choose to accomplish our project is discussed with the help of a spiral model methodology ([76,77]). We include mathematical approaches and algorithms that are implemented throughout in this project which helps in better explaining the kinematics and aerodynamics of the quad-copter.

4.1. Flowchart

The design of the quad-copter is divided into two stages. In the first stage, hardware design and assembling of quad-copter is performed, while in the second stage, software design and its implementation is completed, as shown in Figure 2. For a better explanation, the flowchart of quad-copter design is also shown in the Figure 3.

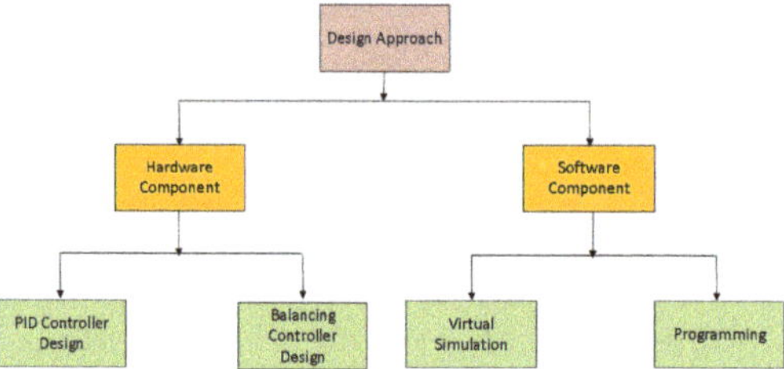

Figure 2. Design Methodology.

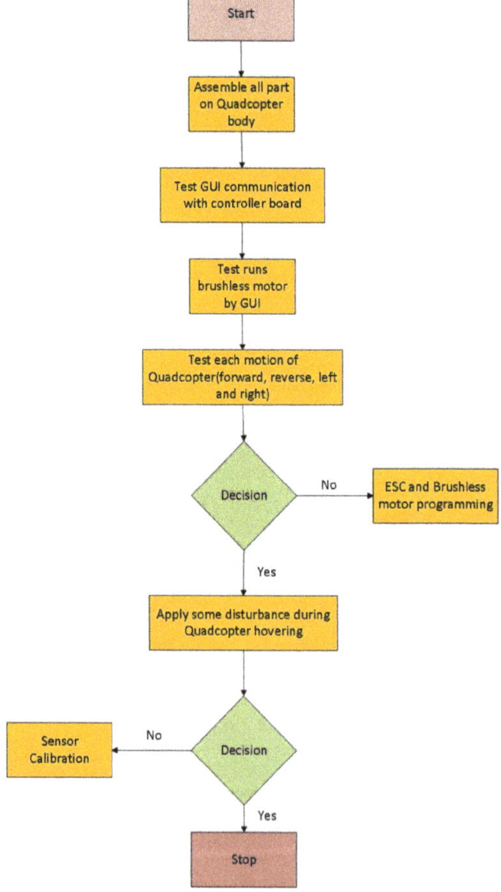

Figure 3. Flow chart of assembling and testing quad-copter.

With the help of the equipment mentioned above, we proceed in the following way. The components are assembled to establish a physical structure of the quad copter, followed by testing the GUI communication with the controller board. Test runs are conducted on the brushless motors by this GUI, checking its forward, reverse, left and right movements, respectively. Afterwards, the ESC and brushless motors are programmed in case there is some issue with the said movement. This is cross-checked by applying some disturbance to the quad copter, and the sensor is calibrated to accommodate the said level of disturbance.

4.2. Mathematical Modelling of Quad-Copter Dynamics

A mathematical model is a description of a system using mathematical concepts and language, and it plays its role in the operation of the device [63–65]. Similarly, the quad-copter's aerodynamics is mathematically modelled with respect to two reference systems which are most important parameters to gain the stable flight and motion [41,42]. The first reference system is the inertial frame of reference which is related to the earth and the body of the quad-copter. The second frame of reference is related to the quad-copter's frame itself, as to how how its translational and rotational motion can be controlled (portraying the frame orientation with respect to the origin of the quad-copter frame, Figure 4).

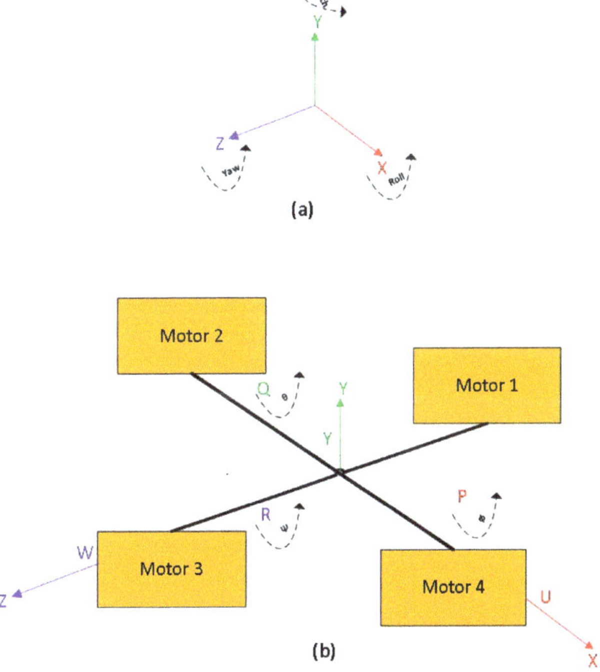

Figure 4. (**a**) Reference system of Quad-copter and (**b**) its movement with respect to the axes.

The position of the quad-copter during flight is given by the parameters Roll (ϕ), Pitch (θ) and Yaw (Ψ) angles, which define its rotation with respect to x-axis, y-axis and z-axis, respectively. The mathematical equation which defines the relationship between the quad-copter with respect to the earth is given by

$$R = \begin{bmatrix} C_\Psi C_\theta & C_\Psi S_\theta S_\phi - S_\Psi C_\phi & C_\Psi S_\theta C_\phi - S_\Psi C_\phi \\ S_\Psi C_\theta & S_\Psi S_\theta S_\phi - C_\Psi C_\phi & S_\Psi S_\theta C_\phi - C_\Psi S_\phi \\ -S_\theta C_\theta S_\phi & & C_\theta C_\phi \end{bmatrix}, \tag{1}$$

where S_ϕ & C_ϕ are notations of $sin(x)$ and $cos(x)$, respectively.

The thrust is the force or pull by which the quad-copter propellers move the air downwards for gaining upward force. Mathematically,

$$T = k \cdot \omega^2. \tag{2}$$

The rotational torque of the quad-copter is proportional to the square of the angular velocity of quad-copter motors which is given as

$$\tau_\psi = (-1)^{i+1} b \cdot \omega_i^2, \tag{3}$$

where motor i in the above equation is positive if the rotation of propeller is clockwise and negative otherwise. The torque is defined as the cross product of applied force and moment arm from the pivot point. Therefore, if L defines the length between a propeller and the copter's mid-point, the total torque of the body of the frame of the quad-copter is given by [78]

$$\tau_B = \begin{bmatrix} L.k.(-\omega_1^2 + \omega_3^2) \\ L.k.(-\omega_2^2 + \omega_4^2) \\ b.(-\omega_1^2 + \omega_2^2 - \omega_3^2 + \omega_4^2) \end{bmatrix}. \tag{4}$$

4.3. Brushless Dc Motor Model

The mathematical modelling of quad-copter motors is implemented by defining the mathematical model [79], which is shown in Figure 5.

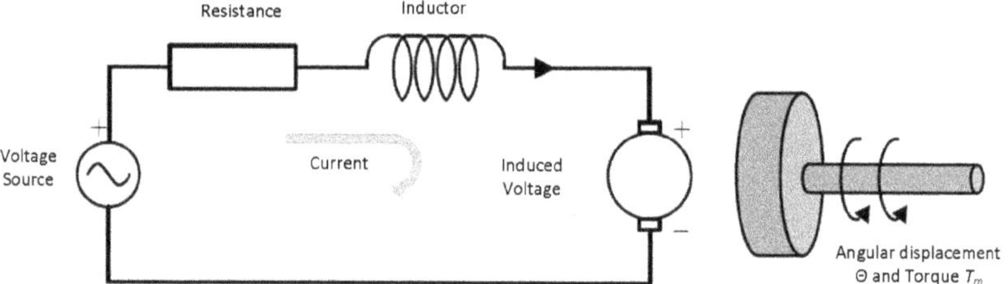

Figure 5. Mathematical Model of BLDC Motor.

Now, to find the transfer function of the model, we use armature voltage V(t) and angular displacement Θ, to define the angular velocity as

$$w(t) = \frac{d\Theta}{dt}. \tag{5}$$

The quad-copter brushless motor mathematical model consist of both electrical and mechanical parts. The equation for the electrical phenomenon is as follows:

$$-e(s) + R_a i_a + L_a \frac{di}{dt} + V_b = 0, \tag{6}$$

where V_b is the Back Electromotive Force which is given by

$$V_b = K_b \omega, \tag{7}$$

where K_b is the motor constant.

Mechanical Characteristics

According to law of conversation of energy, the total sum of torques of motor must be equal to zero. Therefore,

$$T_e - T_{w_r} - T_w - T_L = 0, \tag{8}$$

where T_e is the electromagnetic torque, $T_{\omega'}$ is the torque due to rotational acceleration of the motor, T_ω is the torque generated due to velocity of the motor, and T_L is the torque due to mechanical load across motor. T_e is directly proportional to the armature current i_a and it can be written as

$$T_e = K_t i_a, \tag{9}$$

where K_t is torque constant and it depends on flux density of the stator magnets. $T_{\omega'}$ can be written as

$$T_{\omega'} = J\frac{d\omega_a}{dt}, \tag{10}$$

where J is the inertia of constant.

5. Experimental Results

This section presents the various results and outputs based on real-time processing using a controller. As part of our project, which involves obstacle avoiding, and stable autonomous flight, we wanted to determine whether it was possible to receive correct output from the ultrasonic and other peripheral devices and sensor by given input. Thus, for this purpose, a test bench is designed for simulation. The mathematical simulation was carried out in Matlab (student version, complete details are available on https://www.mathworks.com/help/pdf_doc/matlab/rn.pdf; accessed on 21 December 2021)). For the purpose of our work, no toolboxes were used, as our system did not find any similarity with them. These were conducted in consultation with the technical team, and the simulations were performed at least thrice before proceeding with their interpretations. Test results and transformed useful data after analysis are shown in this section. The results and data obtained from GPS after applying algorithm are also described here, in accordance with Table 2.

Table 2. Description of main parameters.

No.	Parameter	Description
1	Satellites	22 tracking, 66 searching
2	Patch Antenna Size	15 mm × 15 mm × 4 mm
3	Update rate	1 to 10 Hz
4	Position Accuracy	1.8 m
5	Velocity Accuracy	0.1 m/s
6	Warm/cold start	34 s
7	Acquisition sensitivity	-145 dBm
8	Tracking sensitivity	-165 dBm
9	Maximum Velocity	515 m/s
10	Input Voltage range	3.0–5.5 V DC
11	Current drawn during navigation	25 mA tracking, 20 mA
12	Output	NMEA 0183, 9600 baud default
13	Feature	Multi-path detection and compensation

5.1. GPS Simulations

The GPS module used to find the current and desired locations of the quad-copter are presented here. As the purpose of GPS is to find the latitude and longitude, these two parameters are used to track the location of the quad-copter. The specifications of GPS used for autonomous quad-copter are as follows [80].

The Algorithm 1 used for the motion of quad-copter using GPS is *divide and conquer* [81–84], which is described below.

Algorithm 1: Divide and conquer

---

1. Sort points along the x-coordinate.
2. Split the set of points into two equal-sized subsets by a vertical line $x = x_{mid}$.
3. Solve the problem recursively in the left and right subsets. This will give the left-side and right-side minimal distances d_{Lmin} and d_{Rmin}, respectively.
4. Find the minimal distance d_{LRmin} among the pair of points in which one point lies to the left of the dividing vertical and the second point lies to the right.
5. The final answer is the minimum among d_{Lmin}, d_{Rmin}, and d_{LRmin}.

The output results of the GPS module are shown in Figure 6.

Latitude	*33.83514*
Longitude	*-11785384*
Altitude	*70.800 M*
PDOP	*1.1 (1.1)*
HDOP	*0.6 (0.6)*
VDOP	*1.0 (1.0)*
Satellites Tracked	*18*
Satellites in View	*19*

Figure 6. GPS Module output results.

5.2. Proportional (P), Integral (I) and Derivative (D) Controller

A Proportional (P), Integral (I) and Derivative (D) controller is the combination of three different types of controlling devices which are based on the control algorithm, hereby referred to as a PID controller. The mathematical modelling of quad-copter motors is performed by defining the mathematical model which is shown in Figure 7. A complete control mechanism becomes [31,32]

$$u_{sum}(t) = K_p e_{sum}(t) + K_i \int_0^t e_{sum}(t')dt' + K_d \frac{de_{sum}(t)}{dt},$$
$$K_p > 0, K_i > 0, K_d > 0,$$

(11)

where K_p, K_i and K_d are coefficients for the corresponding Proportional (P), Integral (I) and Derivative (D) terms, respectively. Ideally, the quad-copter should directly attain a specific height, after a specific time, without causing any delay in the flight. For this purpose, we take this ideal flight to be 1 arbitrary units (a.u.), after the passage of exactly 1 s. This helps us in obtaining a comparison of the various parameters of the device with this standard (reference) graphical values and the corresponding results.

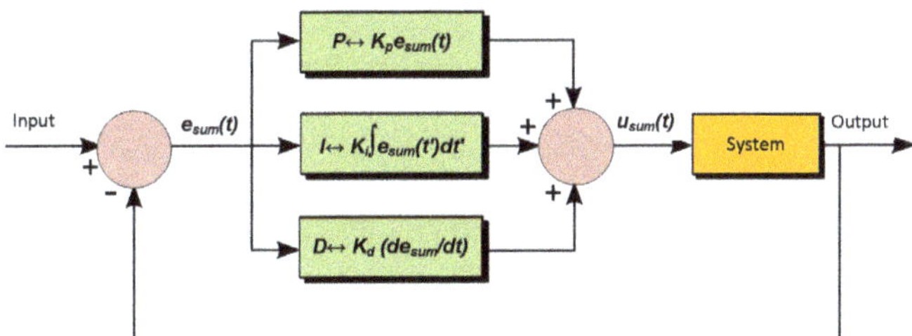

Figure 7. Working principle of a PID controller with feedback mechanism, indicating various components.

5.2.1. Proportional Controller

It is very important to supply the correct amount of current to the quad-copter, as an error could lead to malfunctioning of the motor. For this purpose, the proportional controller is used which causes the motor current of quad-copter set in proportion to the error, as shown in Equation (11) and Figure 7, and the resulting plot is shown in Figure 8. However, the proportional controller fails if the arms of the quad-copter have to pull different weights during flight due to different environment disturbances. This is exactly where the derivative and integral controllers play their part.

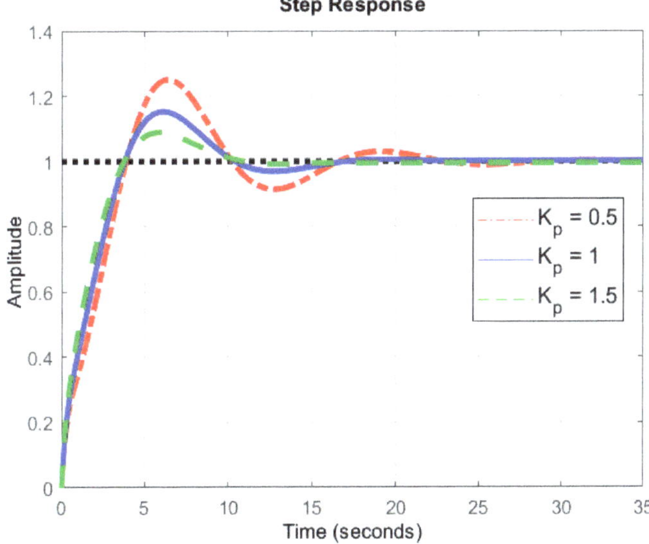

Figure 8. Principle of proportional controller.

5.2.2. Integral Controller

In this controller, the product of integral gain and error of the signal is added to the previous state of the model (Equation (11)), as indicated in Figure 7, with the resulting situation in Figure 9. The integral controller helps to decrease the rise time, thereby minimizing steady-state error. The integral term is proportional to both the magnitude of the error signal and the duration of the error in PID controller response. The different values of integral controller response are shown in Figure 9. A comparison among these values indicates the role of the integral controller in the flight of the device. For instance, when compared with the standard (reference) value, we see that the value of 0.5 shows the best closeness to the reference value, and the value of 2 shows the most different pattern. However, it must be stated that the time obtained by the quad-copter to attain a stable height is a bit more in case of the former value (0.5) as compared to the latter one (2). This could be justified by examining Equation (11), where a value of 2 will take a longer time as compared to 0.5, as the integration term will take time. This, in turn, will fluctuate the values, until a stable height is finally reached. This is exactly what is found in Figure 9.

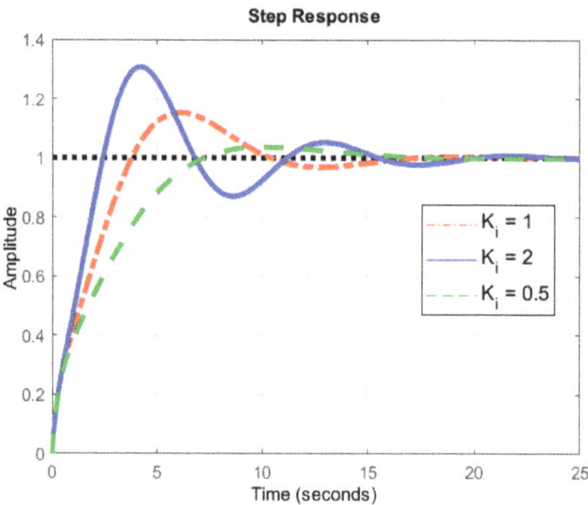

Figure 9. Principle of Integral Controller.

5.2.3. Derivative Controller

The derivative controller is an important part of the PID controller which reacts only when there is a change in error, as per Equation (11). It protects the auto controller from surpassing its limit by generating friction. The derivative of the error signal is calculated by finding the slope of the error signal with time and product of this rate of change with the derivative gain Kd. The response of the derivative controller is shown in Figure 10. An examination of this figure can be correlated with the mathematical interpretation of Equation (11). A value of 2 will take more time to integrate (summing the component values), and this results in an relative fluctuating curve. This will continue until and unless a stable amplitude has been attained, which is desirable for the quad-copter, and mandatory for the transportation of medical items. Therefore, this fact must be kept into account during the operation of the quad-copter, likewise discussed in the former section in detail.

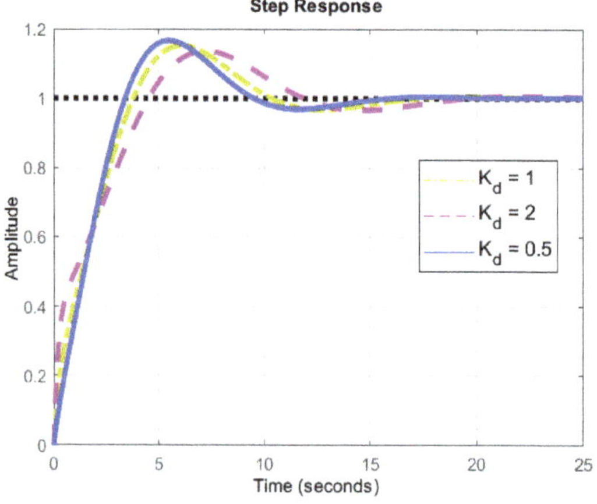

Figure 10. Principle of Derivative Controller.

5.3. Trajectory of the Quad-Copter

In order to maintain a smooth test of the vehicle, some technical aspects have to be brought into focus:

1. With the help of the vehicle's camera, a device can be detected by the vehicle using RF-ID tag on the object. This means that the object can be picked up from the hospital's store where it is located in a certain shelf, and transported to the patient in need;
2. To test the efficiency of the drone's activity, we place the central location of the drone within 300 m of the hospital's store (at furthest);
3. The medicines which have to be transported from the hospital to the patients come in various forms, and are sensitive to environmental variations. At this stage, the vehicle is used to transport only solid medicines and devices, as per recommendations of the physicians of the concerned hospital;
4. Afterwards, the positions of the patients were set at random distances (displacements) from the hospital's store, with the furthest one being at 1.5 km;
5. The maximum weight which the vehicle can lift is 1.5 kg. The maximum speed without any load is 25 km/h, and that with the maximum weight aboard is 21.5 km/h;
6. To carry out the experiments, specific permission was obtained from the local authority, as well as the hospital administration on weekends, as the work was not possible otherwise [43];
7. The weather conditions need to be taken into consideration beforehand. Each measurement was taken on a sunny day, with maximum wind speed of 8 km/h, atmospheric pressure under 1025 hPa, precipitation under 0.5 cm, humidity under under 65%, and visibility under 10.35 km.

In order to check the movement of the quad-copter from source to target destination, we resort to checking its trajectory along various directions. For the sake of convenience, the direction of motion towards the sky (vertical/upward) is nominated as the y-axis, the direction of motion towards right or left is nominated as the z-axis, and the direction towards the destination (horizontal) is nominated as the x-axis. This is conducted in complete accordance with Figure 4. First, we keep the values of x and z axes as constant, and vary the value of y. The motion of the quad-copter was recorded with the help of a video camera. The machine was flown five times along this path, and the average value was taken for each measurement. This value was compared with the simulated results, and the average error between these two values were calculated. These are shown in the respective plot in Figure 11.

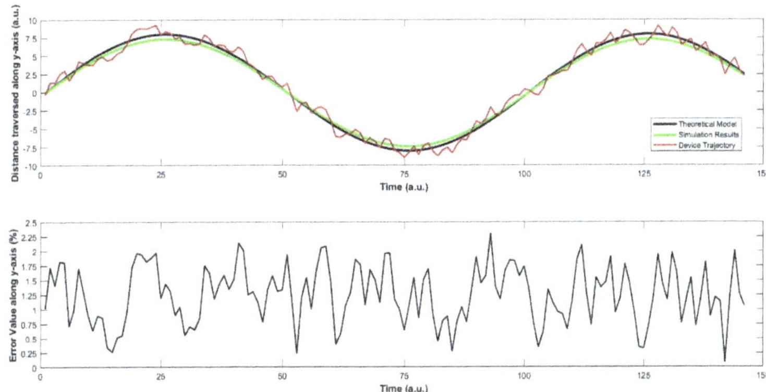

Figure 11. First trajectory scenario: Investigation of change in values of y. (**top**) shows a comparison between the ideal, simulation and actual (trajectory) values during the movement of the quad-copter along the y-axis, and (**bottom**) shows the average error between the simulated results and the actual (trajectory) values, which is very small.

Next, we perform the same test along x-axis, while keeping constant the values along the other two axes. The actual values were compared with the simulation results, and they are shown in Figure 12. Similarly, the case is analysed along the z-axis in Figure 13.

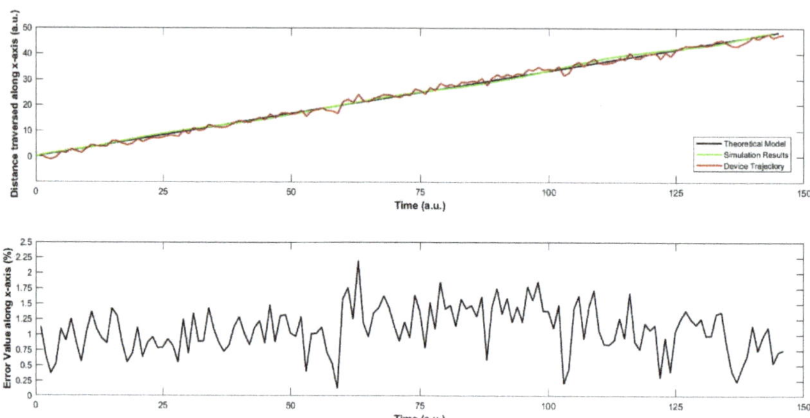

Figure 12. Second trajectory scenario: Investigation of change in values of x. (**top**) shows a comparison between the ideal, simulation and actual (trajectory) values during the movement of the quad-copter along the x-axis, and (**bottom**) shows the average error between the simulated results and the actual (trajectory) values, which is very small.

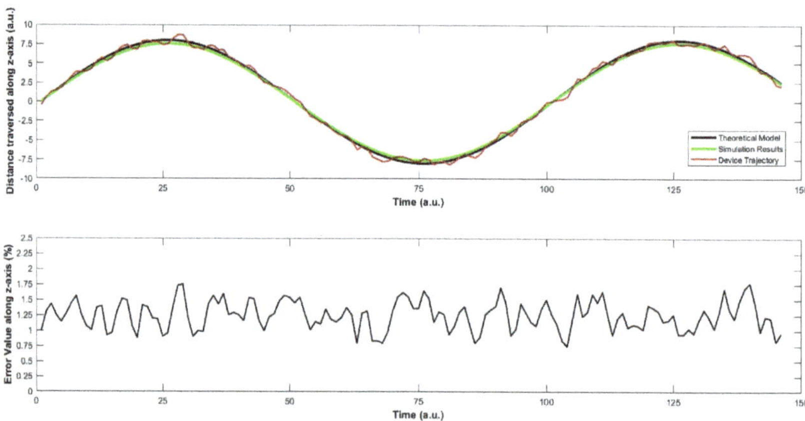

Figure 13. Second trajectory scenario: Investigation of change in values of z. (**top**) shows a comparison between the ideal, simulation and actual (trajectory) values during the movement of the quad-copter along the z-axis, and (**bottom**) shows the average error between the simulated results and the actual (trajectory) values, which is very small.

Observations

1. In Figure 11, simulation results show resemblance with the theoretical ones, as well as the trajectory of the quad-copter. As soon as the device takes off, we see that the error between the simulation and trajectory is less than 1.75%, which increases to a maximum of 2.25% at two occasions, efficiently comparable to the nearest results [21]. First, it is the occasion when the device has consumed about one-third of the travel time. This might be due to the sudden increase in the wind speed at that moment. A similar moment is observed when the device is about to reach the destination.

On average, the error value along the ordinate is 1.17%, which is acceptable for a quad-copter in similar designs [15,21].

2. Regarding Figure 13, a similar trend is observed for the motion of the device along z-axis. The fluctuations in the trajectory are slightly more than those along the y-axis. This is mainly because of two reasons. First, the device is equipped with a sensor that checks its motion along y-axis, but not along the z-axis. After consultation with the local vendors, we could not find a particular solution at that moment. Second, when the air moves along any direction, it has an effect on the motion of the device. This matter was discussed with two pilots of helicopters, who agreed with our stance that the weight of this machine is much smaller than that of a normal helicopter, and this can have an effect on the motion along the z-axis. In addition, they said that this would supposedly not affect any objects loaded on the machine, unless they cross the weight limit of our quad-copter. The average error is found to be 1.28%, which is comparable to recent works on quad-copters with different applications [6,22].

3. Afterwards, the motion of the quad-copter along the x-axis is recorded and compared with the simulation results in Figure 12. When the device travels about half of its distance, some fluctuations are seen in this trajectory which can be interpreted as follows. The quad-copter leaves the store inside the hospital and flies over the ground along its way to the destination which is about half-way. On account of the open area, the air speed is slightly higher as there is less congestion. This again acts as a slight resistance for the device, on its way. Therefore, the device experiences some fluctuations at this point. The average error in the value is 1.04%, which is slightly less than for the y- and z-axes, and no correlated results could be found at this level [9,11].

4. As per the trajectory profile of the device, it is important to note the stability during its movement. At this moment, it is observed that the overall results are within tolerable limits that are the primary focus of the biomedical application. For BCI to further accelerate its progress, the size of the device is important, as discussed in [8,9]. This becomes crucial as the medication becomes sensitive, which is not found in [15]. Although it is successful in imaging issues, the approach in [19] needs to be verified in different weather conditions, streamlining the identical repercussions. This becomes more interesting as there has been a focus on testing and implementation of BCI in virtual environments [21], and much remains to be done for the practical scenario, as we have approached here, with positive prospects in the future. This requires a deep investigation of the device in various dimensions on a continuous basis with BCI, an approach that has been attempted for the first time hitherto.

5. In this manner, our focus in this work was to implement the controller for better control of the quadcopter that can be used to implement the real brain signals as in the literature [11,15–17,22]. The target is to implement it in a biomedical sensor for which various technical aspects have been investigated. The characteristics of the controller were discussed in detail for smooth functioning with the real brain signals that can aid in the prospective design of the said scheme in the future.

6. Conclusions

The proposed framework provided better results in controlling air-vehicles with less error which can be engaged for brain–computer interfaces. In this manner, this research continues in light of recent developments in the areas of UAVs, as there is an urgent need to develop systems that are able to deliver logistics in a reliable and safe way. This becomes more important for the case of biomedical sensors that are designed to be mounted on the human body. In case of an emergency situation, the physician needs to take immediate steps, which might be life-saving in certain situations, and a direct connection between the physician and the patient might certainly not be possible on account of distance. The sensitivity of such a scenario raises a red flag if the locality of the patient is in a rural or distant area, with unspecified transportation options. This was the motivation behind the development of our biomedical sensor that is designed to monitor the health status of elderly patients. Considering the need to transport medical equipment from the hospital to the patient and vice versa, the current work tackles the development of a UAV in the form of a quad-copter. Several issues have been dealt with during the development of this device, after obtaining special permission to check the device in an urban environment. The location of the medical centre and the patient play an important role in the experimentation, for which the investigation was conducted appropriately. The maximum value of average error between simulation and trajectory was found to be at 1.368% in all cases, with a load carrying capacity of 2 kg, under persistent weather conditions that have been mentioned hitherto. This means that not only medication can be supplied, but also some equipment in this range of weight can be moved from the hospital to the affected person which can help in saving precious lives. Another novel aspect is the fact that the equipment which has been used for this purpose is not very expensive, aiding in the physical implementation of the biomedical sensor in terms of its economic parameters. Care has been taken to analyse the device in specific conditions, maintaining the sanctity, precision and accuracy of the work, justified by the physicians of the hospital.

There are numerous ideas that can be explored to extend this work. One of the future work directions of this project is to enhance quad-copter trajectory by implementing machine learning-based scheme to remember the path of trajectory of flight, in order to maximize the logistics of medication. At this time, the device can lift 2 kg, and reviews are being performed extend this range. The experimentation has been satisfactorily done in the prescribed weather conditions (daylight, partly cloudy and windy). Since the weather in the surrounding of the experimental range is like this, we are negotiating with the technical team to extend it to harsher conditions such as rain and snow in the future, and this might require certain permissions from the aviation authority in those areas. We hope that once the results become available, they would lead to supply of medication to the far-off areas requiring a much smaller amount of time, thereby saving precious resources. For this, we are planning to map the vicinity of the medical centre where the UAV is expected to manoeuvre whenever required. Once this is done, the device might be able to fly to the desired location by using the least time and resources, which is one of the main targets of automated vehicles for the future.

Author Contributions: Data curation, U.M.; Formal analysis, U.M.; Funding acquisition, T.S.; Methodology, F.A.; Project administration, T.S. and F.A.; Resources, H.M.; Software, H.M.; Validation, A.A.; Visualization, A.A.; Writing—original draft, U.M.; Writing—review & editing, A.A. All authors have read and agreed to the published version of the manuscript.

Funding: This research received no external funding.

Institutional Review Board Statement: Not applicable.

Informed Consent Statement: Not applicable.

Data Availability Statement: Not applicable.

Acknowledgments: For this work, we are deeply indebted to those physicians in Islamabad and selected areas, who helped us in analysing the environmental scenario, even during their busy schedule during the COVID-19 epidemic. Additionally, the technical conversations with the two pilots are gratefully acknowledged, which helped us in carrying out necessary arrangements during the experimentation phases.

Conflicts of Interest: The authors declare no conflict of interest. The funders had no role in the design of the study; in the collection, analyses, or interpretation of data; in the writing of the manuscript, or in the decision to publish the results.

References

1. Rezeika, A.; Benda, M.; Stawicki, P.; Gembler, F.; Saboor, A.; Volosyak, I. Brain–Computer Interface Spellers: A Review. *Brain Sci.* **2018**, *8*, 57. [CrossRef] [PubMed]
2. Gu, Z.; Chen, Z.; Zhang, J.; Zhang, X.; Yu, Z.L. An Online Interactive Paradigm for P300 Brain-Computer Interface Speller. *IEEE Trans. Neural Syst. Rehabil. Eng.* **2019**, *27*, 152–161. [CrossRef] [PubMed]

3.	Akram, F.; Han, S.M.; Kim, T.-S. An efficient word typing P300-BCI system using a modified T9 interface and random forest classifier. *Comput. Biol. Med.* **2015**, *56*, 30–36. [CrossRef] [PubMed]

4.	Khan, M.J.; Hong, M.J.; Hong, K.-S. Decoding of four movement directions using hybrid NIRS-EEG brain-computer interface. *Front. Hum. Neurosci.* **2014**, *8*, 244. [CrossRef]

5.	Li, Y.; Pan, J.; Wang, F.; Yu, Z. A hybrid BCI system combining P300 and SSVEP and its application to wheelchair control. *IEEE Trans. Biomed. Eng.* **2013**, *60*, 3156–3166.

6.	Galána, F.; Nuttin, M.; Lew, E.; Ferrez, P.W.; Vanacker, G.; Philips, J.; Millánad, J.D.R. A brain-actuated wheelchair: Asynchronous and non-invasive Brain-computer interfaces for continuous control of robots. *Clin. Neurophysiol.* **2008**, *119*, 2159–2169. [CrossRef]

7.	Hochberg, L.R.; Serruya, M.D.; Friehs, G.M.; Mukand, J.A.; Saleh, M.; Caplan, A.H.; Branner, A.; Chen, D.; Penn, R.D.; Donoghue, J.Neuronal ensemble control of prosthetic devices by a human with tetraplegia. *Nature* **2006**, *442*, 164–71. [CrossRef]

8.	Lenhardt, A. A Brain-Computer Interface for Robotic Arm Control. 2011. Available online: http://pub.uni-bielefeld.de/publication/2529157 (accessed on 15 March 2022).

9.	Daly, J.J.; Wolpaw, J.R. Brain–computer interfaces in neurological rehabilitation. *Lancet Neurol.* **2008**, *7*, 1032–1043. [CrossRef]

10.	Masud, U.; Baig, M.I.; Akram, F.; Kim, T.-S. A P300 brain computer interface based intelligent home control system using a random forest classifier. In Proceedings of the 2017 IEEE Symposium Series on Computational Intelligence, SSCI 2017, Honolulu, HI, USA, 27 November–1 December 2017.

11.	Edlinger, G.; Holzner, C.; Guger, C.; Groenegress, C.; Slater, M. Brain-computer interfaces for goal orientated control of a virtual smart home environment. In Proceedings of the 2009 4th International IEEE/EMBS Conference on Neural Engineering, Antalya, Turkey, 29 April–2 May 2009; pp. 463–465.

12.	Finke, A.; Lenhardt, A.; Ritter, H. The MindGame: A P300-based brain-computer interface game. *Neural Netw.* **2009**, *22*, 1329–1333. [CrossRef]

13.	Marshall, D.; Coyle, D.; Wilson, S.; Callaghan, M. Games, Gameplay, and BCI: The State of the Art. *IEEE Trans. Comput. Intell. AI Games* **2013**, *5*, 82–99. [CrossRef]

14.	Khan, M.J.; Hong, K.S. Hybrid EEG-FNIRS-based eight-command decoding for BCI: Application to quadcopter control. *Front. Neurorobot.* **2017**, *11*, 6. [CrossRef] [PubMed]

15.	Lafleur, K.; Cassady, K.; Doud, A.; Shades, K.; Rogin, E.; He, B. Quadcopter control in three-dimensional space using a noninvasive motor imagery-based brain-computer interface. *J. Neural Eng.* **2013**, *10*, 4. [CrossRef] [PubMed]

16.	Naseer, N.; Hong K.-S. fNIRS-based brain-computer interfaces: A review Frontiers in Human. *Neuroscience* **2015**, *9*, 1–15. [CrossRef]

17.	Hong, KS.; Ghafoor, U.; Khan, M.J. Brain–machine interfaces using functional near-infrared spectroscopy: A review. *Artif Life Robot.* **2020**, *25*, 204–218. [CrossRef]

18.	Nicolas-Alonso, L.F.; Gomez-Gil, J. Brain Computer Interfaces, a Review. *Sensors* **2012**, *12*, 1211–1279. [CrossRef]

19.	Duan, X.; Xie, S.; Xie, X.; Meng, Y.; Xu, Z. Quadcopter Flight Control Using a Non-invasive Multi-Modal Brain Computer Interface. *Front. Neurorobot.* **2019**, *13*, 23. [CrossRef]

20.	Masud, U.; Saeed, T.; Malaikah, H.M.; Islam, F.U.; Abbas, G. Smart Assistive System for Visually Impaired People Obstruction Avoidance through Object Detection and Classification. *IEEE Access* **2022**, *10*, 13428–13441. [CrossRef]

21.	Dumitrescu, C.; Costea, I.-M.; Semenescu, A. Using Brain-Computer Interface to Control a Virtual Drone Using Non-Invasive Motor Imagery and Machine Learning. *Appl. Sci.* **2021**, *11*, 11876. [CrossRef]

22.	Chamola, V.; Vineet, A.; Nayyar, A.; Hossain, E. Brain-Computer Interface-Based Humanoid Control: A Review. *Sensors* **2020**, *20*, 3620. [CrossRef]

23.	Hong, K.-S.; Khan, M.J. Hybrid Brain–Computer Interface Techniques for Improved Classification Accuracy and Increased Number of Commands: A Review. *Front. Neurorobot.* **2017**, *11*, 35. [CrossRef]

24.	Akram, F.; Han, H.; Kim, T. A P300-Based Word Typing Brain Computer Interface System Using a Smart Dictionary and Random Forest Classifier. In Proceedings of the ICCGI 2013, The Eighth International Multi-Conference on Computing in the Global Information Technology, Nice, France, 21 July–26 July 2013; pp. 106–109.

25.	Nawaz, H.; Niazi, A.U.; Tahir, A.; Ahmad, N.; Masud, U.; Althobaiti, T.; Alotaibi, A.A.; Ramzan, N. Co-circularly Polarized Planar Antenna with Highly Decoupled Ports for S-Band Full Duplex Applications. *IEEE Access* **2022**, *10*, 16101–16110. [CrossRef]

26.	Amin, F.; Choi, G.S. Hotspots Analysis Using Cyber-Physical-Social System for a Smart City. *IEEE Access* **2020**, *8*, 122197–122209. [CrossRef]

27.	Amin, F.; Ahmad, A.; Sang Choi, G. Towards Trust and Friendliness Approaches in the Social Internet of Things. *Appl. Sci.* **2019**, *9*, 166. [CrossRef]

28.	Kohno, H.; Kubo, T. mKast is dispensable for normal development and sexual maturation of the male European honeybee. *Sci. Rep.* **2018**, *8*, 11877. [CrossRef]

29.	Daugela, I.; Suziedelyte Visockiene, J.; Kumpiene, J.; Suzdalev, I. Measurements of Flammable Gas Concentration in Landfill Areas with a Low-Cost Sensor. *Energies* **2021**, *14*, 3967. [CrossRef]

30.	The Best Drones for 2020. Available online: https://www.pcmag.com/picks/the-best-drones (accessed on 24 October 2020).

31.	Arturo Urquizo. Available online: http://commons.wikimedia.org/wiki/File:PID.svg (accessed on 27 October 2020).

32.	Araki, M. PID Control. Available online: http://www.eolss.net/ebooks/Sample%20Chapters/C18/E6-43-03-03.pdf (accessed on 21 September 2020).

33. Alrayes, Z.O.; Gadalla, M. Development of a Flexible Framework Multi-Design Optimization Scheme for a Hand Launched Fuel Cell-Powered UAV. *Energies* **2021**, *14*, 2951. [CrossRef]
34. Burke, P.J. Demonstration and application of diffusive and ballistic wave propagation for drone-to-ground and drone-to-drone wireless communications. *Sci. Rep.* **2020**, *10*, 14782. [CrossRef]
35. Bindemann, M.; Fysh, M.C.; Sage, S.S.K.; Douglas, K.; Tummon, H.M. Person identification from aerial footage by a remote-controlled drone. *Sci. Rep.* **2017**, *7*, 13629. [CrossRef]
36. Omand, D. The Security Impact of Drones: Challenges and Opportunities. Birmingham Policy Commission. Available online: http://www.birmingham.ac.uk/Documents/research/policycommission/remote-warfare/final-report-october-2014.pdf (accessed on 5 January 2022).
37. Camber, R. Take Off for Police Drones Air Force: Remote-Controlled 'Flying Squad' to Chase Criminals and Hunt for Missing People. *Daily Mail*, 2017. Available online: http://www.dailymail.co.uk/news/article-4329714/Remote-controlled-flying-squad-chase-criminals.html (accessed on 22 October 2020).
38. Masud, U. *Investigations on Highly Sensitive Optical Semiconductor Laser Based Sensorics for Medical and Environmental Applications: The Nanonose*; Kassel University Press: Kassel, Germany, 2015; ISBN 3862195554.
39. Floreano, D.; Wood, R. Science, technology and the future of small autonomous drones. *Nature* **2015**, *521*, 460–466. [CrossRef]
40. Škrinjar, J.P.; Škorput, P.; Furdić, M. Application of Unmanned Aerial Vehicles in Logistic Processes. In *New Technologies, Development and Application*; NT 2018; Lecture Notes in Networks and Systems; Karabegović, I., Ed.; Springer: Cham, Switzerland, 2019; Volume 42. [CrossRef]
41. Mehrer, M.; Moreno, S.; Hartman, D.; Landis, K.; Kim, J. Quadcopter Dynamic Modeling and Simulation. MATLAB and Simulink Student Design Challenge. 2014. Available online: https://de.mathworks.com/academia/student-challenge/spring-2014.html (accessed on 6 January 2022).
42. Gyula Mester Aleksander Rodic. The modeling and simulation of an autonomous quad-rotor microcopter in a virtual outdoor scenario. *Acta Polytech. Hung.* **2011**, *8*, 107–124.
43. Harik, E.H.C.; Guérin, F.; Guinand, F.; Brethé, J.; Pelvillain, H. Towards an autonomous warehouse inventory scheme. In Proceedings of the 2016 IEEE Symposium Series on Computational Intelligence, Athens, Greece, 6–9 December 2016.
44. Masud, U.; Baig, M.I. Investigation of Cavity Length and Mode Spacing Effects in Dual-Mode Sensor. *IEEE Sens. J.* **2018**, *18*, 2737–2743. [CrossRef]
45. Pakistani Tech Company Develops Drones to Plant Thousands of Trees in a Day. Available online: https://propakistani.pk/2020/10/23/pakistani-tech-company-develops-drones-to-plant-thousands-of-trees-in-a-day/?fbclid=IwAR0Hvnkkz8CtDM168It33Abh3VMJA6onCJwHYH4DmahFVhUMxiFXK4lgGos (accessed on 24 October 2020).
46. Masud, U.; Ali, M.; Ikram, M. Calibration and stability of highly sensitive fibre based laser through relative intensity noise. *Phys. Scr.* **2020**, *95*, 055505. [CrossRef]
47. Masud, U.; Jeribi, F.; Zeeshan, A.; Tahir, A.; Ali, M. Highly Sensitive Microsensor Based on Absorption Spectroscopy: Design Considerations for Optical Receiver. *IEEE Access* **2020**, *8*, 100212–100225. [CrossRef]
48. Usman, M.; Mudassar, A.; Farhan, Q.; Ahmed, Z.; Momna, I. Dual mode spectroscopic biomedical sensor: Technical considerations for the wireless testbed. *Phys. Scr.* **2020**, *95*, 105206.
49. Han, S.S.; Ghafoor, U.; Saeed, T.; Elahi, H.; Masud, U.; Kumar, L.; Selvaraj, J.; Ahmad, M.S. Silicon Particles/Black Paint Coating for Performance Enhancement of Solar Absorbers. *Energies* **2021**, *14*, 7140. [CrossRef]
50. Butt, O.M.; Saeed, T.; Elahi, H.; Masud, U.; Ghafoor, U.; Che, H.S.; Rahim, N.A.; Ahmad, M.S. A Predictive Approach to Optimize a HHO Generator Coupled with Solar PV as a Standalone System. *Sustainability* **2021**, *13*, 12110. [CrossRef]
51. Kim, Y.K.; Wang, H.; Mahmud, M.S. 9—Wearable body sensor network for health care applications. In *Woodhead Publishing Series in Textiles, Smart Textiles and their Applications*; Koncar, V., Ed.; Woodhead Publishing: Cambridge, UK, 2016; pp. 161–184, ISBN 9780081005743. [CrossRef]
52. Bradke, B.S.; Miller, T.A.; Everman, B. Photoplethysmography behind the Ear Outperforms Electrocardiogram for Cardiovascular Monitoring in Dynamic Environments. *Sensors* **2021**, *21*, 4543. [CrossRef]
53. Batista, A.D.; Silva, W.R.; Mizaikoff, B. Molecularly imprinted materials for biomedical sensing. *Med. Devices Sens.* **2021**, *4*, e10166. [CrossRef]
54. Zhou, G.; Wang, Y.; Cui, L. *Biomedical Sensor, Device and Measurement Systems, Advances in Bioengineering, Pier Andrea Serra*; Intech Open: Rijeka, Croatia, 2015. [CrossRef]
55. Masud, U.; Jeribi, F.; Alhameed, M.; Akram, F.; Tahir, A.; Naudhani, M.Y. Two-Mode Biomedical Sensor Build-up: Characterization of Optical Amplifier. *CMC-Comput. Mater. Contin.* **2022**, *70*, 5487–5489. [CrossRef]
56. Harsányi, G. Sensors in biomedical applications. *Sens. Rev.* **2001**, *21*, 4. [CrossRef]
57. Wang, P.; Liu, Q. *Biomedical Sensors and Measurement*; Springer: Berlin/Heidelberg, Germany, 2011. [CrossRef]
58. Maza, I.; Caballero, F.; Capitán, J.; Martínez-de-Dios, J.R.; Ollero, A. Experimental results in multi-UAV coordination for disaster management and civil security applications. *J. Intell. Rob. Syst.* **2011**, *61*, 563–585. [CrossRef]
59. Kim, M.; Matson, E.T. A cost-optimization model in multi-agent system routing for drone delivery. In *Highlights of Practical Applications of Cyber-Physical Multi-Agent Systems, PAAMS 2017*; Communications in Computer and Information Science; Springer: Cham, Switzerland, 2017; p. 722.

60. Masud, U.; Jeribi, F.; Alhameed, M.; Tahir, A.; Javaid, Q.; Akram, F. Traffic Congestion Avoidance System Using Foreground Estimation and Cascade Classifier. *IEEE Access* **2020**, *8*, 178859–178869. [CrossRef]
61. The Top 100 Drone Companies to Watch in 2020. Available online: https://uavcoach.com/drone-companies/ (accessed on 21 July 2020).
62. Unmanned Systems Technology. Available online: https://www.unmannedsystemstechnology.com/supplier-directory/ (accessed on 17 September 2020).
63. On Line Store of Saravana Electronics. Available online: http://www.alselectro.com/frame-f450.html (accessed on 13 January 2020).
64. Zeeshan, A.; Masud, U.; Saeed, T.; Hobiny, A. On the effects of chemical reaction on controlled heat and mass transfer in magnetized non-Newtonian biofluid through a long rectangular tunnel. *J. Therm. Anal. Calorim.* **2021**, *143*, 2637–2646. [CrossRef]
65. Hassan, M.; Faisal, A.; Bhatti, M.M. Interaction of aluminum oxide nanoparticles with flow of polyvinyl alcohol solutions base nanofluids over a wedge. *Appl. Nanosci.* **2018**, *8*, 53–60. [CrossRef]
66. Ellahi, R.; Hassan, M.; Zeeshan, A. Study of natural convection MHD nanofluid by means of single and multi-walled carbon nanotubes suspended in a salt-water solution. *IEEE Trans. Nanotechnol.* **2015**, *14*, 726–734. [CrossRef]
67. Sheikholeslami, M.; Ellahi, R.; Ashorynejad, H.; Domairry, G.; Hayat, T. Effects of heat transfer in flow of nanofluids over a permeable stretching wall in a porous medium. *J. Comput. Theor. Nanosci.* **2014**, *11*, 486–496. [CrossRef]
68. Barzkar, A.; Ghassemi, M. Electric Power Systems in More and All Electric Aircraft: A Review. *IEEE Access* **2020**, *8*, 169314–169332. [CrossRef]
69. Naus, K.; Szymak, P.; Piskur, P.; Niedziela, M.; Nowak, A. Methodology for the Correction of the Spatial Orientation Angles of the Unmanned Aerial Vehicle Using Real Time GNSS, a Shoreline Image and an Electronic Navigational Chart. *Energies* **2021**, *14*, 2810. [CrossRef]
70. What is DSHOT ESC Protocol. Available online: https://oscarliang.com/dshot/ (accessed on 18 September 2020).
71. Marco, M.; Kris, S. Design and Implementation of an Electronic Speed Controller for Brushless DC Motors. Ph.D. Thesis, Malta College of Arts, Science & Technology, Paola, Malta, 2017. [CrossRef]
72. Dixit, K.R.; Krishna, P.P.; Antony, R. Design and development of H frame quadcopter for control system with obstacle detection using ultrasound sensors. In Proceedings of the 2017 International Conference on Circuits, Controls, and Communications (CCUBE), Bangalore, India, 15–16 December 2017; pp. 100–104. [CrossRef]
73. Vedder, B.; Eriksson, H.; Skarin, D.; Vinter, J.; Jonsson, M. Towards Collision Avoidance for Commodity Hardware Quadcopters with Ultrasound Localization. In Proceedings of the 2015 International Conference on Unmanned Aircraft Systems (ICUAS) Denver Marriott Tech Center Denver, Denver, CO, USA, 9–12 June 2015.
74. Koumaras, H.; Makropoulos, G.; Batistatos, M.; Kolometsos, S.; Gogos, A.; Xilouris, G.; Sarlas, A.; Kourtis, M.-A. 5G-Enabled UAVs with Command and Control Software Component at the Edge for Supporting Energy Efficient Opportunistic Networks. *Energies* **2021**, *14*, 1480. [CrossRef]
75. Electronic Speed Controller Reference Design for Drones. Available online: https://eepower.com/new-industry-products/electronic-speed-controller-reference-design-for-drones/# (accessed on 12 September 2020).
76. Nugraha, A.T.; Agustinah, T. Quadcopter path following control design using output feedback with command generator tracker LOS based at square path. *J. Phys. Conf. Ser.* **2018**, *947*, 012074. [CrossRef]
77. Scher, C.L.; Griffoul, E.; Cannon, C.H. Drone-based photogrammetry for the construction of high-resolution models of individual trees. *Trees* **2019**, *33*, 1385–1397. [CrossRef]
78. Quadcopter Dynamics and Simulation. Available online: https://andrew.gibiansky.com/blog/physics/quadcopter-dynamics/ (accessed on 12 September 2018).
79. Pérez Gordillo, A.M.; Villegas Santos, J.S.; Lo Mejia, O.D.; Suárez Collazos, L.J.; Escobar, J.A. Numerical and Experimental Estimation of the Efficiency of a Quadcopter Rotor Operating at Hover. *Energies* **2019**, *12*, 261. [CrossRef]
80. Adafruit Ultimate GPS. Available online: https://learn.adafruit.com/adafruit-ultimate-gps (accessed on 12 September 2021).
81. Cormen, T.H.; Leiserson, C.E.; Rivest, R.L.; Stein, C. *Introduction to Algorithms*; MIT Press: Cambridge, MA, USA, 2009; ISBN 978-0-262-53305-8.
82. Levitin, A.V. *Introduction to the Design and Analysis of Algorithms*; Addison Wesley: Reading, MA, USA, 2002.
83. Xiang, C.; Wang, X.; Ma, Y.; Xu, B. Practical Modeling and Comprehensive System Identification of a BLDC Motor. *Math. Probl. Eng.* **2015**, *2015*, 879581. [CrossRef]
84. Mondal, S.; Mitra, A.; Chattopadhyay, M. Mathematical modeling and simulation of Brushless DC motor with ideal Back EMF for a precision speed control. In Proceedings of the 2015 IEEE International Conference on Electrical, Computer and Communication Technologies (ICECCT), Coimbatore, India, 5–7 March 2015; pp. 1–5. [CrossRef]

Article

LASSO Homotopy-Based Sparse Representation Classification for fNIRS-BCI

Asma Gulraiz [1], Noman Naseer [1,*], Hammad Nazeer [1], Muhammad Jawad Khan [2], Rayyan Azam Khan [3] and Umar Shahbaz Khan [4,5]

[1] Department of Mechatronics and Biomedical Engineering, Air University, Islamabad 44000, Pakistan; asmagulraizkiani@gmail.com (A.G.); hammad@mail.au.edu.pk (H.N.)
[2] School of Mechanical and Manufacturing Engineering, National University of Science and Technology, Islamabad 44000, Pakistan; jawad.khan@smme.nust.edu.pk
[3] Department of Mechanical Engineering, University of Saskatchewan, Saskatoon, SK S7N 5A9, Canada; rayyan.khan@usask.ca
[4] Department of Mechatronics Engineering, National University of Sciences and Technology, H-12, Islamabad 44000, Pakistan; u.shahbaz@ceme.nust.edu.pk
[5] National Centre of Robotics and Automation (NCRA), Rawalpindi 46000, Pakistan
* Correspondence: noman.naseer@mail.au.edu.pk

Abstract: Brain-computer interface (BCI) systems based on functional near-infrared spectroscopy (fNIRS) have been used as a way of facilitating communication between the brain and peripheral devices. The BCI provides an option to improve the walking pattern of people with poor walking dysfunction, by applying a rehabilitation process. A state-of-the-art step-wise BCI system includes data acquisition, pre-processing, channel selection, feature extraction, and classification. In fNIRS-based BCI (fNIRS-BCI), channel selection plays a vital role in enhancing the classification accuracy of the BCI problem. In this study, the concentration of blood oxygenation (HbO) in a resting state and in a walking state was used to decode the walking activity and the resting state of the subject, using channel selection by Least Absolute Shrinkage and Selection Operator (LASSO) homotopy-based sparse representation classification. The fNIRS signals of nine subjects were collected from the left hemisphere of the primary motor cortex. The subjects performed the task of walking on a treadmill for 10 s, followed by a 20 s rest. Appropriate filters were applied to the collected signals to remove motion artifacts and physiological noises. LASSO homotopy-based sparse representation was used to select the most significant channels, and then classification was performed to identify walking and resting states. For comparison, the statistical spatial features of mean, peak, variance, and skewness, and their combination, were used for classification. The classification results after channel selection were then compared with the classification based on the extracted features. The classifiers used for both methods were linear discrimination analysis (LDA), support vector machine (SVM), and logistic regression (LR). The study found that LASSO homotopy-based sparse representation classification successfully discriminated between the walking and resting states, with a better average classification accuracy ($p < 0.016$) of 91.32%. This research provides a step forward in improving the classification accuracy of fNIRS-BCI systems. The proposed methodology may also be used for rehabilitation purposes, such as controlling wheelchairs and prostheses, as well as an active rehabilitation training technique for patients with motor dysfunction.

Keywords: BCI; fNIRS; SRC; channel selection; classification

Citation: Gulraiz, A.; Naseer, N.; Nazeer, H.; Khan, M.J.; Khan, R.A.; Shahbaz Khan, U. LASSO Homotopy-Based Sparse Representation Classification for fNIRS-BCI. *Sensors* **2022**, *22*, 2575. https://doi.org/10.3390/s22072575

Academic Editor: Sung-Phil Kim

Received: 28 January 2022
Accepted: 23 March 2022
Published: 28 March 2022

1. Introduction

Nowadays, many elderly people have motor dysfunction and joint problems because of age factors, stroke, and spinal cord injuries. Due to this, they face many problems when walking, which strongly influences their lives [1]. According to WHO data, mental illnesses and neurological disorders are major sources of morbidity, death, and disability. Mental, neurological, and behavioral diseases have a major impact on the world's population, impacting more than 450 million individuals. According to the Global Burden of Disease Report, neurological and mental disorders account for four out of the six primary causes of years lived with disability, accounting for 33 percent of the years lived with disability and 13 percent of disability-adjusted life years (DALYs) [2]. People with walking disabilities need to improve their walking patterns or capability by using rehabilitation and assistive devices [3]. The brain-computer interface (BCI) is the best way to accommodate the neuro-rehabilitation process, by providing a communication pathway between the brain and the peripheral devices [4]. The field of perceptible neuroscience scrutinizes itself with calibrating information processing models of the brain with the operational and structural (e.g., hemodynamic, metabolic, and electrical) features of the brain [5]. In the last few years, the development of BCI has played an important role in the analysis of brain dysfunction disorders and musculoskeletal gait. A typical BCI system consists of the following five major stages: signal acquisition, pre-processing, feature extraction, classification, and control commands to peripheral devices. In the signal acquisition stage, brain signals are acquired using brain signal acquisition modalities. The acquired signals contain noises such as physiological, instrumentation, and motion artifacts. These errors can be removed with the help of appropriate filters in the pre-processing stage. After obtaining filtered and processed data, some useful features can be extracted. These extracted features contain intrinsic information related to brain activity. In the classification stage, the most suitable classifier is employed for the extracted features, to predict the response to a particular class. Furthermore, the brain activity discriminated by the classifier is used as a command to control the external devices. The general flow diagram of functional near-infrared spectroscopy (fNIRS)-BCI is shown in Figure 1. BCI systems provide the end-user with full control over the channels used to communicate with the brain and external devices, regardless of the level of dependence on the output channel [1].

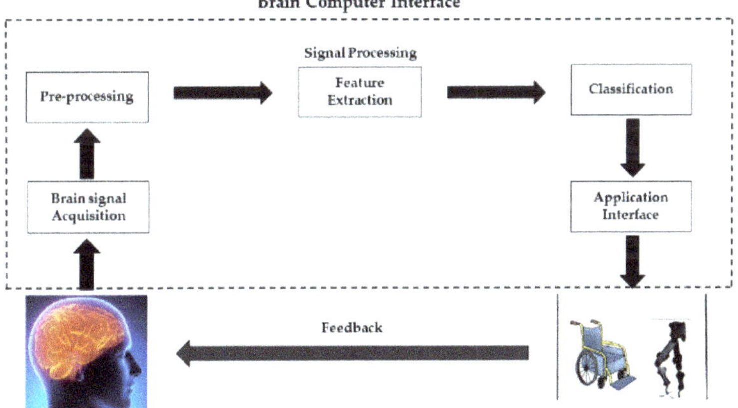

Figure 1. Block diagram of BCI system.

For improving mild cognitive impairment (MCI), BCI based on functional near-infrared spectroscopy (fNIRS) had a positive result [3]. It has been widely used in the rehabilitation process [6]. fNIRS is utilized to concentrate on the brain areas of interest in eleven sicknesses, including stroke, MCI, traumatic brain injury, and harm recognition [7]. There are several modalities used to acquire brain signals for rehabilitation, such as magnetic field measurement using magnetoencephalography (MEG) [8,9], electroencephalography (EEG), radioactive tracer-based positron emission tomography (PET) [10,11], functional magnetic resonance imaging (fMRI) [12,13], gamma emission-based single-photon emission computed tomography (SPECT) [14,15], and fNIRS. fNIRS is widely used due to its advantages of mobility and ease of use compared to other neuroimaging modalities when researching the brain basis of cognitive inputs during gait [16,17]. The fNIRS modality has been most commonly used over the recent decades, because of its portability and high spatial resolution. fNIRS is operated in wavelengths between 650 and 1000 nm; in this range, the blood oxygenation concentration (*HbO*) and the blood deoxygenation concentration (HbR) are more clear [7]. Several classifiers and techniques are applied to fNIRS signals [6,18], to improve the accuracy and efficiency of BCI systems, to help disabled and elderly people in their daily life [7,19].

For the classification of different brain activity, fNIRS-based BCI mostly extracted features such as mean, peak, variance, skewness, kurtosis values, etc., from the obtained data [20]. In the literature, studies have been performed using single, multiple or a combination of features to classify two- or multiple-class fNIRS-BCI problems [21]. Support vector machine (SVM) and linear discrimination analysis (LDA) are mainly used to classify walking and resting states, but the classification accuracy is low and needs to be improved [22].

To improve the classification accuracy, it is important to introduce some new methods and technologies in the field of fNIRS-BCI. In this study, a new classification method is discussed, which is sparse representation-based classification (SRC). SRC has been used in the compressed sensing (CS) theory; the core concept of CS is that we can represent a huge amount of data with a few data points [23]. Weighted SRC was applied to EEG-BCI to classify motor imagery, and achieved good classification accuracy results [24]. Sparse representation-based classification was used to translate the motor imagery of a single index finger classification, with an accuracy of 81.32%; the results were used to construct a BCI-enhanced finger rehabilitation system [25]. Optimization features, such as spatial-frequency-temporal, were calculated from the public dataset of EEG, and were used as predictors for SRC. The classification accuracy achieved was higher than on the original basis [26]. Shin et al. classified motor images using SRC and compared the results with SVM. They discovered that SRC had better results than SVM and LDA, in terms of classification accuracy, testing duration, and noise robustness [27]. This study includes the use of LASSO homotopy-based SRC for channel selection for the fNIRS-BCI system, to identify walking and resting states, Figure 2.

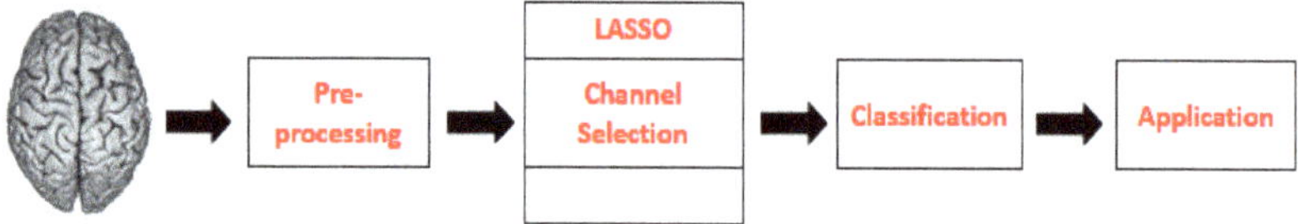

Figure 2. BCI system with LASSO-based sparse representation classification for channel selection.

2. Materials and Methods

2.1. Experimental Design

The raw optical signals from the brain during activity and resting states were collected by dynamic near-infrared optical tomography (DYNOT; NIRx Medical Technologies, New York, NY, USA). For signal acquisition, the sampling frequency was set to 1.81 Hz, with operating wavelengths of 760 and 830 nm. A total of nine healthy male subjects, aged approximately 30 ± 3, were called up for the study. All the subjects were right-handed and had no neurological disorders. The experiments were conducted in accordance with the latest Declaration of Helsinki, and verbal consent from the subjects was collected before experimentation.

2.2. Experimental Paradigm

The subjects were asked to take an initial rest in a quiet room for 30 s before the start of the activity. After the initial rest, subjects were asked to start walking with their right leg on the treadmill for 10 s, followed by a 20 s rest while standing on the treadmill. Ten trials were performed for each subject. For baseline correction, a 30 s rest was given at the end of each experiment. The length of the experiment for each subject was 300 s. The experimental paradigm is shown in Figure 3.

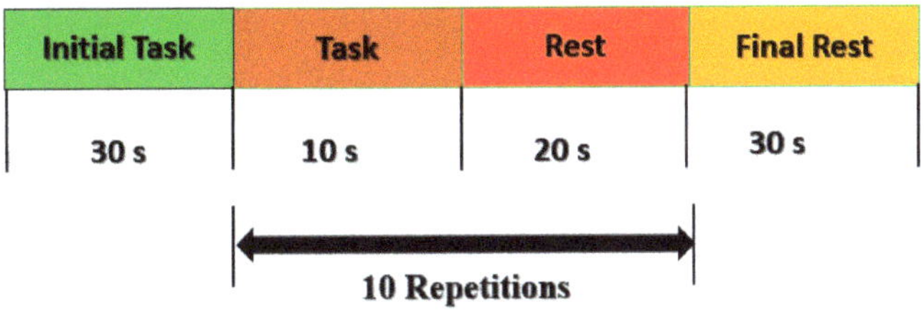

<table>
<tr><td>Initial Task</td><td>Task</td><td>Rest</td><td>Final Rest</td></tr>
<tr><td>30 s</td><td>10 s</td><td>20 s</td><td>30 s</td></tr>
</table>

10 Repetitions

Figure 3. Experimental paradigm for data acquisition: after an initial 30 s rest, a single trial consisted of a 10 s period of walking followed by a 20 s rest.

2.3. Experimental Configuration

In accordance with the literature [28], the twelve-channel configuration maintained a minimum distance distribution of 3 cm between the source and the detector. Brain signals from the left hemisphere of the primary motor cortex (M1) were acquired. There were nine optodes, out of which five were sources and four were light detectors. The configuration of the source and detector, with channels, is shown in Figure 4.

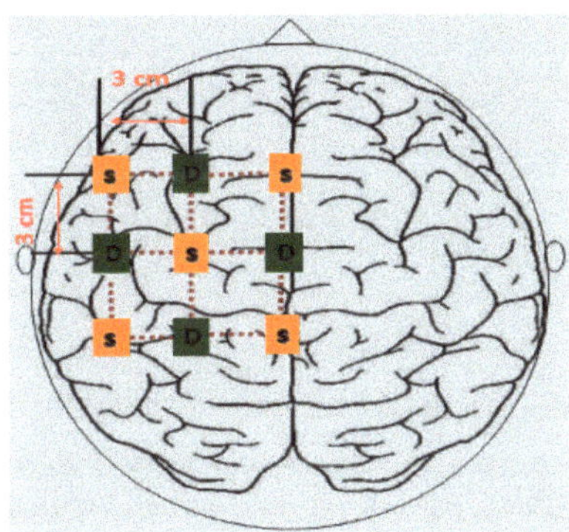

Figure 4. Position of source and detectors on the left hemisphere of the motor cortex. D represents the detectors and S represents sources.

2.4. Data Acquisition

Raw optical density signals were converted into oxy and deoxyhemoglobin concentration changes ($\Delta C_{HbO}(t)$ and $\Delta C_{HbR}(t)$ by using the modified Beer–Lambert law (MBLL) shown in Equation (1) [29].

$$\begin{bmatrix} \Delta C_{HbO}(t) \\ \Delta C_{HbR}(t) \end{bmatrix} = \frac{\begin{bmatrix} \alpha_{HbO}(\lambda_1) & \alpha_{HbR}(\lambda_1) \\ \alpha_{HbO}(\lambda_2) & \alpha_{HbR}(\lambda_2) \end{bmatrix}^{-1} \begin{bmatrix} \Delta A(t,\lambda_1) \\ \Delta A(t,\lambda_2) \end{bmatrix}}{d \times l}, \tag{1}$$

where $\alpha_{HbR}(\lambda_{1,2})$ and $\alpha_{HbO}(\lambda_{1,2})$ are the extinction coefficients of HbO and HbR in μM^{-1} cm^{-1}, respectively, and $\Delta C_{HbR}(t)$ and $\Delta C_{HbO}(t)$ are the concentration changes in HbR and HbO in μM, respectively. Furthermore, l is the source and detector distance, d is the curved path length factor, and $A(t, \lambda_1)$ and $A(t, \lambda_2)$ are the absorption coefficients at two different instants.

2.5. Signal Processing

In this study, we only used the *HbO* response of brain activity for further processing. Noises including respiration between 1 and 1.5 Hz, heartbeat 0.5 Hz, and instrumental noise are present in the signals. These noises were removed using high-pass and low-pass filters with cut-off frequencies of 0.01 and 0.5 Hz [7]. The Hemodynamic Response filter and Gaussian filter were applied to the acquired signal for the removal of drift noise, using the NIRS-SPM toolbox [30]. For the motion artifacts, a hemodynamic response filter and discrete cosine transform were applied using the NIRS-SPM toolbox. Figure 5 shows the average trial $\Delta C_{HbR}(t)$ signals of subject four for channels 9–12.

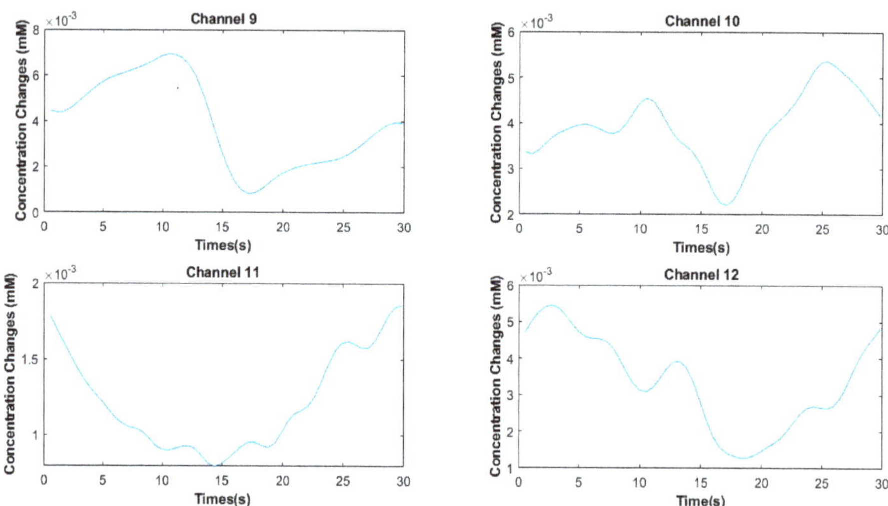

Figure 5. Average trial ΔC_{HbO} signals of subject four for channels 9–12.

2.6. Feature Extraction

A prior study explained several combinations of statistical features, with the goal of finding an effective filter for a given cortical region [31]. In this paper, spatial features were extracted from *HbO* data of all the active channels. The features were calculated for the entire task and rest session. The signal mean was calculated as follows:

$$\text{mean} = \frac{1}{N} \sum_{i=0}^{N} X_i, \tag{2}$$

where the total number of observations is represented as N, and X_i represents the $\Delta C_{HbO}(t)$ across each observation. The variance was calculated as follows:

$$Var = \frac{\sum (X_i - X)^2}{n-1}, \tag{3}$$

where X_i represents the $\Delta C_{HbO}(t)$ across each observation, X is the mean value of observations, and N is the total number of observations. The *Skewness* was calculated as follows:

$$Skewness = \frac{\sum (X_i - \mu)^2}{N \times \sigma}, \tag{4}$$

where X_i is each observation, μ is the mean of each observation, σ is the standard deviation of data, and N is the total number of observations. The peak values were calculated using the max function in MATLAB.

2.7. Channel Selection

Selecting channels of interest (COI) or a region of interest (ROI) in BCI can save processing time, reduce dimensionality, improve performance, and provide adequate brain region identification with low noise signals. In the literature, the z-score approach, which uses cross-correlation and z-scores for ROI/COI selection, was utilized to improve the performance of the fNIRS-BCI system [21]. The hemodynamic responses with positive t-values were selected by using the t-value method [32]. For pain-related cortical activations, the cross-correlation approach was employed to identify potentially dominating channels in both hemispheres. The response delay was calculated after a visual check, to identify probable dominating channels. The active channels that were next to each other were chosen [33]. In this paper, the LASSO homotopy-based sparse representation method is used for channel selection.

2.7.1. Sparse Representation Classification

The basic idea of the SRC method is to recognize the true class of new signals by learning the sparsest representation (fewest significant coefficients) of the test signals, in terms of training signals [34]. A principle that a signal can be approximated by, using a linear combination of dictionary atoms, is formulated as follows [35]:

$$\left(\frac{b}{A}, x, k\right) = x_1 a_1 + \ldots + x_k a_k + \varepsilon, \tag{5}$$

where the dictionary is represented as $A = [a_1, \cdots, a_k]$, dictionary atom is represented as a_i, x is a sparse coefficient vector, and ε is an error term. A, x, and k are the model parameters. In general, the SRC algorithm produces a dictionary before solving the optimization problem, reconstructing, and calculating the residual.

For a certain category, when the residual is very small and the other categories are very large, the unknown category of the object belongs to that category [3]. The simplest sparse representation classification model is shown in Figure 6.

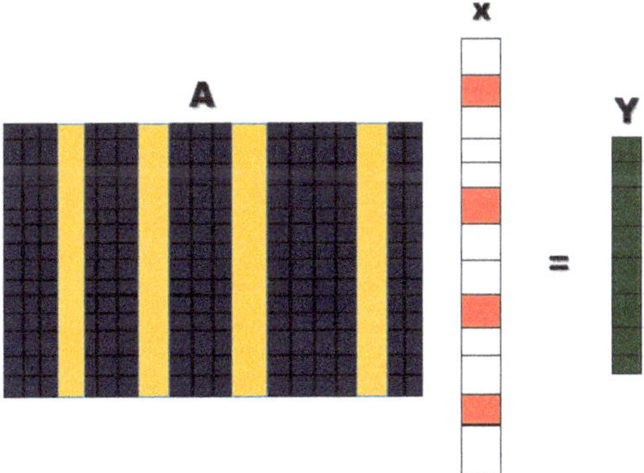

Figure 6. Sparse representation model. The dictionary is represented as $A = [a_1, \cdots, a_k]$, dictionary atom is represented as a_i, x is a sparse coefficient vector and Y is the output signal result as combination of $A \times x$.

2.7.2. LASSO Homotopy

The notion of homotopy comes from topology, and the homotopy technique is mostly used to solve problems involving nonlinear systems of equations. The homotopy approach was first developed to tackle the l_1 penalty least squares problem [36]. Least absolute selection and shrinkage operator are representative approaches that use the homotopy-based strategy to tackle the sparse representation problem with l_1-norm regularization (LASSO) [36]. Regularization is a crucial concept for avoiding data overfitting, especially when the learned and test data differ significantly. Regularization is implemented by adding a penalty term to the best fit produced from the trained data, in order to attain lower variance with the tested data, as well as by compressing the coefficients of the least important predictor or channel variable over the output variable. L_1 regularization forces the weights of uninformative features and channels to be zero, by subtracting a small amount from the weight at each iteration, and, thus, making the weight of each channel or predictor equal to zero. LASSO homotopy starts optimization at a large value of λ parameter along the solution path and terminates at a point of λ, which is approximately zero, giving an optimal solution. The mathematical model of LASSO homotopy is represented as follows:

$$\frac{1}{2N}\left(y - X\beta'\right)'\left(y - X\beta'\right) + \lambda \sum_{j=1}^{P}|\beta_j|, \tag{6}$$

In the first term, y is the prediction value or test sample, X is the feature vector or trained sample, and β' is the vector of coefficients (weights on the basis of significance). The first term in the equation is the residual sum of squares (error term) and the second is product of $\lambda \times$ sum of the absolute values of the magnitude of coefficients (penalty term). λ denotes the amount of shrinkage. $\lambda = 0$ implies that all the features are considered and is equivalent to the linear regression, where the only residual square is considered to build a predictive model. $\lambda = \infty$ implies that no features are considered (i.e., as λ approaches infinity, it eliminates more and more features and channels).

2.8. Classification Algorithms

K-fold cross-validation is used to estimate classification performance. To ensure data separation for training and testing of classifiers for each channel selection method and activity utilized, the dataset was separated into training and testing sets, and the value of k was set to five-fold cross-validation.

In MATLAB®, the classification learner app was used for classification and validation of data. Several classifiers were selected and employed on the data, on the basis of prediction speed and training time. Following the literature [22], the following classifiers were used: linear discrimination analysis (LDA), logistic regression (LR), and support vector machine (SVM). The following settings were made during classification: covariance structure for LDA was set to diagonal covariance, and the kernel function for SVM was the Gaussian function.

3. Results

In this study, the LASSO homotopy method was employed for the channel selection of *HbO* signals with significant information; Table 1 shows the channels selected for each subject. From Table 1, we observe that the maximum and minimum channels selected by the LASSO homotopy method are nine and two for distinct subjects, respectively. The classification was performed using LDA, LR, and SVM on the data of the selected channels. The subject-wise average classification accuracies of all the classifiers used are given in Table 2. For comparison purpose, classification accuracies were calculated using conventional statistical features. Tables 3 and 4 show the subject-wise classification accuracies of three- and four-feature combinations of statistical features. A comparison of the overall average classification accuracies of all the classifiers after channel selection using LASSO homotopy, and without channel selection, is shown in Table 5. In Table 6, the results of the *t*-test are shown [37]. A comparative bar graph is shown in Figure 7, for the average classification accuracies of all the classifiers.

Table 1. Subject-wise channel selection using LASSO homotopy-based spare representation.

Subjects	Selected Channels
1	1, 2, 3, 4, 7, 8, 9, 10, 11
2	2, 3, 4, 5, 6, 7, 9, 11
3	2, 6, 8, 9, 10, 11
4	8, 9, 12
5	1, 2, 5, 6, 7, 8, 12
6	1, 5, 8, 11, 12
7	2, 4, 5, 6, 8, 9, 11, 12
8	6, 10
9	1, 2, 3, 4, 6, 7, 8, 9

Table 2. Subject-wise classification accuracies of all subjects (%) were obtained by implementing LASSO homotopy for channel selection of *HbO* signals and classification using SVM, LDA, and LR of the walking and resting states (binary classification) of 9 subjects.

Subjects	LDA	LR	SVM
1	72.6%	69.1%	95.7%
2	75.7%	76.7%	95.9%
3	74.6%	83%	95.2%
4	68%	67.4%	85.4%
5	71.9%	72.4%	91.3%
6	68%	70.4%	95.2%
7	75.9%	74.6%	95.4%
8	62.6%	62.2%	75.9%
9	69.8%	69.8%	91.3%

Table 3. Subject-wise classification accuracies of all subjects (%) were obtained by extracting features (i.e., SM. SP, and SV) of *HbO* signals and classification using SVM, LDA, and LR of the walking and resting states (binary classification) of 9 subjects.

Subjects	LDA	LR	SVM
1	65.5%	63.9%	75.5%
2	66.5%	65.2%	72.4%
3	63.9%	62.8%	70.4%
4	66.9%	68.1%	68.9%
5	66.7%	66.7%	71.5%
6	61.9%	65.7%	71.3%
7	63.9%	64.8%	71.7%
8	66.5%	66.5%	71.7%
9	68.1%	67.4%	81.5%

Table 4. Subject-wise classification accuracies of all subjects (%) were obtained by extracting features (i.e., SM. SP, SV, and SK) of *HbO* signals and classification using SVM, LDA, and LR of the walking and resting states (binary classification) of 9 subjects.

Subjects	LDA	LR	SVM
1	65.4%	65.2%	78.1%
2	66.5%	69.4%	78.5%
3	64.6%	63%	71.9%
4	65%	65.9%	73%
5	66.1%	65.4%	74.8%
6	61.5%	65.9%	73.5%
7	62.8%	64.1%	72.6%
8	66.3%	68%	85.2%
9	67.6%	68%	85.2%

Table 5. Average classification accuracies of all subjects (%) were obtained by extracting features and selecting channels of *HbO* signals and classification using SVM, LDA, and LR of the walking and resting states (binary classification) of 9 subjects.

	LDA	LR	SVM
After LASSO Homotopy	71.01%	71.6%	91.32%
Mean, Peak and Variance	65.54%	65.67%	72.7%
Mean, Peak, Variance and Skewness	65.08%	65.9%	76.2%

Table 6. Statistical significance of the LASSO homotopy-based sparse representation method.

Bonferroni Correction Applied ($p < 0.0167$)			
SVM	**vs.**	LDA	1.0886×10^{-6}
		LR	6.8421×10^{-6}

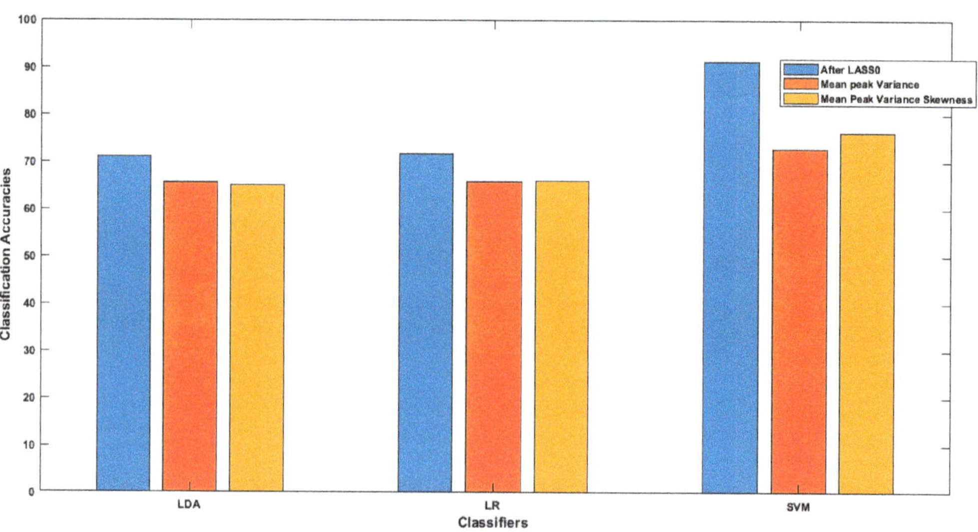

Figure 7. This figure shows a bar chart comparison of average classification accuracies for walking and resting states of all classifiers using both methods.

4. Discussion

In the literature, recent studies have focused on enhancing the classification accuracies of fNIRS-BCI systems using the optimal classification technique [22], general linear model [38], vector-based phase analysis [38–41], optimal feature selection [31,38], optimal feature combination [42], *t*-value method [43,44], cross-correlation [45], and dominant channel selection [46]. An accurate and reliable fNIRS-BCI performance may lead to producing applications in neuro-robotics, rehabilitation, clinical BCI, for monitoring and analysis, and neuro-ergonomics.

In the present study, a new method for selecting channels on the basis of the strong influence of individual input variables on the output response was introduced to increase fNIRS-BCI performance, especially in terms of classification accuracy. In the literature, there were many optimization techniques used to enhance the classification accuracy of the fNIRS-BCI system, to make it more robust and reliable. A comparative analysis between classifications of fNIRS-BCI, based on two methods, was conducted. The classification accuracies based on the proposed method were compared with the accuracies based on the conventional method of excessively used feature extractions, without channel selection,

using all the channel data. In the first method, we observed that by using two different combinations of spatial features, we achieved average classification accuracies, for LDA, LR, and SVM, of 65 ± 1.34%, 65 ± 1.6%, and 72 ± 4.9%, respectively. After the implementation of the other method, LASSO homotopy-based sparse representation for channel selection, the classification accuracies of LDA, LR, and SVM improved to 71.01, 71.6, and 91.32%, respectively. This study shows that selecting the channels with intrinsic brain information as features for classification improves the classification accuracy of fNIRS-BCI. LASSO homotopy-based SRC enhances both the prediction accuracy and model interpretability. It lowers the variability of the system estimations, by precisely decreasing some of the coefficients, and making models that are easy to understand, produce, and interpret [47]. For the channel selection method used for EEG-BCI, the classification accuracy was 93.08%, by selecting only eight channels out of 64 when classifying motor imagery tasks [48]. A similar study was performed to select cortical activation-based channel selection using the z-score method for fNIRS-BCI problems, achieving a classification accuracy of 88% [21]. LASSO homotopy-based SRC autonomously selects the most significant channels for the fNIRS-BCI system, thus greatly improving the overall classification accuracy.

This study has a few limitations, including the fact that it only applies to a single activity at a time, because specific tasks are linked to certain brain regions, and subject-based channels were selected due to the different brain sizes. LASSO homotopy-based SRC selects channels with the minimum residual sum of error. Furthermore, the offline study is performed and analyzed, while the online study may be conducted for other cognitive activities. Moreover, several machine learning algorithms are applied in this study to analyze performance. Further deep learning algorithms may be implemented with LASSO homotopy-based SRC for analysis, and may perform better.

5. Conclusions

This study attempts to apply LASSO homotopy-based sparse representation to fNIRS to identify the following two binary classes of data: walking state and resting state. The average classification accuracies are 71.01, 71.6, and 91.32% for LDA, LR, and SVM, respectively. The results show that LASSO homotopy-based SRC can effectively identify classes with significantly ($p < 0.0167$) improved classification accuracies. This study shows the better performance of LASSO homotopy-based SRC as a step to improve the classification performance of state-of-the-art fNIRS-BCI problems.

Author Contributions: Conceptualization, A.G. and N.N.; methodology, A.G.; software, R.A.K.; validation, M.J.K. and H.N.; formal analysis, A.G. and R.A.K.; investigation, H.N.; resources, U.S.K.; data curation, M.J.K. and H.N.; writing—original draft preparation, A.G. and H.N.; writing—review and editing, N.N.; visualization, A.G. and H.N.; supervision, N.N.; project administration, N.N.; funding acquisition, U.S.K. All authors have read and agreed to the published version of the manuscript.

Funding: This research was funded by the National Centre of Robotics and Automation Rawalpindi, Pakistan, grant No. NCRA-RF-027.

Institutional Review Board Statement: The study was conducted according to the guidelines of the Declaration of Helsinki and approved by the Institutional Review Board of Pusan National University, Busan, Republic of Korea.

Informed Consent Statement: Informed consent was obtained from all subjects involved in the study.

Data Availability Statement: The original data used for this study can be shared upon reasonable request by the associate author.

Acknowledgments: We would like to acknowledge the National Centre of Robotics and Automation (NCRA), Rawalpindi, Pakistan, for providing the necessary support, laboratory equipment, and facilities to conduct this study.

Conflicts of Interest: The authors declare that there are no conflict of interest regarding the publication of this paper.

References

1. Naseer, N.; Hong, K.-S. fNIRS-based brain-computer interfaces: A review. *Front. Hum. Neurosci.* **2015**, *9*, 3. [CrossRef]
2. Hussain, G.; Rasul, A.; Anwar, H.; Sohail, M.U.; Kamran, S.K.S.; Baig, S.M.; Shabbir, A.; Iqbal, J. Epidemiological Data of Neurological Disorders in Pakistan and Neighboring Countries: A Review. *Pak. J. Neurol. Sci.* **2017**, *12*, 52–70.
3. Li, H.; Gong, A.; Zhao, L.; Zhang, W.; Wang, F.; Fu, Y. Decoding of Walking Imagery and Idle State Using Sparse Representation Based on fNIRS. *Intell. Neurosci.* **2021**, *2021*, 6614112. [CrossRef]
4. Shih, J.J.; Krusienski, D.J.; Wolpaw, J. Brain-Computer Interfaces in Medicine. *Mayo Clin. Proc.* **2012**, *87*, 268–279. [CrossRef] [PubMed]
5. Pinti, P.; Aichelburg, C.; Gilbert, S.; Hamilton, A.; Hirsch, J.; Burgess, P.; Tachtsidis, I. A Review on the Use of Wearable Functional Near-Infrared Spectroscopy in Naturalistic Environments. *Jpn. Psychol. Res.* **2018**, *60*, 347–373. [CrossRef]
6. Ghaffar, M.S.B.A.; Khan, U.S.; Iqbal, J.; Rashid, N.; Hamza, A.; Qureshi, W.S.; Tiwana, M.I.; Izhar, U. Improving classification performance of four class FNIRS-BCI using Mel Frequency Cepstral Coefficients (MFCC). *Infrared Phys. Technol.* **2020**, *112*, 103589. [CrossRef]
7. Hong, K.-S.; Yaqub, M.A. Application of functional near-infrared spectroscopy in the healthcare industry: A review. *J. Innov. Opt. Health Sci.* **2019**, *12*, 91. [CrossRef]
8. James, L.M.; Enghdal, B.E.; Leuthold, A.C.; Georgopoulos, A.P. Classification of Trauma-Related Outcomes in US Veterans Using Magnetoencephalography (MEG). *J. Neurol. Neuromed.* **2021**, *6*, 13–20. [CrossRef]
9. Thorpe, D.R.; Engdahl, B.E.; Leuthold, A.; Georgopoulos, A.P. Assessing Recovery from Mild Traumatic Brain Injury (Mtbi) using Magnetoencephalography (MEG): An Application of the Synchronous Neural Interactions (SNI) Test. *J. Neurol. Neuromed.* **2020**, *5*, 28–34. [CrossRef]
10. Bokhari, S.; Schneider, R.H.; Salerno, J.W.; Rainforth, M.V.; Gaylord-King, C.; Nidich, S.I. Effects of cardiac rehabilitation with and without meditation on myocardial blood flow using quantitative positron emission tomography: A pilot study. *J. Nucl. Cardiol.* **2019**, *28*, 1596–1607. [CrossRef]
11. Nelles, G.; Jentzen, W.; Jueptner, M.; Müller, S.; Diener, H.C. Arm Training Induced Brain Plasticity in Stroke Studied with Serial Positron Emission Tomography. *NeuroImage* **2001**, *13*, 1146–1154. [CrossRef] [PubMed]
12. Nishida, D.; Mizuno, K.; Yamada, E.; Hanakawa, T.; Liu, M.; Tsuji, T. The neural correlates of gait improvement by rhythmic sound stimulation in adults with Parkinson's disease–A functional magnetic resonance imaging study. *Parkinsonism Relat. Disord.* **2021**, *84*, 91–97. [CrossRef] [PubMed]
13. Boyne, P.; Doren, S.; Scholl, V.; Staggs, E.; Whitesel, D.; Maloney, T.; Awosika, O.; Kissela, B.; Dunning, K.; Vannest, J. Functional magnetic resonance brain imaging of imagined walking to study locomotor function after stroke. *Clin. Neurophysiol.* **2020**, *132*, 167–177. [CrossRef] [PubMed]
14. Du, Y.; Zaidi, H. Single-Photon Emission Computed Tomography: Principles and Applications. In *Encyclopedia of Biomedical Engineering*; Narayan, R., Ed.; Elsevier: Oxford, UK, 2019; pp. 493–506. [CrossRef]
15. Dorraji, E.S.; Oteiza, A.; Kuttner, S.; Martin-Armas, M.; Kanapathippillai, P.; Garbarino, S.; Kalda, G.; Scussolini, M.; Piana, M.; Fenton, K.A. Positron emission tomography and single photon emission computed tomography imaging of tertiary lymphoid structures during the development of lupus nephritis. *Int. J. Immunopathol. Pharmacol.* **2021**, *35*, e005772. [CrossRef]
16. Khan, H.; Nazeer, H.; Engell, H.; Naseer, N.; Korostynska, O.; Mirtaheri, P. Prefrontal Cortex Activation Measured during Different Footwear and Ground Conditions Using fNIRS—A Case Study. In Proceedings of the 2021 International Conference on Artificial Intelligence and Mechatronics Systems (AIMS), Delft, The Netherlands, 12–16 July 2021.
17. Khan, H.; Naseer, N.; Yazidi, A.; Eide, P.K.; Hassan, H.W.; Mirtaheri, P. Analysis of Human Gait Using Hybrid EEG-fNIRS-Based BCI System: A Review. *Front. Hum. Neurosci.* **2021**, *14*, 613254. [CrossRef]
18. Hamid, H.; Naseer, N.; Nazeer, H.; Khan, M.J.; Khan, R.A.; Khan, U.S. Analyzing Classification Performance of fNIRS-BCI for Gait Rehabilitation Using Deep Neural Networks. *Sensors* **2022**, *22*, 1932. [CrossRef]
19. Khan, R.A.; Naseer, N.; Qureshi, N.K.; Noori, F.M.; Nazeer, H.; Khan, M.U. fNIRS-based Neurorobotic Interface for gait rehabilitation. *J. Neuroeng. Rehabil.* **2018**, *15*, 7. [CrossRef]
20. Abdalmalak, A.; Milej, D.; Yip, L.; Khan, A.R.; Diop, M.; Owen, A.M.; Lawrence, K.S. Assessing Time-Resolved fNIRS for Brain-Computer Interface Applications of Mental Communication. *Front. Neurosci.* **2020**, *14*, 105. [CrossRef]
21. Nazeer, H.; Naseer, N.; Mehboob, A.; Khan, M.J.; Khan, R.A.; Khan, U.S.; Ayaz, Y. Enhancing classification performance of fNIRS-BCI by identifying cortically active channels using the z-score method. *Sensors* **2020**, *20*, 6995. [CrossRef]
22. Naseer, N.; Qureshi, N.K.; Noori, F.M.; Hong, K.-S. Analysis of different classification techniques for two-class functional near-infrared spectroscopy-based brain-computer interface. *Comput. Intell. Neurosci.* **2016**, *2016*, 5480760. [CrossRef]
23. Zhu, Z.; Yin, H.; Chai, Y.; Li, Y.; Qi, G. A novel multi-modality image fusion method based on image decomposition and sparse representation. *Inf. Sci.* **2018**, *432*, 516–529. [CrossRef]
24. Sreeja, S.; Samanta, D. Distance-based weighted sparse representation to classify motor imagery EEG signals for BCI applications. *Multimed. Tools Appl.* **2020**, *79*, 13775–13793. [CrossRef]
25. Miao, M.; Zhao, F.; Liu, F.; Zeng, H.; Wang, A. Index finger motor imagery EEG pattern recognition in BCI applications using dictionary cleaned sparse representation-based classification for healthy people. *Rev. Sci. Instrum.* **2017**, *88*, 094305. [CrossRef] [PubMed]

26. Miao, M.; Wang, A.; Liu, F. A spatial-frequency-temporal optimized feature sparse representation-based classification method for motor imagery EEG pattern recognition. *Med. Biol. Eng. Comput.* **2017**, *55*, 1589–1603. [CrossRef] [PubMed]
27. Shin, Y.; Lee, S.; Ahn, M.; Cho, H.; Jun, S.C.; Lee, H.-N. Noise robustness analysis of sparse representation based classification method for non-stationary EEG signal classification. *Biomed. Signal Process. Control* **2015**, *21*, 8–18. [CrossRef]
28. Gratton, G.; Brumback, C.R.; Gordon, B.; Pearson, M.A.; Low, K.A.; Fabiani, M. Effects of measurement method, wavelength, and source-detector distance on the fast optical signal. *NeuroImage* **2006**, *32*, 1576–1590. [CrossRef] [PubMed]
29. Delpy, D.T.; Cope, M.; van der Zee, P.; Arridge, S.; Wray, S.; Wyatt, J. Estimation of optical pathlength through tissue from direct time of flight measurement. *Phys. Med. Biol.* **1988**, *33*, 1433–1442. [CrossRef]
30. Ye, J.C.; Tak, S.; Jang, K.E.; Jung, J.; Jang, J. NIRS-SPM: Statistical parametric mapping for near-infrared spectroscopy. *NeuroImage* **2009**, *44*, 428–447. [CrossRef]
31. Noori, F.M.; Naseer, N.; Qureshi, N.K.; Nazeer, H.; Khan, R. Optimal feature selection from fNIRS signals using genetic algorithms for BCI. *Neurosci. Lett.* **2017**, *647*, 61–66. [CrossRef]
32. Santosa, H.; Hong, M.J.; Hong, K.-S. Lateralization of music processing with noises in the auditory cortex: An fNIRS study. *Front. Behav. Neurosci.* **2014**, *8*, 418. [CrossRef]
33. Rojas, R.F.; Huang, X.; Ou, K.L.; Lopez-Aparicio, J. Cross Correlation Analysis of Multi-Channel Near Infrared Spectroscopy. *Comput. Sci. Inf. Technol.* **2016**, *6*, 23–33. [CrossRef]
34. Zhang, H.; Patel, V.M. Sparse representation-based open set recognition. *IEEE Trans. Pattern Anal. Mach. Intell.* **2016**, *39*, 1690–1696. [CrossRef]
35. Li, Y.; Ngom, A. Sparse representation approaches for the classification of high-dimensional biological data. *BMC Syst. Biol.* **2013**, *7*, S6. [CrossRef] [PubMed]
36. Zhang, Z.; Xuelong, L.; Yang, J.; Li, X.; Zhang, D. A Survey of Sparse Representation: Algorithms and Applications. *IEEE Access* **2015**, *3*, 490–530. [CrossRef]
37. Cleophas, T.J.; Zwinderman, A.H. *Statistical Analysis of Clinical Data on a Pocket Calculator*; Springer: Berlin/Heidelberg, Germany, 2011.
38. Qureshi, N.K.; Naseer, N.; Noori, F.M.; Nazeer, H.; Khan, R.; Saleem, S. Enhancing Classification Performance of Functional Near-Infrared Spectroscopy- Brain–Computer Interface Using Adaptive Estimation of General Linear Model Coefficients. *Front. Neurorobotics* **2017**, *11*, 33. [CrossRef] [PubMed]
39. Zafar, A.; Hong, K.-S. Neuronal Activation Detection Using Vector Phase Analysis with Dual Threshold Circles: A Functional Near-Infrared Spectroscopy Study. *Int. J. Neural Syst.* **2018**, *28*, 1850031. [CrossRef]
40. Zafar, A.; Ghafoor, U.; Yaqub, M.; Hong, K.-S. Initial-dip-based classification for fNIRS BCI. In *Neural Imaging and Sensing*; SPIE: Bellingham, WA, USA, 2019.
41. Nazeer, H.; Naseer, N.; Khan, R.A.; Noori, F.M.; Qureshi, N.K.; Khan, U.S.; Khan, M.J. Enhancing classification accuracy of fNIRS-BCI using features acquired from vector-based phase analysis. *J. Neural Eng.* **2020**, *17*, 056025. [CrossRef]
42. Naseer, N.; Noori, F.M.; Qureshi, N.K.; Hong, K.-S. Determining Optimal Feature-Combination for LDA Classification of Functional Near-Infrared Spectroscopy Signals in Brain-Computer Interface Application. *Front. Hum. Neurosci.* **2016**, *10*, 237. [CrossRef]
43. Hong, K.-S.; Santosa, H. Decoding four different sound-categories in the auditory cortex using functional near-infrared spectroscopy. *Hear. Res.* **2016**, *333*, 157–166. [CrossRef]
44. Nguyen, H.-D.; Hong, K.-S. Bundled-optode implementation for 3D imaging in functional near-infrared spectroscopy. *Biomed. Opt. Express* **2016**, *7*, 3491–3507. [CrossRef]
45. Petrantonakis, P.C.; Kompatsiaris, I. Single-Trial NIRS Data Classification for Brain–Computer Interfaces Using Graph Signal Processing. *IEEE Trans. Neural Syst. Rehabil. Eng.* **2018**, *26*, 1700–1709. [CrossRef] [PubMed]
46. Muthukrishnan, R.; Rohini, R. LASSO: A feature selection technique in predictive modeling for machine learning. In Proceedings of the 2016 IEEE International Conference on Advances in Computer Applications (ICACA), Coimbatore, India, 24 October 2016.
47. Varsehi, H.; Firoozabadi, S.M.P. An EEG channel selection method for motor imagery based brain–computer interface and neurofeedback using Granger causality. *Neural Netw.* **2020**, *133*, 193–206. [CrossRef] [PubMed]
48. Naseer, N.; Hong, K.-S.; Bhutta, M.R.; Khan, M.J. Improving classification accuracy of covert yes/no response decoding using support vector machines: An fNIRS study. In Proceedings of the 2014 International Conference on Robotics and Emerging Allied Technologies in Engineering (iCREATE), Islamabad, Pakistan, 22–24 April 2014.

Article

Analyzing Classification Performance of fNIRS-BCI for Gait Rehabilitation Using Deep Neural Networks

Huma Hamid [1], Noman Naseer [1,*], Hammad Nazeer [1], Muhammad Jawad Khan [2], Rayyan Azam Khan [3] and Umar Shahbaz Khan [4,5]

[1]   Department of Mechatronics and Biomedical Engineering, Air University, Islamabad 44000, Pakistan; humahamid244@gmail.com (H.H.); hammad@mail.au.edu.pk (H.N.)
[2]   School of Mechanical and Manufacturing Engineering, National University of Science and Technology, Islamabad 44000, Pakistan; jawad.khan@smme.nust.edu.pk
[3]   Department of Mechanical Engineering, University of Saskatchewan, Saskatoon, SK S7N 5A9, Canada; rayyan.khan@usask.ca
[4]   Department of Mechatronics Engineering, National University of Sciences and Technology, Islamabad 44000, Pakistan; u.shahbaz@ceme.nust.edu.pk
[5]   National Centre of Robotics and Automation (NCRA), Rawalpindi 46000, Pakistan
*   Correspondence: noman.naseer@mail.au.edu.pk

Abstract: This research presents a brain-computer interface (BCI) framework for brain signal classification using deep learning (DL) and machine learning (ML) approaches on functional near-infrared spectroscopy (fNIRS) signals. fNIRS signals of motor execution for walking and rest tasks are acquired from the primary motor cortex in the brain's left hemisphere for nine subjects. DL algorithms, including convolutional neural networks (CNNs), long short-term memory (LSTM), and bidirectional LSTM (Bi-LSTM) are used to achieve average classification accuracies of 88.50%, 84.24%, and 85.13%, respectively. For comparison purposes, three conventional ML algorithms, support vector machine (SVM), k-nearest neighbor (k-NN), and linear discriminant analysis (LDA) are also used for classification, resulting in average classification accuracies of 73.91%, 74.24%, and 65.85%, respectively. This study successfully demonstrates that the enhanced performance of fNIRS-BCI can be achieved in terms of classification accuracy using DL approaches compared to conventional ML approaches. Furthermore, the control commands generated by these classifiers can be used to initiate and stop the gait cycle of the lower limb exoskeleton for gait rehabilitation.

Keywords: functional near-infrared spectroscopy; brain-computer interface; convolutional neural network; long short-term memory; neurorehabilitation

Citation: Hamid, H.; Naseer, N.; Nazeer, H.; Khan, M.J.; Khan, R.A.; Shahbaz Khan, U. Analyzing Classification Performance of fNIRS-BCI for Gait Rehabilitation Using Deep Neural Networks. *Sensors* **2022**, *22*, 1932. https://doi.org/10.3390/s22051932

Academic Editor: Ki H. Chon

Received: 28 January 2022
Accepted: 25 February 2022
Published: 1 March 2022

1. Introduction

The world has been striving to create a communication channel based on signals obtained from the brain. A brain-computer interface (BCI) is a communication system that provides its users with control channels independent of the brain's output channel to control external devices using brain activity [1,2]. The BCI system was first introduced by Vidal in 1973 in which he proposed three assumptions regarding BCI, including analysis of complex data in the form of small wavelets [3]. A typical BCI system consists of five stages, as shown in Figure 1. The first stage is the brain-signal acquisition using a neuroimaging modality. The second is preprocessing those signals as they contain physiological noises and motion artefacts [4]. The third stage is feature extraction in which meaningful features are selected [5]. These features are then classified using suitable classifiers. The final stage is the application interface in which the classified BCI signals are given to an external device as a control command [6].

Researchers have been using different techniques to acquire brain signals [7]. These techniques include electroencephalography (EEG), functional near-infrared spectroscopy

(fNIRS), magnetoencephalography (MEG), and functional magnetic resonance imaging (fMRI) [8]. EEG is a technique used to analyze brain activity by measuring changes in the electrical activity of the active neurons in the brain [9], while MEG measures the changes in magnetic fields associated with the brain's electrical activity changes [10]. fMRI is another modality for analyzing brain function by measuring the localized changes in cerebral blood flow stimulated by cognitive, sensory, or motor tasks [11,12]. In this study, we will only be dealing with fNIRS-BCI. fNIRS is a non-invasive optical neuroimaging technique that measures the concentration changes of oxy-hemoglobin (ΔHbO) and deoxy-hemoglobin (ΔHbR) that are associated with the brain activity stimulated, when participants perform experiments, such as arithmetic tasks, motor imagery, motor execution, etc. [13]. It is a non-invasive, portable and easy-to-use brain imaging technique that helps study the brain's functions in the laboratory, naturalistic, and real-world settings [14]. fNIRS consists of near-infrared light emitter detector pairs. The emitter emits light with several distinct wavelengths absorbed in the subject's scalp, consequently causing scattered photons; while some of them are absorbed, the others disperse and pass through the cortical areas where HbO and HbR chromophores absorb the light and have different absorption coefficients. The concentration of HbO and HbR changes along the photon path in consideration of the modified Beer-Lambert law [15].

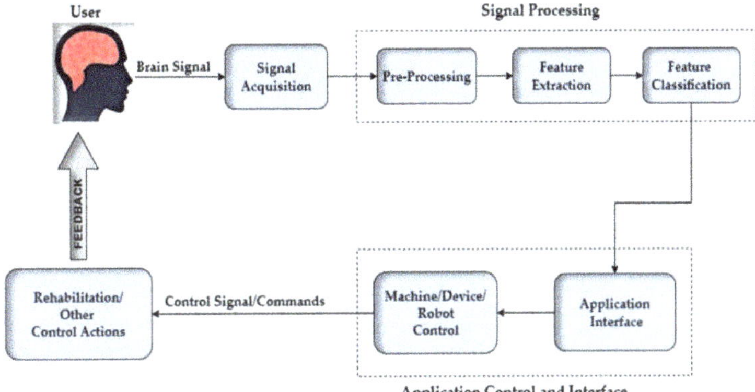

Figure 1. Basic design and operation of the brain-computer interface (BCI)-based control.

In the recent decade, the research on fNIRS-BCI has increased enormously, and new diverse techniques, particularly in its applications, may grow exponentially over the following years [16]. One of the significant fields of fNIRS application is cognitive neuroscience, particularly in real-world cognition [17], neuroergonomics [18], neurodevelopment [19], neurorehabilitation [20], and in social interactions. fNIRS-BCI can provide a modest input for BCI systems in the real time, but improvements are required for this system as there is the difficulty faced with the collection and interpretation of the data for classifiers due to noise in the data and subject variations [21].

Well-designed wearable assistive devices for rehabilitation and performance augmentation purposes have been developed that run independently of physical or muscular interventions [22–24]. Recent studies focus on acquiring the user's intent through brain signals to control these devices/limbs [25–27]. In assistive technologies, the fNIRS-BCI system is a suitable technique for controlling exoskeletons and wearable robots by using intuitive brain signals instead of being controlled manually by various buttons in order to get the desired postures [28]. In addition, it has a better spatial resolution, fewer artefacts, and acceptable temporal resolution, which makes it a suitable choice for rehabilitation and mental task applications [29].

To find the best machine-learning (ML) method for fNIRS-based active-walking classification, a series of experiments with various ML algorithms and configurations were conducted; the classification accuracy achieved was above 95% [30] for classifying 1000 samples using different ML algorithms, such as random forest, decision tree, logistic regression, support vector machine (SVM) and k-nearest neighbor (k-NN). In order to achieve minimum execution delay and minimum computation cost for an online BCI system, linear discriminant analysis (LDA) was used with combinations of six features for walking intention and rest tasks [31].

The traditional method of extracting and selecting acceptable features can result in performance degradation. In contrast, deep neural networks (DNNs) can extract different features from raw fNIRS signals, nullifying the need for a manual feature extraction stage in the BCI system development, but limited studies are available so far [32,33].

Convolutional neural network (CNN) is a deep-learning (DL) approach that can automatically learn relevant features from the input data [34]. In a study, CNN architecture was compared to conventional ML algorithms, and CNNs performed better in terms of classification with an average classification accuracy of 72.35 $\pm$ 4.4% for the four-class motor imagery fNIRS signals [35]. CNN-based time series classification (TSC) methods to classify fNIRS-BCI are compared with ML methods, such as SVM. The results showed that the CNN-based methods performed better in terms of classification accuracy for left-handed and right-handed motor imagery tasks and achieved up to 98.6% accuracy [36].

Time-series data can be handled more precisely using long short-term memory (LSTMs) modules. In a study, four DL models were evaluated including multilayer perceptron (MLP), forward and backward long short-term memory (LSTMs), and bidirectional LSTM (Bi-LSTM) for the assessment of human pain in nonverbal patients, and Bi-LSTM model achieved the highest classification accuracy of 90.6% [37]. Using the LSTM network, large time scale connectivity can be determined with the help of the InceptionTime neural network, which is an attention LSTM neural network utilized for the brain activations of mood disorders [38]. A recent study assessed four-level mental workload states using DNNs, such as CNNs and LSTM for fNIRS-BCI system, with average classification accuracies of 87.45% and 89.31% [39].

This study contributes to the development of a neurorehablitation tool in the gait training of elderly and disabled people. The main objective of this study is to compare two classification approaches, ML and DL, to achieve high performance in terms of classification accuracy on the time-series fNIRS data. The complete summary of the methods employed in this research is depicted in Figure 2.

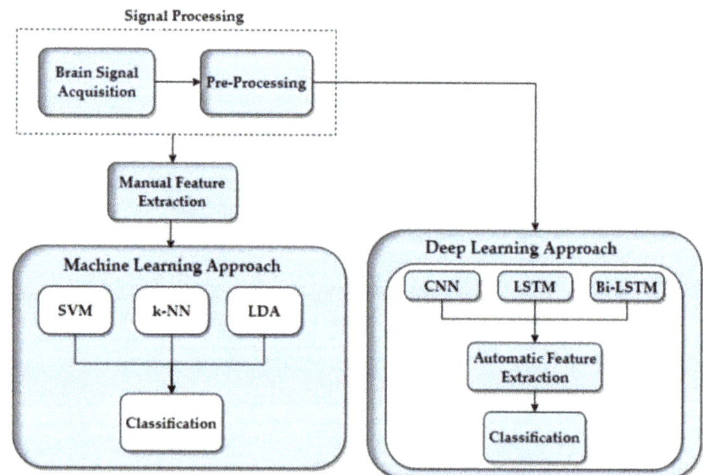

Figure 2. Time-series functional near-infrared spectroscopy (fNIRS) signal classification for walking and rest tasks using conventional machine-learning (ML) and DL algorithms. Signal processing and feature engineering followed by classification using ML and DL approaches.

2. Materials and Methods

2.1. Experimental Paradigm

Nine healthy right-handed male subjects of 27 ± 5 median age were selected. The subjects had no history of motor disability or any visual neurological disorders affecting the experimental results. fNIRS-based BCI signals were acquired from the primary motor cortex (M1) in the left hemisphere for self-paced walking [40]. Before the start of each experiment, the subjects were asked to take a rest for 30 s in a quiet room to establish the baseline condition; it was followed by 10 s of walking on a treadmill, followed by 20 s of rest while standing on the treadmill. At the end of each experiment, a 30 s rest was also given for baseline correction of the signals. Each subject performed 10 trials, as shown in Figure 3. Excluding the initial (30 s) and final (30 s) of rest, the total length of each experiment was 300 s for each subject. All the experiments were conducted in accordance with the latest Declaration of Helsinki and approved by the Institutional Review Board of Pusan National University, Busan, Republic of Korea [41].

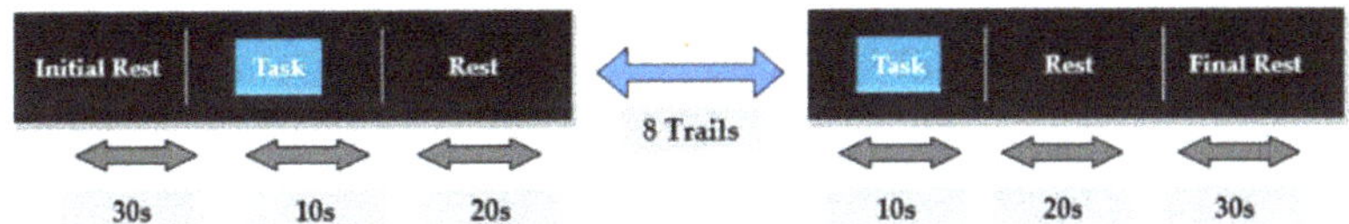

Figure 3. Experimental paradigm with experimental 10 trials with initial and final 30 s rest.

2.2. Experimental Configuration

In this study a multi-channel continuous-wave imaging system (DYNOT: Dynamic Near-infrared Optical Tomography; NIRx Medical Technologies, NY, USA) was used to acquire the brain signals, which operate on two wavelengths, 760 and 830 nm, with a 1.81 Hz sampling frequency. Four near-infrared light detectors and five sources (total of nine optodes) were placed on the left hemisphere of the M1 with 3 cm of distance between a source and a detector [42]. A total of twelve channels were formed in accordance with the defined configuration, as shown in Figure 4.

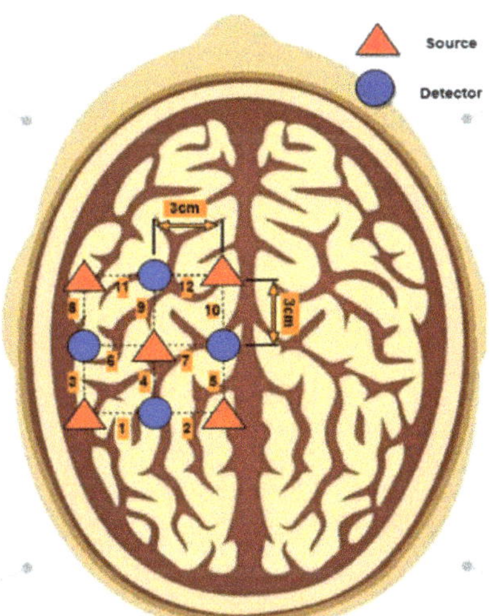

Figure 4. Optode placement on the left hemisphere of the motor cortex with channel configuration using 10–20 international system.

2.3. Signal Acquisition

The acquired light intensity signals from the left hemisphere of the M1 were first converted into oxy- and deoxy-hemoglobin concentration changes ($\Delta c_{HbO}(t)$, and $\Delta c_{HbR}(t)$) using the modified Beer-Lambert law [43].

$$\begin{bmatrix} \Delta c_{HbO}(t) \\ \Delta c_{HbR}(t) \end{bmatrix} = \frac{\begin{bmatrix} \alpha_{HbO}(\lambda_1) & \alpha_{HbR}(\lambda_1) \\ \alpha_{HbO}(\lambda_2) & \alpha_{HbR}(\lambda_2) \end{bmatrix}^{-1} \begin{bmatrix} \Delta A(t,\lambda_1) \\ \Delta A(t,\lambda_2) \end{bmatrix}}{d * l} \tag{1}$$

where $\Delta c_{HbO}(t)$ and $\Delta c_{HbR}(t)$ are the concentration changes of HbO and HbR in [μM], $A(t, \lambda_1)$ and $A(t, \lambda_2)$ are the absorptions at two different instants, l is the emitter–detector distance (in millimeters), d is the unitless differential path length factor (DPF), and $\alpha_{HbO}(\lambda)$ and $\alpha_{HbR}(\lambda)$ are the extinction coefficients of HbO and HbR in [$\mu M^{-1}\ cm^{-1}$].

2.4. Signal Processing

After obtaining oxy-hemoglobin concentration changes ($\Delta c_{HbO}(t)$ and $\Delta c_{HbR}(t)$), the brain signals acquired were filtered with suitable filters using the modified Beer-Lambert law. In order to minimize the physiological or instrumental noises, such as heartbeat (1–1.5 Hz), respiration (~0.5 Hz), Mayer waves (blood pressure), artefacts, and others, the signals were low-pass filtered at a cut-off frequency of 0.5 Hz and a high-pass filter with cut-off frequency of 0.01 Hz [44]. The filter used for $\Delta c_{HbO}(t)$ signals were hemodynamic response (hrf) using NIRS-SPM toolbox [45]. The averaged $\Delta c_{HbO}(t)$ signal for task and rest of subject 1 after filtering is shown in Figure 5.

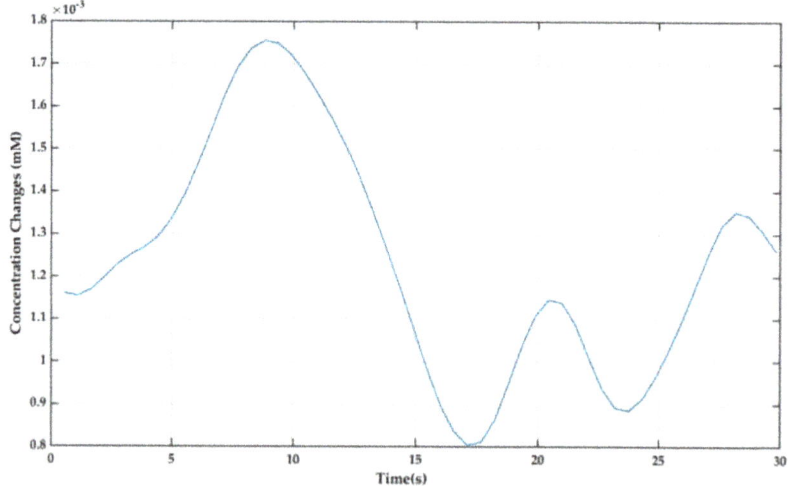

Figure 5. Averaged $\Delta c_{HbO}(t)$ signal for task and rest of subject 1.

2.5. Feature Extraction

For the conventional ML algorithms, five different features of filtered $\Delta c_{HbO}(t)$ signals were extracted using the spatial average for all 12 channels. Five statistical properties (mean, variance, skewness, kurtosis, and peak) of the averaged signals were calculated for the entire task and rest sessions. In this study, a feature combination of signal mean, signal peak, and signal variance a was used for the ML classifiers. This specific combination was selected based on the higher classification accuracies that were obtained using these features [46,47].

For the mean, the equation was as follows:

$$\overline{X} = \frac{1}{N} \sum_{i=1}^{N} (Z_i) \tag{2}$$

where N is the total number of observations, and Z_i is the $\Delta c_{HbO}(t)$ across each observation. For signal variance, the calculation was as follows:

$$\sigma^2 = \frac{\sum_{i=1}^{N} (X_i - \overline{X})}{n - 1} \tag{3}$$

where n is the sample size, X_i is the ith element of the sample, and $\overline{X}$ is the mean of the sample. To calculate signal peak, the max function in MATLAB$^{\circledR}$ was used.

3. Classification Using Machine-Learning Algorithms

3.1. Support Vector Machine (SVM)

SVM is a commonly used classification technique suitable for fNIRS-BCI systems for handling high-dimensional data [48,49]. In supervised learning, the SVM classifier creates hyperplanes to maximize the distance between the separating hyperplanes and the closest training points [50]. The hyperplanes, known as the class vectors, are called support vectors. The separating hyperplane in the two-dimension features space is given by:

$$f(x) = r \cdot x + b \tag{4}$$

where b is a scaling factor, and r, $x \in R^2$ and $b \in R^1$. The optimal solution, r^*, that is the distance between the hyperplane and the nearest training point(s) is maximized by minimizing the cost function. The optimal solution, r^* is given by the equation.

Minimize

$$\frac{1}{2} ||w||^2 + C \sum_{i=1}^{n} \xi_i \tag{5}$$

Subject to

$$y_i \left(w^T x_i + b \right) \geq 1 - \xi_i, \quad \xi_i \geq 0 \tag{6}$$

where y_i represents the class label for the ith sample, T is the transpose, and n is the total number of samples, $||w||^2 = w^T w$. where w^T and $x_i \in R^2$, $b \in R^1$, C is the trade-off parameter between the margin and error, and ξ_i is the training error.

3.2. k-Nearest Neighbor (k-NN)

k-NN predicts the test sample's category; the k value represents the number of neighbors considered and classifies it in the same class as its nearest neighbor based upon the largest category probability [51]. Euclidean distance is the distance between the trained and the test object given by the equation.

$$D_E(p,q) = \sqrt{\sum_{i=1}^{n} (p_i - q_i)^2} \tag{7}$$

where n is the n-space, p and q are two points in the Euclidean n-space, and p_i, q_i are the two vectors, stating from the origin of the space.

k-NN is a widely used efficient classification method for BCI applications because of its low computational requirements and simple implementation [52,53].

3.3. Linear Discriminant Analysis (LDA)

LDA has discriminant hyperplanes to separate classes from each other. LDA performs well in various BCI systems because of its simplicity and execution speed [54]. It minimizes the variance of the class and maximizes the separation between the mean of the class by maximizing the Fisher's criterion [55]. The equation for Fisher's criterion is given by:

$$J\left(v\right) = \frac{v^T S_b v}{v^T S_w v} \tag{8}$$

where S_b and S_w are the between-class and within-class scatter matrices given as:

$$S_b = (m_1 - m_2)(m_1 - m_2)^T, \tag{9}$$

$$S_w = \sum_{x_n \in C_1} (x_n - m_1)(x - m_2)^T + \sum_{x_n \in C_2} (x_n - m_1)(x - m_2)^T$$

where x_n denotes samples, m_1 and m_2 are the group means of classes C_1 and C_2, respectively.

4. Classification Using Deep-Learning Algorithms

fNIRS signal classification with conventional ML methods is composed of local and global feature extraction, e.g., independent component analysis (ICA) and principal component analysis (PCA), selection of possible features, their combinations, and dimensionality reduction, which leads to the biasness and overfitting of the data [56]. It is because of these limitations a large amount of time is consumed in the mining and preprocessing of the data [57].

The BCI classification task can avoid local filtering, noise removal, and manual local feature extraction by utilizing DL algorithms as an alternative to avoid the need for manual feature engineering, data cleaning, analysis, transformation, and dimensionality reduction before feeding it to the learning machines [58]. Extracting and selecting appropriate features is critical with fNIRS-BCI signals, and the multi-dimensionality and complexity of fNIRS signals make it ideal for DL algorithms to work with.

4.1. Convolutional Neural Networks (CNNs)

CNNs are a type of neural networks that are capable of automatically learning appropriate features from the input fNIRS time-series data. CNNs consist of several layers, such as the convolutional layer, pooling layer, fully connected layer, and output layer [59]. The input fNIRS time-series data (the changes in the HbO concentrations) across all the channels are passed through CNN layers. The convolutional layer contains filters that are known as convolution kernels to extract features. CNN minimizes the classification errors by adjusting the weight parameters of each filter using forward and backward propagation.

The convolution operation is the sliding of a filter over the time series, which results in activation maps also known as feature maps that stores the features and patterns of the fNIRS data [60]. Convolution operation for time stamp t is given by the equation:

$$C_t = f\left(\omega * X_{t-l/2:t+l/2} + b\right) \mid \forall\, t \in [1,\, T] \tag{10}$$

where C is the output of a convolution (dot product) on a time series, X, of length, T, with a filter, ω, of length, l, b is a bias parameter, and f is a non-linear function, such as the rectified linear unit (ReLU).

After the convolutional layer, we have a pooling layer that downsamples the spatial size of the data and also reduces the number of parameters in the network [61]. The pooling layer is followed by a fully connected layer, as shown in Figure 6 in which each data point is treated as a single neuron that outputs the class scores, and each neuron is connected to all the neurons in the previous layer [62].

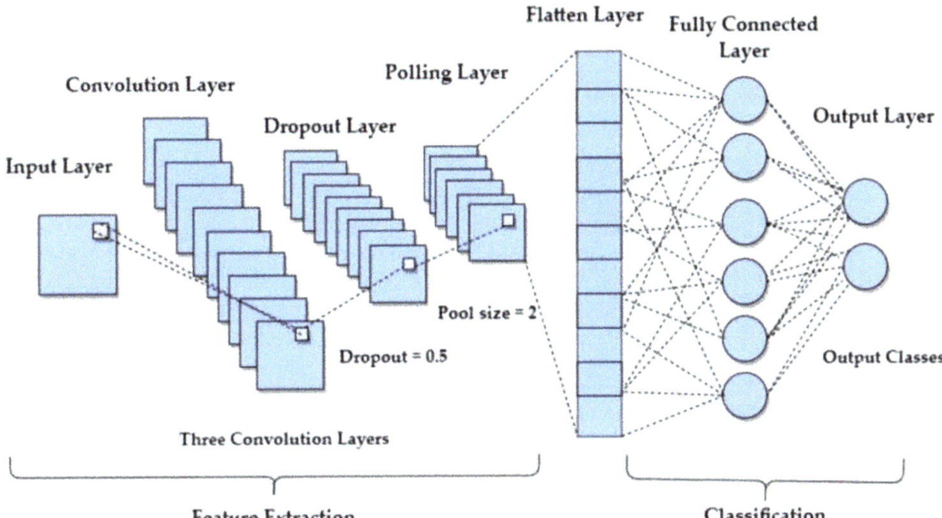

Figure 6. Convolutional neural network (CNN) model with convolutional layer, dropout layer, pooling layer, flatten layer, fully connected layer, and output layer.

The proposed CNN model consists of the input layer, three one-dimensional convolutional layers, max-pooling layers, dropout layers, a fully connected layer, and an output layer. The three convolutional layers contain filters 128, 64, 32 with a kernel size of 3, 5, 11, respectively. A dropout layer of 0.5 'dropout ratio' was added after each convolutional layer to avoid overfitting, followed by a pooling layer with a stride of two. The input fNIRS time-series data after passing through a number of convolutional, max-pooling, and dropout layers is flattened and fed into the fully connected layers for the purpose of classification. The fully connected layer of 100 units is incorporated with ReLU activation. The output layer consists of two neurons corresponding to the two classes with Softmax activation. The optimization technique used was Adam with a batch size of 150 and 500 number of epochs.

4.2. Long Short-Term Memory (LSTM) and Bi-LSTM

LSTM is a DL algorithm that can achieve high accuracies in terms of classification, processing, and forecasting predictions on the time-series fNIRS data. LSTMs have internal mechanisms called gates that can regulate the flow of information [63]. These gates, such as forget gate, input gate, and output gate, can learn which data in a sequence are necessary to keep or throw away [64]. By doing that, it can pass relevant information down the long chain of sequences to make predictions. The equations for forget gate (f_t), input gate (i_t) and output gate (o_t) are given by:

$$f_t = \sigma\left(W_f \cdot [h_{t-1}, x_t] + b_f\right) \tag{11}$$

$$i_t = \sigma(W_i \cdot [h_{t-1}, x_t] + b_i) \tag{12}$$

$$o_t = \sigma(W_o \cdot [h_{t-1}, x_t] + b_o) \tag{13}$$

where W_f, W_i, and W_o are the weight matrices of forget, input, and output gates, respectively, and h_{t-1} is the hidden state.

These gates contain sigmoid and Tanh activations to help regulate the values flowing through the network [65]. General sigmoid function is given by:

$$f(x) = \frac{1}{1 + e^{-k(x - x_0)}}$$ (14)

where x_0 is the x-value of the sigmoid midpoint, e is the natural logarithm base, and k is the growth rate.

For LSTM the data has to be reshaped because it expects the data in the form of $(m \times k \times n)$, where m is the number of samples, n refers to the number of fNIRS channels (12 $\Delta c_{HbO}(t)$ channels), and k refers to the time steps. The proposed LSTM model consisted of an input layer, four LSTM layers, a fully connected layer, and an output layer, as shown in Figure 7. A dropout layer of 0.5 'dropout ratio' was added after the LSTM layers to avoid overfitting. The output from the dropout layer is flattened and fed to the fully connected layer of 64 units, also known as the dense layer, and incorporated with ReLU activation. Finally, an output layer consists of two neurons corresponding to the two classes with Softmax activation. The early-stopping technique was used to avoid overfitting with the patience of 50; a batch size of 150 over 500 epochs with Adam optimization technique.

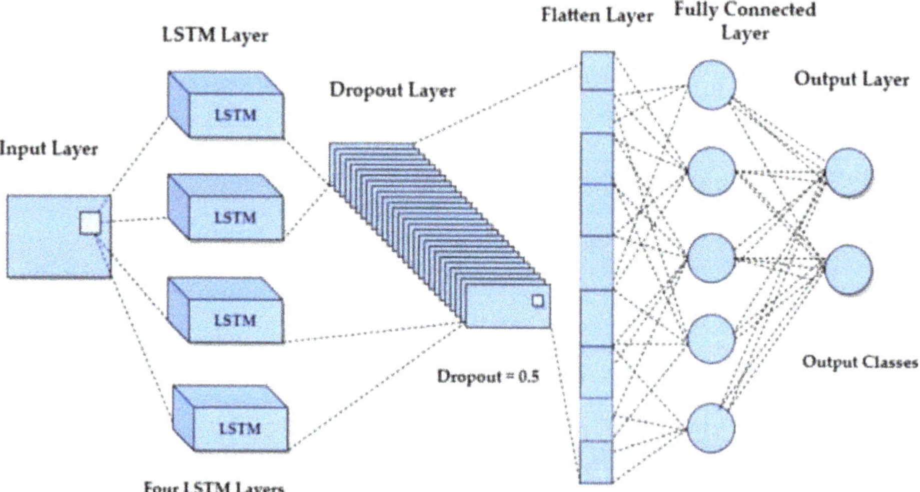

Figure 7. Long short-term memory (LSTM) model with four LSTM layers, dropout layer, flatten layer, fully connected layer, and output layer.

Bi-LSTM models are a combination of both forward and backward LSTMs [66]. These models can run inputs in two ways, from past to future and from future to past and have both forward and backward information about the sequence at every time step [67]. Bi-LSTM differs from conventional LSTMs as they are unidirectional, and with bidirectional, we are able at any point in time to preserve information from both past and future, which is why they perform better than conventional LSTMs [68].

The proposed Bi-LSTM model consisted of two Bi-LSTM layers with 64 hidden units, a fully connected layer, and an output layer, as shown in Figure 8. The fully connected layer of 64 units is employed with ReLU activation, and the output layer consists of two neurons corresponding to the two classes with Softmax activation.

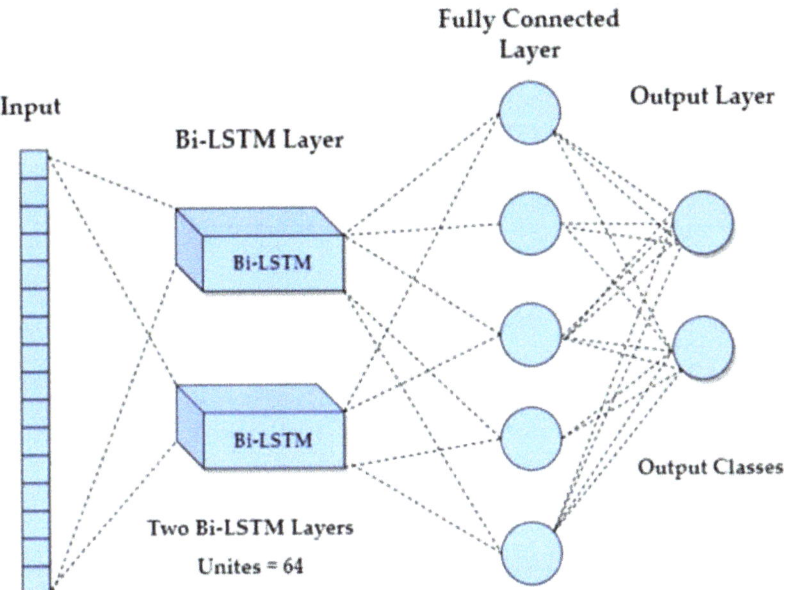

Figure 8. Bidirectional LSTM (Bi-LSTM) model with two Bi-LSTM layers with 64 units, fully connected layer, and output layer.

5. Results

The results evaluated for all the methods used in this study are presented in this section, including the validation of the methods. ML algorithms (SVM, k-NN, and LDA) were performed on MATLAB® 2020a Classification Learner App, whereas DL algorithms (CNN, LSTM, and Bi-LSTM) were performed on Python 3.7.12 on Google Colab using Keras models with TensorFlow.

5.1. Classification Accuracy of Machine-Learning Algorithms

For ML algorithms, five features (signal mean, signal variance, signal skewness, signal kurtosis, and signal peak) across all 12 channels of filtered $\Delta c_{HbO}(t)$ signals were spatially calculated. Three feature combinations that were signal mean, signal variance, and signal peak yielded the best results. The manually extracted features from fNIRS data of walking and rest states of nine subjects are fed to the three conventional ML algorithms, SVM, k-NN, and linear LDA, and the highest accuracies obtained were 78.90%, 77.01%, and 66.70% across 12 channels, respectively, as given in Table 1.

Table 1. Accuracy of conventional machine-learning (ML) algorithms, k-nearest neighbor (k-NN), support vector machine (SVM), and linear discriminant analysis (LDA) for all nine subjects.

Classifier	Sub1	Sub2	Sub3	Sub4	Sub5	Sub6	Sub7	Sub8	Sub9
SVM	78.90%	76.70%	66.70%	71.50%	72.00%	72.80%	73.50%	75.70%	77.40%
k-NN	77.01%	74.40%	68.30%	70.60%	73.50%	74.10%	72.02%	73.50%	84.80%
LDA	64.30%	66.30%	63.70%	66.30%	66.70%	65.20%	65.60%	67%	67.60%

To curb overfitting, 10-fold cross-validation was used for the training of the models. The accuracy of conventional ML algorithms for all nine subjects is shown in a bar graph in Figure 9.

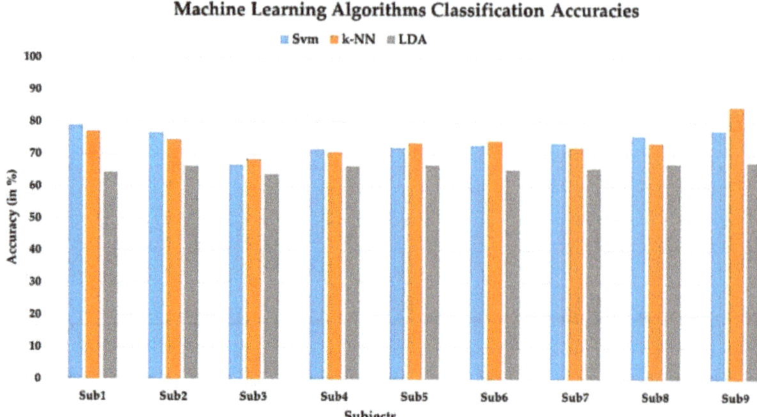

Figure 9. Machine-learning (ML) classifier accuracies (in %) for nine subjects. The ML classifiers are support vector machine (SVM), k-nearest neighbor (k-NN), and linear discriminant analysis (LDA).

5.2. Classification Accuracy of Deep Learning Algorithms

To evaluate the deep-learning models, the dataset was initially split into an 80:20 ratio, the training set (80%) and the testing set (20%). The training set used for DL methods in this study has 12 feature dimensions. The learning performance of the algorithm is affected by the size of the training set, which is why 20% of the validation set were employed for each subject since the larger training set usually provides higher classification performance [69]. Although, for CNN, a smaller number of samples after the 30%validation set also attained classification accuracy reaching 90%. The pre-processed fNIRS data of walking and rest states of nine subjects is fed to the three DL algorithms, CNN, LSTM, and Bi-LSTM; the highest classification accuracies obtained were 95.47%, 95.35%, and 95.54% across 12 channels, respectively. The classification accuracy of DL algorithms for all nine subjects is shown in a bar graph in Figure 10.

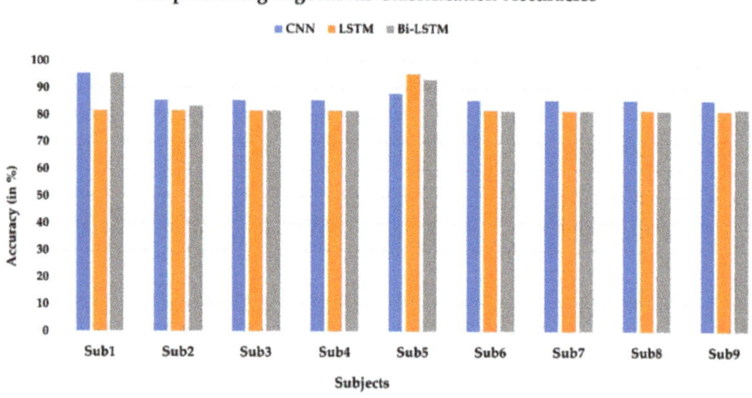

Figure 10. Deep-learning (DL) classifier accuracies (in %) for nine subjects. The DL classifiers are convolutional neural network (CNN), long short-term memory (LSTM), and bidirectional LSTM (Bi-LSTM).

All the DL models (CNN, LSTM, and Bi-LSTM) were compiled with the metric "accuracy", which is the measure of the number of correct predictions from all the predictions that were made. To further evaluate the effectiveness of the model, model "precision", which is the number of positive predictions divided by the total number of positive predicted values and model "recall", which is the number of actual positives divided by the total number of positive values were also calculated. Accuracy, precision, and recall of DL algorithms are summarized in Tables 2–4. The loss function used for the models was "categorical cross-entropy" which is a measure of prediction error, and the optimization technique used was "Adam optimizer". In order to avoid overfitting, early-stopping technique was used to halt the training when the error during the last 50 epochs is not reduced. Learning rate of 0.001 and decay factor of 0.5 were used in all DL models.

Table 2. Accuracy, precision, and recall of deep-learning (DL) algorithm, convolutional neural network (CNN) for all nine subjects.

CNN	Sub1	Sub2	Sub3	Sub4	Sub5	Sub6	Sub7	Sub8	Sub9
Accuracy	95.47%	88.10%	85.71%	87.72%	95.29%	85.63%	85.70%	87.37%	85.52%
Precision	90.78%	86.65%	88.28%	82.94%	93.72%	86.18%	79.32%	85.23%	83.79%
Recall	87.88%	80.74%	84.37%	85.63%	90.49%	82.87%	82.60%	88.06%	81.63%

Table 3. Accuracy, precision and recall of deep learning (DL) algorithm, long short-term memory (LSTM) for all nine subjects.

LSTM	Sub1	Sub2	Sub3	Sub4	Sub5	Sub6	Sub7	Sub8	Sub9
Accuracy	83.81%	82.84%	82.72%	81.83%	95.35%	83.04%	81.72%	82.00%	84.81%
Precision	78.24%	83.36%	80.92%	80.83%	90.76%	85.49%	80.29%	81.43%	82.45%
Recall	80.04%	82.32%	81.75%	81.25%	93.45%	84.35%	81.82%	83.63%	79.83%

Table 4. Accuracy, precision and recall of deep learning (DL) algorithm, bidirectional LSTM (Bi-LSTM) for all nine subjects.

Bi-LSTM	Sub1	Sub2	Sub3	Sub4	Sub5	Sub6	Sub7	Sub8	Sub9
Accuracy	95.54%	83.55%	81.81%	82.42%	93.28%	81.67%	81.85%	82.62%	83.42%
Precision	90.74%	80.23%	82.45%	81.72%	95.56%	80.48%	84.90%	80.53%	85.37%
Recall	92.38%	82.08%	80.76%	83.62%	91.49%	82.43%	83.73%	84.29%	80.97%

To evaluate the statistical significance of ML and DL methods, a *t*-test was performed for the best performing DL method (CNN) and all the other five classifiers accuracies, as shown in Table 5 The results of these statistical tests showed significant improvement of classification accuracy for the proposed DL method ($p < 0.008$) and the null hypothesis, meaning no statistical significance was rejected.

Table 5. Statistical significance between CNN and all other five classifiers accuracies.

Classifiers	Bonferroni Correction Applied ($p < 0.008$)
CNN vs. SVM	1.42×10^{-5}
CNN vs. k-NN	8.63×10^{-5}
CNN vs. LDA	4.01×10^{-12}
CNN vs. LSTM	5.35×10^{-9}
CNN vs. Bi-LSTM	2.19×10^{-8}

5.3. Validation

For the purpose of validation of the proposed methods, the analysis was also performed on an open-access database containing fNIRS brain signals ($\Delta c_{HbO}(t)$ and $\Delta c_{HbR}(t)$)

for the dominant foot tapping vs. rest [70]. The analysis was performed for 20 subjects with 25 trials for each subject. By applying the ML methods (SVM, k-NN, and LDA) on the fNIRS dataset for the dominant foot tapping vs. rest tasks, the average classification accuracies were estimated at 66.63%, 68.38%, and 65.96%, respectively, while for DL methods (CNN, LSTM, and Bi-LSTM) the average classification accuracies were estimated at 79.726%, 77.21%, and 78.97%, respectively. The students' t-test showed significantly higher performance ($p < 0.008$) for the proposed DL method. The results obtained from the validation dataset also confirmed the high performance of the proposed DL methods over the conventional ML methods.

6. Discussion

Around the world, there are a substantial number of people that have gait impairment and permanent disability in their lower limbs [71]. In the recent decade, the development of wearable/portable assistive devices for mobility rehabilitation and performance augmentation focuses on acquiring the user's intent through brain signals to control these devices/limbs [72]. In the field of assistive technologies, the fNIRS-BCI system is a relatively suitable technique for the control of exoskeletons and wearable robots by using intuitive brain signals instead of being controlled manually by various buttons to get the desired postures [28,31]. It has a better spatial resolution, fewer artefacts, and acceptable temporal resolution, which makes it a suitable choice for rehabilitation and mental task applications [29,73]. High accuracy BCI systems are to be designed in order to improve the quality of life of people with gait impairment since any misclassification can probably result in a serious accident [56]. To achieve this, the proposed DL and conventional ML methods are investigated for a state-of-the-art fNIRS-BCI system. The control commands generated through these models can be used to initiate and stop the gait cycle of the lower limb exoskeleton for gait rehabilitation.

Researchers have been exploring different ways to improve the classification accuracies by using different feature extraction techniques, feature combinations, or by using different machine-learning algorithms [30]. In a study, six feature combinations, signal mean (SM), signal slope (SS), signal variance (SV), slope kurtosis (KR), and signal peak (SP) have been used for walking and rest data, and the highest average classification accuracy of 75% was obtained from SVM using the hrf filter [31]. In this study, we used three feature combinations of the signal mean, signal variance, and signal peak, and the accuracy obtained from SVM using these features were 73.91%. In a recent study, four-level, mental workload states were assessed using the fNIRS-BCI system by utilizing DNNs, such as CNN and LSTM, and the average accuracy obtained using CNN was 87.45% [39]. Our study achieved almost the same average classification accuracy for CNN with 87.06% for two-class motor execution of walking and rest tasks.

CNN generally refers to a two-dimensional CNN used for image classification, in which the kernel slides along two dimensions on the image data. Recently, researchers have started using deep learning for fNIRS-BCI and bioinformatics problems and have achieved good performances using 2-D CNNs [35,74]. However, in this study, we have considered the one-dimensional CNN for time series fNIRS signals of motor execution tasks and reached a satisfactory classification accuracy. The highest average classification accuracy obtained in this study is 88.50%. For time-series fNIRS data, LSTMs and Bi-LSTMs can achieve high accuracy in terms of classification, processing, and forecasting predictions. In a study for assessing human pain in nonverbal patients, LSTM and Bi-LSTM models were evaluated, and the Bi-LSTM model achieved the highest classification accuracy of 90.6% [37]. In another study, the LSTM based conditional generative adversarial network (CGAN) system was analyzed to determine whether the subject's task is left-hand finger tapping, right-hand finger tapping, or foot tapping based on the fNIRS data patterns, and the classification accuracy obtained was 90.2%. In the present study, the highest accuracy achieved on any subject with LSTM and Bi-LSTM is 95.35% and 95.54%, respectively, across all 12 channels. The comparison of the average accuracies obtained using ML and DL approaches is shown in a bar graph in Figure 11.

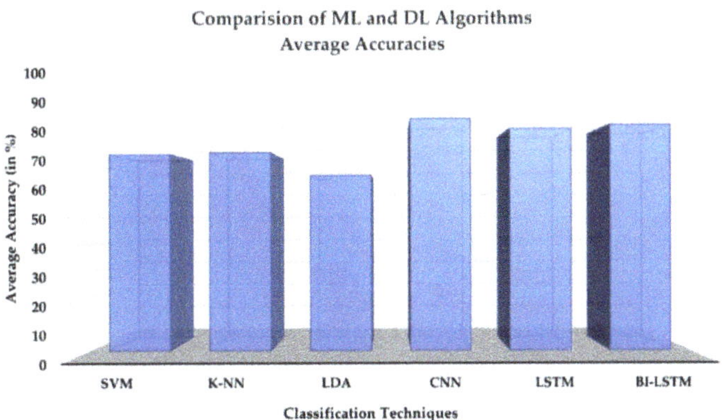

Figure 11. Comparison between machine-learning (ML) classifiers (support vector machine (SVM), k-nearest neighbor (k-NN), and linear discriminant analysis (LDA)) and deep-learning (DL) classifiers (convolutional neural network (CNN), long short-term memory (LSTM), and bidirectional LSTM (Bi-LSTM)) based on average accuracies.

7. Conclusions

In this study, two approaches, ML and DL, are investigated to decode two-class data of walking and rest tasks to obtain maximum classification accuracy. The DL approaches proposed in this study, CNN, LSTM, and Bi-LSTM, attained enhanced performance of the fNIRS-BCI system in terms of classification accuracy as compared to conventional ML algorithms across all nine subjects. The highest average classification accuracy of 88.50% was obtained using CNN. CNN showed significantly ($p < 0.008$) better performance as compared to all other ML and DL algorithms. The control commands generated by the classifiers can be used to start and stop the gait cycle of the lower limb exoskeleton which can effectively assist elderly and disabled people in the gait training.

Author Contributions: Conceptualization, H.H. and N.N.; methodology, H.H.; software, R.A.K.; validation, M.J.K., and H.N.; formal analysis, H.H. and R.A.K.; investigation, H.N.; resources, U.S.K.; data curation, M.J.K. and H.N.; writing—original draft preparation, H.H. and H.N.; writing—review and editing, N.N.; visualization, H.H. and H.N.; supervision, N.N.; project administration, N.N.; funding acquisition, U.S.K. All authors have read and agreed to the published version of the manuscript.

Funding: This study was funded by Higher Education Commission through the National Centre of Robotics and Automation, Rawalpindi, Pakistan, grant number NCRA-RF-027.

Institutional Review Board Statement: The study was conducted according to the guidelines of the Declaration of Helsinki and approved by the Institutional Review Board of Pusan National University, Busan, Republic of Korea.

Informed Consent Statement: Informed consent was obtained from all subjects involved in the study.

Data Availability Statement: The dataset presented in this study is available upon the request from the corresponding author.

Acknowledgments: The authors extend their appreciation to the National Centre of Robotics and Automation (NCRA), Rawalpindi, Pakistan, for the funding and for providing the necessary support and facilities to conduct this study. The authors would also like to acknowledge Keum-Shik Hong from Pusan National University, Busan, the Republic of Korea, for providing the necessary support and the opportunity to use the equipment in his lab.

Conflicts of Interest: The authors declare no conflict of interest.

References

1. Nicolas-Alonso, L.F.; Gomez-Gil, J. Brain Computer Interfaces, a Review. *Sensors* **2012**, *12*, 1211–1279. [CrossRef]
2. Nijboer, F.; Furdea, A.; Gunst, I.; Mellinger, J.; McFarland, D.J.; Birbaumer, N.; Kübler, A. An auditory brain-computer interface (BCI). *J. Neurosci. Methods* **2008**, *167*, 43–50. [CrossRef]
3. Vidal, J.J. Toward Direct Brain-Computer Communication. *Annu. Rev. Biophys. Bioeng.* **1973**, *2*, 157–180. [CrossRef]
4. Pinti, P.; Scholkmann, F.; Hamilton, A.; Burgess, P.; Tachtsidis, I. Current Status and Issues Regarding Pre-processing of fNIRS Neuroimaging Data: An Investigation of Diverse Signal Filtering Methods within a General Linear Model Framework. *Front. Hum. Neurosci.* **2019**, *12*, 505. [CrossRef]
5. Nazeer, H.; Naseer, N.; Khan, R.A.; Noori, F.M.; Qureshi, N.K.; Khan, U.S.; Khan, M.J. Enhancing classification accuracy of fNIRS-BCI using features acquired from vector-based phase analysis. *J. Neural Eng.* **2020**, *17*, 056025. [CrossRef]
6. Moore, M. Real-world applications for brain-computer interface technology. *IEEE Trans. Neural Syst. Rehabil. Eng.* **2003**, *11*, 162–165. [CrossRef]
7. Paszkiel, S.; Szpulak, P. Methods of Acquisition, Archiving and Biomedical Data Analysis of Brain Functioning. *Adv. Intell. Syst. Comput.* **2018**, *720*, 158–171. [CrossRef]
8. Crosson, B.; Ford, A.; McGregor, K.; Meinzer, M.; Cheshkov, S.; Li, X.; Walker-Batson, D.; Briggs, R.W. Functional imaging and related techniques: An introduction for rehabilitation researchers. *J. Rehabil. Res. Dev.* **2010**, *47*, vii–xxxiv. [CrossRef]
9. Kumar, J.S.; Bhuvaneswari, P. Analysis of Electroencephalography (EEG) Signals and Its Categorization–A Study. *Procedia Eng.* **2012**, *38*, 2525–2536. [CrossRef]
10. Cohen, D. Magnetoencephalography: Detection of the Brain's Electrical Activity with a Superconducting Magnetometer. *Science* **1972**, *175*, 664–666. [CrossRef]
11. DeYoe, E.A.; Bandettini, P.; Neitz, J.; Miller, D.; Winans, P. Functional magnetic resonance imaging (FMRI) of the human brain. *J. Neurosci. Methods* **1994**, *54*, 171–187. [CrossRef]
12. Hay, L.; Duffy, A.; Gilbert, S.; Grealy, M. Functional magnetic resonance imaging (fMRI) in design studies: Methodological considerations, challenges, and recommendations. *Des. Stud.* **2022**, *78*, 101078. [CrossRef]
13. Naseer, N.; Hong, K.-S. fNIRS-based brain-computer interfaces: A review. *Front. Hum. Neurosci.* **2015**, *9*, 3. [CrossRef]
14. Yücel, M.A.; Lühmann, A.V.; Scholkmann, F.; Gervain, J.; Dan, I.; Ayaz, H.; Boas, D.; Cooper, R.J.; Culver, J.; Elwell, C.E.; et al. Best practices for fNIRS publications. *Neurophotonics* **2021**, *8*, 012101. [CrossRef]
15. Hong, K.-S.; Naseer, N. Reduction of Delay in Detecting Initial Dips from Functional Near-Infrared Spectroscopy Signals Using Vector-Based Phase Analysis. *Int. J. Neural Syst.* **2016**, *26*, 1650012. [CrossRef]
16. Zephaniah, P.V.; Kim, J.G. Recent functional near infrared spectroscopy based brain computer interface systems: Developments, applications and challenges. *Biomed. Eng. Lett.* **2014**, *4*, 223–230. [CrossRef]
17. Pinti, P.; Tachtsidis, I.; Hamilton, A.; Hirsch, J.; Aichelburg, C.; Gilbert, S.; Burgess, P.W. The present and future use of functional near-infrared spectroscopy (fNIRS) for cognitive neuroscience. *Ann. N. Y. Acad. Sci.* **2020**, *1464*, 5–29. [CrossRef]
18. Dehais, F.; Karwowski, W.; Ayaz, H. Brain at Work and in Everyday Life as the Next Frontier: Grand Field Challenges for Neuroergonomics. *Front. Neuroergonomics* **2020**, *1*. [CrossRef]
19. Zhang, F.; Roeyers, H. Exploring brain functions in autism spectrum disorder: A systematic review on functional near-infrared spectroscopy (fNIRS) studies. *Int. J. Psychophysiol.* **2019**, *137*, 41–53. [CrossRef]
20. Han, C.-H.; Hwang, H.-J.; Lim, J.-H.; Im, C.-H. Assessment of user voluntary engagement during neurorehabilitation using functional near-infrared spectroscopy: A preliminary study. *J. Neuroeng. Rehabil.* **2018**, *15*, 1–10. [CrossRef]
21. Karunakaran, K.D.; Peng, K.; Berry, D.; Green, S.; Labadie, R.; Kussman, B.; Borsook, D. NIRS measures in pain and analgesia: Fundamentals, features, and function. *Neurosci. Biobehav. Rev.* **2021**, *120*, 335–353. [CrossRef]
22. Berger, A.; Horst, F.; Müller, S.; Steinberg, F.; Doppelmayr, M. Current State and Future Prospects of EEG and fNIRS in Robot-Assisted Gait Rehabilitation: A Brief Review. *Front. Hum. Neurosci.* **2019**, *13*, 172. [CrossRef]
23. Rea, M.; Rana, M.; Lugato, N.; Terekhin, P.; Gizzi, L.; Brötz, D.; Fallgatter, A.; Birbaumer, N.; Sitaram, R.; Caria, A. Lower limb movement preparation in chronic stroke: A pilot study toward an fnirs-bci for gait rehabilitation. *Neurorehabilit. Neural Repair* **2014**, *28*, 564–575. [CrossRef]
24. Khan, H.; Naseer, N.; Yazidi, A.; Eide, P.K.; Hassan, H.W.; Mirtaheri, P. Analysis of Human Gait Using Hybrid EEG-fNIRS-Based BCI System: A Review. *Front. Hum. Neurosci.* **2021**, *14*, 605. [CrossRef]
25. Li, D.; Fan, Y.; Lü, N.; Chen, G.; Wang, Z.; Chi, W. Safety Protection Method of Rehabilitation Robot Based on fNIRS and RGB-D Information Fusion. *J. Shanghai Jiaotong Univ.* **2021**, *27*, 45–54. [CrossRef]
26. Afonin, A.N.; Asadullayev, R.G.; Sitnikova, M.A.; Shamrayev, A.A. A Rehabilitation Device for Paralyzed Disabled People Based on an Eye Tracker and fNIRS. *Stud. Comput. Intell.* **2020**, *925*, 65–70. [CrossRef]
27. Khan, M.A.; Das, R.; Iversen, H.K.; Puthusserypady, S. Review on motor imagery based BCI systems for upper limb post-stroke neurorehabilitation: From designing to application. *Comput. Biol. Med.* **2020**, *123*, 103843. [CrossRef]
28. Liu, D.; Chen, W.; Pei, Z.; Wang, J. A brain-controlled lower-limb exoskeleton for human gait training. *Rev. Sci. Instrum.* **2017**, *88*, 104302. [CrossRef]

29. Hong, K.-S.; Naseer, N.; Kim, Y.-H. Classification of prefrontal and motor cortex signals for three-class fNIRS–BCI. *Neurosci. Lett.* **2015**, *587*, 87–92. [CrossRef]

30. Ma, D.; Izzetoglu, M.; Holtzer, R.; Jiao, X. Machine Learning-based Classification of Active Walking Tasks in Older Adults using fNIRS. *arXiv* **2021**, arXiv:2102.03987.

31. Khan, R.A.; Naseer, N.; Qureshi, N.K.; Noori, F.M.; Nazeer, H.; Khan, M.U. fNIRS-based Neurorobotic Interface for gait rehabilitation. *J. Neuroeng. Rehabil.* **2018**, *15*, 1–17. [CrossRef] [PubMed]

32. Wang, Z.; Yan, W.; Oates, T. Time series classification from scratch with deep neural networks: A strong baseline. In Proceedings of the 2017 International Joint Conference on Neural Networks (IJCNN), Anchorage, AK, USA, 14–19 May 2017; pp. 1578–1585.

33. Ho, T.K.K.; Gwak, J.; Park, C.M.; Song, J.-I. Discrimination of Mental Workload Levels from Multi-Channel fNIRS Using Deep Leaning-Based Approaches. *IEEE Access* **2019**, *7*, 24392–24403. [CrossRef]

34. Zheng, Y.; Liu, Q.; Chen, E.; Ge, Y.; Zhao, J.L. Time Series Classification Using Multi-Channels Deep Convolutional Neural Networks. In *Web-Age Information Management. WAIM 2014*; Lecture Notes in Computer Science; Li, F., Li, G., Hwang, S., Yao, B., Zhang, Z., Eds.; Springer: Cham, Switzerland, 2014; Volume 8485. [CrossRef]

35. Janani, A.; Sasikala, M.; Chhabra, H.; Shajil, N.; Venkatasubramanian, G. Investigation of deep convolutional neural network for classification of motor imagery fNIRS signals for BCI applications. *Biomed. Signal Process. Control* **2020**, *62*, 102133. [CrossRef]

36. Ma, T.; Wang, S.; Xia, Y.; Zhu, X.; Evans, J.; Sun, Y.; He, S. CNN-based classification of fNIRS signals in motor imagery BCI system. *J. Neural Eng.* **2021**, *18*, 056019. [CrossRef] [PubMed]

37. Rojas, R.F.; Romero, J.; Lopez-Aparicio, J.; Ou, K.-L. Pain Assessment based on fNIRS using Bi-LSTM RNNs. In Proceedings of the International IEEE/EMBS Conference on Neural Engineering (NER), Virtual Event, Italy, 4–6 May 2021; IEEE: Piscataway, NJ, USA, 2021; pp. 399–402. [CrossRef]

38. Ma, T.; Lyu, H.; Liu, J.; Xia, Y.; Qian, C.; Evans, J.; Xu, W.; Hu, J.; Hu, S.; He, S. Distinguishing bipolar depression from major depressive disorder using fnirs and deep neural network. *Prog. Electromagn. Res.* **2020**, *169*, 73–86. [CrossRef]

39. Asgher, U.; Khalil, K.; Khan, M.J.; Ahmad, R.; Butt, S.I.; Ayaz, Y.; Naseer, N.; Nazir, S. Enhanced Accuracy for Multiclass Mental Workload Detection Using Long Short-Term Memory for Brain-Computer Interface. *Front. Neurosci.* **2020**, *14*, 584. [CrossRef]

40. Mihara, M.; Miyai, I.; Hatakenaka, M.; Kubota, K.; Sakoda, S. Role of the prefrontal cortex in human balance control. *NeuroImage* **2008**, *43*, 329–336. [CrossRef]

41. Christie, B. Doctors revise Declaration of Helsinki. *BMJ* **2000**, *321*, 913. [CrossRef]

42. Okada, E.; Firbank, M.; Schweiger, M.; Arridge, S.; Cope, M.; Delpy, D.T. Theoretical and experimental investigation of near-infrared light propagation in a model of the adult head. *Appl. Opt.* **1997**, *36*, 21–31. [CrossRef]

43. Delpy, D.T.; Cope, M.; Van Der Zee, P.; Arridge, S.; Wray, S.; Wyatt, J. Estimation of optical pathlength through tissue from direct time of flight measurement. *Phys. Med. Biol.* **1988**, *33*, 1433–1442. [CrossRef]

44. Santosa, H.; Hong, M.J.; Kim, S.-P.; Hong, K.-S. Noise reduction in functional near-infrared spectroscopy signals by independent component analysis. *Rev. Sci. Instrum.* **2013**, *84*, 073106. [CrossRef] [PubMed]

45. Ye, J.C.; Tak, S.; Jang, K.E.; Jung, J.; Jang, J. NIRS-SPM: Statistical parametric mapping for near-infrared spectroscopy. *NeuroImage* **2009**, *44*, 428–447. [CrossRef]

46. Qureshi, N.K.; Naseer, N.; Noori, F.M.; Nazeer, H.; Khan, R.; Saleem, S. Enhancing Classification Performance of Functional Near-Infrared Spectroscopy- Brain-Computer Interface Using Adaptive Estimation of General Linear Model Coefficients. *Front. Neurorobot.* **2017**, *11*, 33. [CrossRef] [PubMed]

47. Nazeer, H.; Naseer, N.; Mehboob, A.; Khan, M.J.; Khan, R.A.; Khan, U.S.; Ayaz, Y. Enhancing Classification Performance of fNIRS-BCI by Identifying Cortically Active Channels Using the z-Score Method. *Sensors* **2020**, *20*, 6995. [CrossRef] [PubMed]

48. Naseer, N.; Hong, M.J.; Hong, K.-S. Online binary decision decoding using functional near-infrared spectroscopy for the development of brain-computer interface. *Exp. Brain Res.* **2014**, *232*, 555–564. [CrossRef] [PubMed]

49. Noori, F.M.; Naseer, N.; Qureshi, N.K.; Nazeer, H.; Khan, R. Optimal feature selection from fNIRS signals using genetic algorithms for BCI. *Neurosci. Lett.* **2017**, *647*, 61–66. [CrossRef] [PubMed]

50. Cortes, C.; Vapnik, V. Support-vector networks. *Mach. Learn.* **1995**, *20*, 273–297. [CrossRef]

51. Guo, G.; Wang, H.; Bell, D.; Bi, Y.; Greer, K. KNN Model-Based Approach in Classification. In *On The Move to Meaningful Internet Systems 2003: CoopIS, DOA, and ODBASE. OTM 2003*; Lecture Notes in Computer Science; Meersman, R., Tari, Z., Schmidt, D.C., Eds.; Springer: Berlin, Heidelberg, 2003; Volume 2888. [CrossRef]

52. Sumantri, A.F.; Wijayanto, I.; Patmasari, R.; Ibrahim, N. Motion Artifact Contaminated Functional Near-infrared Spectroscopy Signals Classification using K-Nearest Neighbor (KNN). *J. Phys. Conf. Ser.* **2019**, *1201*, 012062. [CrossRef]

53. Zhang, S.; Li, X.; Zong, M.; Zhu, X.; Wang, R. Efficient kNN Classification with Different Numbers of Nearest Neighbors. *IEEE Trans. Neural Netw. Learn. Syst.* **2018**, *29*, 1774–1785. [CrossRef]

54. Naseer, N.; Noori, F.M.; Qureshi, N.K.; Hong, K.-S. Determining Optimal Feature-Combination for LDA Classification of Functional Near-Infrared Spectroscopy Signals in Brain-Computer Interface Application. *Front. Hum. Neurosci.* **2016**, *10*, 237. [CrossRef]

55. Xanthopoulos, P.; Pardalos, P.M.; Trafalis, T.B. Linear Discriminant Analysis. In *Modern Multivariate Statistical Techniques*; Springer: New York, NY, USA, 2012; pp. 27–33. [CrossRef]

56. Trakoolwilaiwan, T.; Behboodi, B.; Lee, J.; Kim, K.; Choi, J.-W. Convolutional neural network for high-accuracy functional near-infrared spectroscopy in a brain-computer interface: Three-class classification of rest, right-, and left-hand motor execution. *Neurophotonics* **2017**, *5*, 011008. [CrossRef] [PubMed]
57. Saadati, M.; Nelson, J.; Ayaz, H. Multimodal fNIRS-EEG Classification Using Deep Learning Algorithms for Brain-Computer Interfaces Purposes. In *Advances in Intelligent Systems and Computing*; Springer: Berlin/Heidelberg, Germany, 2019; Volume 953, pp. 209–220. [CrossRef]
58. Saadati, M.; Nelson, J.; Ayaz, H. Convolutional Neural Network for Hybrid fNIRS-EEG Mental Workload Classification. In *Advances in Intelligent Systems and Computing*; Springer: Berlin/Heidelberg, Germany, 2019; Volume 953, pp. 221–232. [CrossRef]
59. Albawi, S.; Mohammed, T.A.; Al-Zawi, S. Understanding of a convolutional neural network. In Proceedings of the 2017 International Conference on Engineering and Technology (ICET), Antalya, Turkey, 21–23 August 2017; pp. 1–6.
60. Borovykh, A.; Bohte, S.; Oosterlee, C.W. Conditional Time Series Forecasting with Convolutional Neural Networks. *arXiv* **2017**, arXiv:1703.04691.
61. Zhou, B.; Khosla, A.; Lapedriza, A.; Oliva, A.; Torralba, A. Learning deep features for discriminative localization. In Proceedings of the IEEE Conference on Computer Vision and Pattern Recognition, Las Vegas, NV, USA, 27–30 June 2016; pp. 2921–2929.
62. Ismail Fawaz, H.; Forestier, G.; Weber, J.; Idoumghar, L.; Muller, P.-A. Deep learning for time series classification: A review. *Data Min. Knowl. Discov.* **2019**, *33*, 917–963. [CrossRef]
63. Hochreiter, S.; Schmidhuber, J. Long short-term memory. *Neural Comput.* **1997**, *9*, 1735–1780. [CrossRef]
64. Gers, F.A.; Schmidhuber, J.; Cummins, F. Learning to Forget: Continual Prediction with LSTM. *Neural Comput.* **2000**, *12*, 2451–2471. [CrossRef]
65. Siami-Namini, S.; Tavakoli, N.; Namin, A.S. The Performance of LSTM and BiLSTM in Forecasting Time Series. In Proceedings of the 2019 IEEE International Conference on Big Data, Los Angeles, CA, USA, 9–12 December 2019; pp. 3285–3292. [CrossRef]
66. Schuster, M.; Paliwal, K.K. Bidirectional recurrent neural networks. *IEEE Trans. Signal Process* **1997**, *45*, 2673–2681. [CrossRef]
67. Sun, Q.; Jankovic, M.V.; Bally, L.; Mougiakakou, S.G. Predicting Blood Glucose with an LSTM and Bi-LSTM Based Deep Neural Network. In Proceedings of the 2018 14th Symposium on Neural Networks and Applications, NEUREL 2018, Belgrade, Serbia, 20–21 November 2018; pp. 1–5. [CrossRef]
68. Murad, A.; Pyun, J.-Y. Deep Recurrent Neural Networks for Human Activity Recognition. *Sensors* **2017**, *17*, 2556. [CrossRef]
69. Kwon, J.; Im, C.-H. Subject-Independent Functional Near-Infrared Spectroscopy-Based Brain-Computer Interfaces Based on Convolutional Neural Networks. *Front. Hum. Neurosci.* **2021**, *15*, 646915. [CrossRef]
70. Open Access fNIRS Dataset for Classification of the Unilateral Finger- and Foot-Tapping. Available online: https://figshare.com/articles/dataset/Open_access_fNIRS_dataset_for_classification_of_the_unilateral_finger-_and_foot-tapping/9783755/1 (accessed on 21 January 2022).
71. Pamungkas, D.S.; Caesarendra, W.; Soebakti, H.; Analia, R.; Susanto, S. Overview: Types of Lower Limb Exoskeletons. *Electronics* **2019**, *8*, 1283. [CrossRef]
72. Remsik, A.; Young, B.; Vermilyea, R.; Kiekhoefer, L.; Abrams, J.; Elmore, S.E.; Schultz, P.; Nair, V.; Edwards, D.; Williams, J.; et al. A review of the progression and future implications of brain-computer interface therapies for restoration of distal upper extremity motor function after stroke. *Expert Rev. Med. Devices* **2016**, *13*, 445–454. [CrossRef]
73. Huppert, T.J. Commentary on the statistical properties of noise and its implication on general linear models in functional near-infrared spectroscopy. *Neurophotonics* **2016**, *3*, 010401. [CrossRef] [PubMed]
74. Le, N.-Q.; Nguyen, B.P. Prediction of FMN Binding Sites in Electron Transport Chains Based on 2-D CNN and PSSM Profiles. *IEEE/ACM Trans. Comput. Biol. Bioinform.* **2019**, *18*, 2189–2197. [CrossRef] [PubMed]

MDPI

Article

Single-Trial Classification of Error-Related Potentials in People with Motor Disabilities: A Study in Cerebral Palsy, Stroke, and Amputees

Nayab Usama [1], Imran Khan Niazi [1,2,3], Kim Dremstrup [1] and Mads Jochumsen [1,*]

[1] Department of Health Science and Technology, Aalborg University, 9000 Aalborg, Denmark; nu@hst.aau.dk (N.U.); imrankn@hst.aau.dk (I.K.N.); kdn@hst.aau.dk (K.D.)

[2] Centre for Chiropractic Research, New Zealand College of Chiropractic, Auckland 1060, New Zealand

[3] Health and Rehabilitation Research Institute, AUT University, Auckland 0627, New Zealand

* Correspondence: mj@hst.aau.dk

Abstract: Brain-computer interface performance may be reduced over time, but adapting the classifier could reduce this problem. Error-related potentials (ErrPs) could label data for continuous adaptation. However, this has scarcely been investigated in populations with severe motor impairments. The aim of this study was to detect ErrPs from single-trial EEG in offline analysis in participants with cerebral palsy, an amputation, or stroke, and determine how much discriminative information different brain regions hold. Ten participants with cerebral palsy, eight with an amputation, and 25 with a stroke attempted to perform 300–400 wrist and ankle movements while a sham BCI provided feedback on their performance for eliciting ErrPs. Pre-processed EEG epochs were inputted in a multi-layer perceptron artificial neural network. Each brain region was used as input individually (Frontal, Central, Temporal Right, Temporal Left, Parietal, and Occipital), the combination of the Central region with each of the adjacent regions, and all regions combined. The Frontal and Central regions were most important, and adding additional regions only improved performance slightly. The average classification accuracies were 84 ± 4%, 87± 4%, and 85 ± 3% for cerebral palsy, amputation, and stroke participants. In conclusion, ErrPs can be detected in participants with motor impairments; this may have implications for developing adaptive BCIs or automatic error correction.

Keywords: error-related potentials; brain-computer interface; cerebral palsy; amputation; stroke; neurorehabilitation; artificial neural network

Citation: Usama, N.; Niazi, I.K.; Dremstrup, K.; Jochumsen, M. Single-Trial Classification of Error-Related Potentials in People with Motor Disabilities: A Study in Cerebral Palsy, Stroke, and Amputees. *Sensors* **2022**, *22*, 1676. https://doi.org/10.3390/s22041676

Academic Editor: Sung-Phil Kim

Received: 19 January 2022
Accepted: 18 February 2022
Published: 21 February 2022

1. Introduction

Brain-computer interfaces (BCIs) provide individuals with severe motor impairments the possibility to control external devices using only brain activity [1–3]. Examples of such devices could be wheelchairs and robotic manipulators for mobility restoration, speller devices for communication, and electrical stimulators or rehabilitation robots for motor rehabilitation after e.g., stroke [1,4,5]. Various control signals can be used to control BCIs, such as steady-state visually evoked potentials [2,6], P300, movement-related cortical potentials, and sensorimotor rhythms. These control signals are recorded from the electrical activity of the brain and processed to enhance the signal-to-noise ratio, after which they are detected/classified and translated into device commands. To ensure good performance of the BCI, several factors need to be attended to such as proper electrode montage and impedances for recording the brain activity and good calibration data for the classifier [7,8]. The calibration data that often are recorded prior to the actual use of the BCI may not represent the actual brain activity well after the BCI has been used for some time, e.g., due to changes in electrode impedance or if the user starts to fatigue. This problem could be accounted for if the classifier in the BCI is continuously updated. Error-related potentials (ErrPs) have been proposed as a means for this [9]. An ErrP is elicited when a person realizes an error e.g., the output of the BCI is different than expected. With proper ErrP detection, erroneously classified data can be correctly labelled and the classifier in the BCI can be correctly updated based on the most recent data. Another application of ErrPs within BCI is error correction, where an erroneous action of the BCI, e.g., in a P300 speller or movement of a robotic arm, can be detected and the incorrect action can be reverted automatically [9,10]. If the ErrPs are properly detected, the performance of the BCI can be improved, since the potential errors do not need to be corrected manually (see e.g., [11–13]). It has been shown in several studies that ErrPs can be detected from single-trial EEG [9,10], primarily from able-bodied individuals, but only a few studies have investigated the detection of ErrPs in individuals with movement disabilities. It has previously been shown how various factors modulate the ErrP in stroke patients [14,15], and that ErrPs in stroke patients can be detected from single-trial EEG [16]. In addition, it has been reported that ErrPs can be elicited and detected in individuals with spinal cord injury [13,17,18], amyotrophic lateral sclerosis [19], and epilepsy [20,21]. Lastly, error processing has been investigated in individuals with Parkinson's disease [22] and cerebral palsy [23], but detection of ErrPs in these conditions has not been performed. ErrPs have generally been detected using temporal waveform features of a bandpass filtered epoch (~0–1 s after the feedback of the outcome) from electrodes on the scalp in the proximity of the anterior cingulate cortex amongst other neural generators (roughly around FCz according to the 10–20 EEG system) [9,10]. ErrPs can be detected from a single or few electrodes around FCz [24,25], but studies have reported that additional discriminative information can be obtained from using more electrodes covering other parts of the brain [24,26–35]. The aim of this study was twofold; first it was investigated whether ErrPs could be detected in individuals with motor disabilities after cerebral palsy, an amputation, or stroke in offline analysis, and secondly, how much discriminative information different brain regions bring to the detection of ErrPs.

2. Materials and Methods

2.1. Participants

In this study, ten participants with cerebral palsy (for clinical characteristics see Table 1), eight amputees (see Table 2), and 25 participants with a stroke (see Table 3) were recruited. The experiments were conducted at Allied Hospital, Faisalabad, Pakistan. All participants or their parents provided written informed consent before the experiment. The procedures were approved by the local ethical committee at Allied Hospital and were conducted according to the Helsinki Declaration. The cerebral palsy participants were recruited through the Department of Pediatrics, amputees were enlisted through the Department of Orthopedics, and stroke participants were recruited at the Department of Neurology at Allied Hospital Faisalabad. All the cerebral palsy participants were diagnosed between ages of 1–3 years. The cerebral palsy participants' motor abilities were assessed by a pediatrician at Allied Hospital in terms of the gross motor function classification system (GMFCS), (I = ambulatory, II = some limitations in motor functions, III = dependent on others or some assistive devices). The motor abilities of the stroke participants were assessed by a Neurologist in terms of the Bruunstrom Stage classification. The data from the stroke patients have been presented in a recent study, but the experimental details are described in detail in the following sections [16].

2.2. Data Recording

EEG was recorded from 64 channels with active electrodes with a sampling rate of 1200 Hz (g.HIamp and g.GAMMASYS, G.Tec, Graz, Austria). The electrodes were placed according to the 10-10 system and were grounded to AFz and referenced to a linked ear reference. During the experiment the electrode impedances were below 10 kΩ. The EEG was synchronized to visual cues through an Arduino controller from a custom-made MATLAB script (MathWorks®, Natick, MA, USA) which sent a trigger to the EEG amplifier.

The external triggers were used to divide the continuous EEG into epochs containing error and correct responses.

2.3. Experimental Details

The participants were seated in a comfortable chair in front of a computer screen which displayed the visual cues throughout the experiment. The experiment consisted of 15 runs for the participants with cerebral palsy and 20 runs for the amputees and stroke participants, where each run consisted of 20 trials. Each trial started with an idle phase lasting five seconds where the participant could relax, this was followed by a preparation phase lasting three seconds where the participant was cued to bring the attention back to the screen and prepare to attempt to perform a movement. In the movement phase a picture of the hand or foot was shown pointing to the left or right indicating that a wrist extension of the right or left hand should be performed or a dorsiflexion of the right or left foot. The movement phase lasted three seconds for the amputees and stroke participants and five seconds for the participants with cerebral palsy to allow them more time to process what intended movement that should be attempted. The amputees were instructed to imagine the movement of their amputated limb. A single movement attempt was performed in each movement phase. After the movement phase, visual feedback with a ratio of 70/30 (for correct/ incorrect) was presented as a green coloured tick mark or a red coloured cross sign indicating whether the movement was correctly or incorrectly detected based on the brain activity. No actual detection of movements was performed, but it was conveyed to the participant that a BCI classified the movements [16,36]. During the feedback monitoring the participant was instructed to avoid unnecessary movements and eye blinks. An equal number of 75, 100, and 100 movements were performed for the left/right hand and foot for the cerebral palsy, amputees, and stroke participants respectively, i.e., 300, 400, and 400 movements in total. At the end of each run a break was given until the participant was comfortable with resuming the experiment. The experiments were completed in approximately 100–180 min.

Table 1. Characteristics of the participants with cerebral palsy. Gender, age, and diagnosis (diplegia or hemiplegia) as well as the affected side, and gross motor function classification system (GMFCS) score are presented.

Participant	Gender	Age (Years)	Diagnose	GMFCS
01	F	12	Diplegia	II
02	F	10	Diplegia	II
03	F	10	Hemiplegia-right	II
04	M	16	Hemiplegia-left	II
05	M	11	Hemiplegia-left	I
06	M	9	Hemiplegia-right	II
07	F	14	Hemiplegia-right	II
08	M	12	Hemiplegia-right	III
09	M	13	Diplegia	III
10	M	15	Diplegia	III

2.4. Signal Processing

2.4.1. Pre-Processing

The continuous EEG data were bandpass filtered between 1–10 Hz with an 8th order zero phase-shift Butterworth filter. After the filtering, bad channels and epochs were rejected from the analysis. Channels having a mean amplitude more than three standard deviations above the overall mean amplitude across all channels were defined as bad channels and removed. Next, the filtered data were divided into 0.7 s epochs (starting from 0.1 s after the presentation of the visual feedback, i.e., green tick mark or red plus sign, until 0.8 s after) to capture the negative and positive peaks of the error and correct responses. Bad epochs were defined as an epoch with peak-peak amplitude exceeding ± 150 µV. To have an equal number of error and correct responses, random correct responses were selected for the further data analysis to match the number of error responses. The data analysis was performed in MATLAB (MathWorks®).

Table 2. Characteristics of the participants with an amputation. Gender, age, time since amputation, the level of amputation, and affected side are presented.

Participant	Gender	Age (Years)	Time Since Amputation (Years)	Amputation Level	Amputation Side
01	M	13	3	Hip disarticulation	Left
02	F	45	2	Transfemoral	Left
03	M	32	5	Wrist disarticulation	Right
04	M	27	1	Transfemoral	Right
05	M	30	2	Shoulder disarticulation	Left
06	M	32	5	Transcredial	Right
07	M	53	7	Knee disarticulation	Right
08	M	12	5	Hip disarticluation	Left

Table 3. Characteristics of the participants with an amputation. Gender, age, time since amputation, the level of amputation and affected side are presented.

Participant	Gender	Age (Years)	Affected Side	Type of Stroke	Time Since Injury (Days)	Bruunstrom Stage
01	M	48	Right	Haemorrhage	91	II
02	M	55	Right	Ischemic	172	V
03	M	41	Left	Ischemic	70	III
04	M	50	Left	Haemorrhage	90	III
05	M	57	Right	Haemorrhage	52	V
06	M	52	Right	Ischemic	188	V
07	M	24	Left	Haemorrhage	180	IV
08	F	32	Left	Ischemic	25	II
09	F	26	Left	Haemorrhage	20	I
10	M	60	Right	Ischemic	87	IV
11	M	54	Left	Ischemic	220	VII
12	M	46	Left	Ischemic	42	III
13	M	58	Right	Ischemic	84	III
14	M	37	Right	Haemorrhage	36	II
15	M	42	Left	Haemorrhage	118	V
16	M	24	Left	Haemorrhage	45	IV
17	F	26	Right	Ischemic	12	I
18	M	62	Right	Haemorrhage	118	III
19	M	30	Right	Ischemic	60	III
20	F	53	Left	Ischemic	93	IV
21	F	38	Right	Haemorrhage	45	VI
22	F	28	Left	Ischemic	27	V
23	M	45	Left	Ischemic	90	IV
24	M	35	Left	Haemorrhage	17	II
25	M	45	Right	Haemorrhage	280	VI

2.4.2. Classification

For the classification of error and correct responses a multi-layer perceptron artificial neural network (MLP ANN) was used. In a recent study [16], we found that MLP ANN performed better than a linear discriminant analysis classifier, which is often used for classifying ErrPs. Therefore, we chose to perform the classification with MLP ANN.

The input for the MLP ANN was the entire pre-processed epoch. The MLP ANN had 5 layers where the input, layer 1, was the data points in the epoch for the channels of interest (dimension: number of channels x number of samples in epoch). Three hidden layers were used, they had a size of 100-50-25, whereas the output layer was of size 1 with a sigmoid activation function. The MLP ANN was trained using the scaled conjugate gradient descent method, and the performance of the MLP ANN was validation checked with cross-entropy.

The classification was performed with different electrode configurations to investigate how well error responses can be discriminated from correct responses. The electrodes were divided into six specific brain regions: Frontal, Central, Parietal, Occipital, Temporal Left, and Temporal Right (see Figure 1). Initially, each region was used as input for the MLP ANN individually. We had a hypothesis about ErrP classification being highest at the Central brain region based on existing ErrP literature and the proximity of the anterior cingulate cortex. Thus, the classification was performed again with the Central region as input combined with each of the adjacent brain areas (i.e., Frontal, Parietal, Temporal Left, and Temporal Right). Next, the classification was performed with all brain regions except the Occipital region, and lastly, the classification was performed with all brain regions combined. In all the classification scenarios, 10-fold cross-validation was used, the same folds were used across the different classification scenarios. The analyses were performed using MATLAB (MathWorks®).

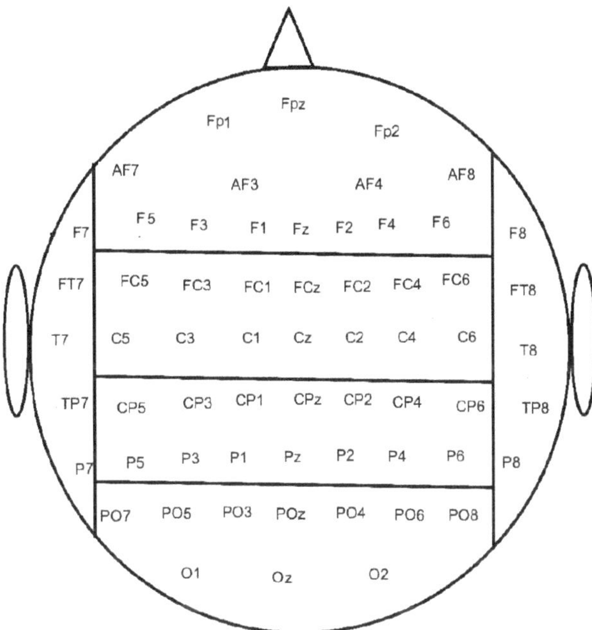

Figure 1. The electrodes were divided into Frontal, Central, Parietal, Occipital, Temporal Left, and Temporal Right brain regions. Note that there is a different number of electrodes in the different brain regions.

3. Results

On average, 0.6 ± 0.7 channels (range: 0–2) and 65 ± 56 epochs (range: 9–141) were excluded for the cerebral palsy participants, 0.5 ± 1.4 channels (range: 0–4) and 59 ± 71 epochs (range: 0–191) were excluded for the amputees, and 0.3 ± 0.5 channels (range: 0–1) and 72 ± 64 epochs (range: 2–228) were excluded for the stroke participants.

The average error and correct responses are presented in Figure 2 for participants with cerebral palsy, participants with an amputation, and participants with a stroke. From the averages it can be seen that there was a consistent negative peak between 0.3 and 0.4 s after the presentation of the feedback and a positive peak 0.1 s after the negative peak. Based on the grand averages, there was a slightly higher peak-peak amplitude between the negative and positive peaks for the error responses compared to the correct responses. In Figure 3, topographical plots are shown from representative participants. It can be seen that the Central and Frontal channels show the most negative and positive peaks, although most other channels also show a similar negative or positive peak, but with smaller amplitudes. In the following sections, the classification accuracies are presented as mean ± standard error across participants.

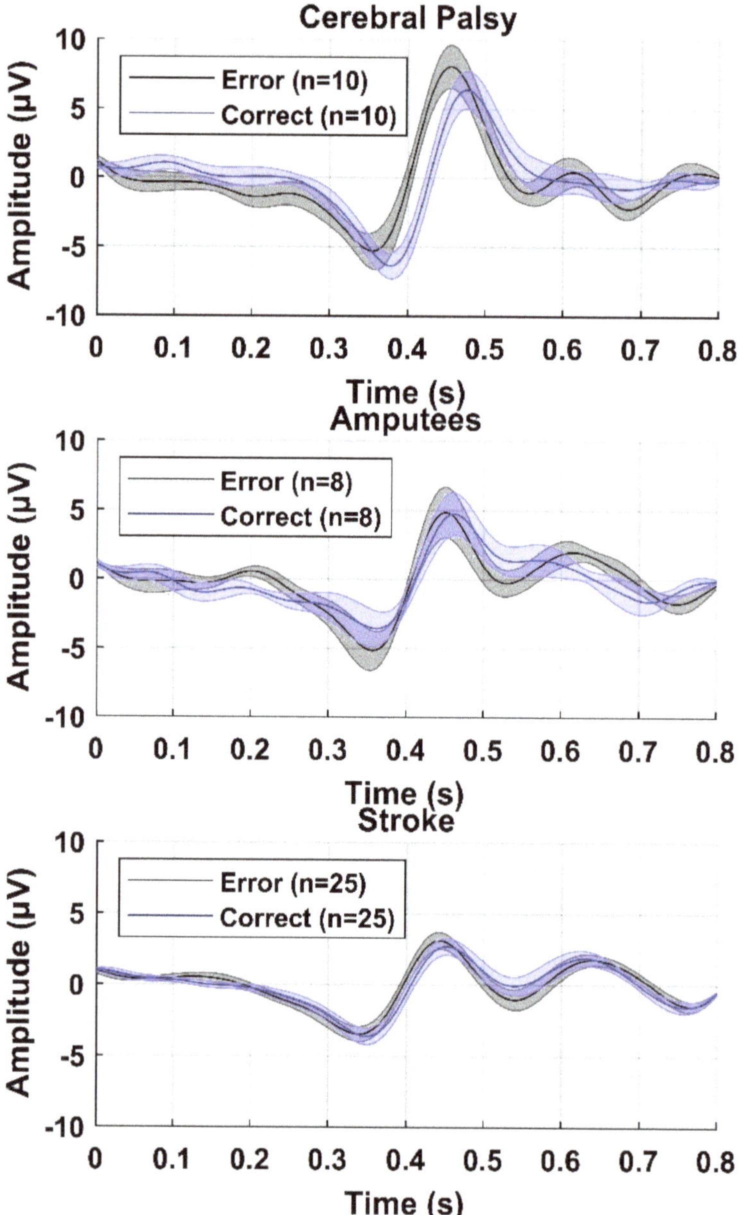

Figure 2. Grand average of error and correct responses from FCz across the ten participants with cerebral palsy (**top**), eight participants with an amputation (**middle**), and 25 participants with a stroke (**bottom**). The solid line is the mean across participants and the shaded area represents the standard error.

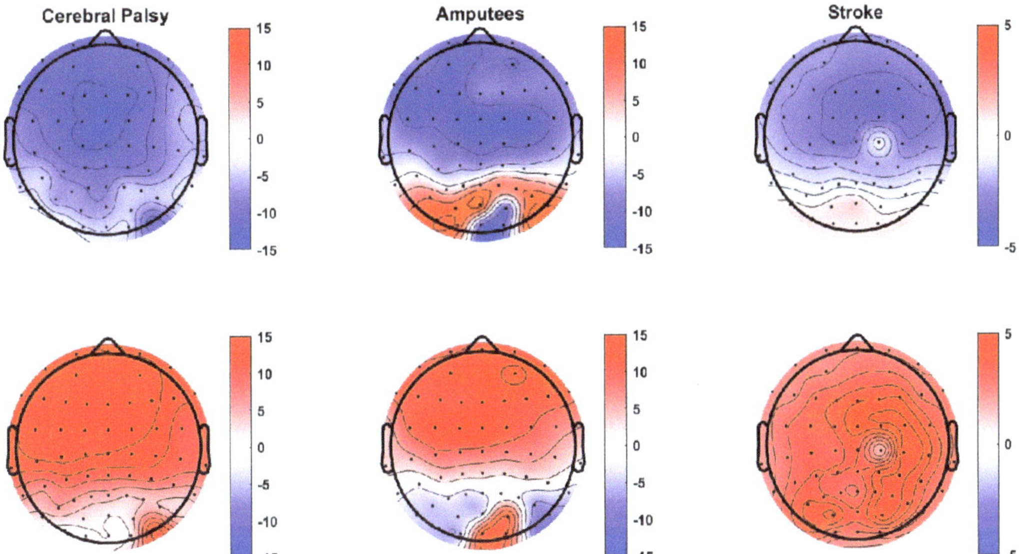

Figure 3. Topographical plot of the negative (**top**) and positive (**bottom**) peaks of the error responses for a representative participant with cerebral palsy (**left column**), a participant with an amputation (**middle column**), and a participant with a stroke (**right column**). The unit of the color bar is in μV.

The classification results for participants with cerebral palsy are presented in Figure 4. With a single brain region as input, the highest classification accuracies between error and correct responses were obtained with the electrodes from the Frontal region ($87 \pm 3\%$) followed by the Central region ($84 \pm 3\%$). The lowest classification accuracies were obtained from the Occipital region ($78 \pm 5\%$). The classification accuracies did not increase when combining different regions.

The classification results for participants with a stroke are presented in Figure 6. Like the cerebral palsy and amputation participants, the highest classification accuracies with a single brain region as input were obtained with the electrodes from the Frontal ($84 \pm 2\%$) and Central region ($84 \pm 3\%$). The lowest classification accuracies were obtained from the Parietal region ($79 \pm 3\%$). The classification accuracies increased slightly when combining the Frontal and Central region ($85 \pm 3\%$).

The classification results for participants with an amputation are presented in Figure 5. As with the cerebral palsy patients, the highest classification accuracies with a single brain region as input were obtained with the electrodes from the Frontal region ($84 \pm 5\%$) followed by the Central region ($83 \pm 5\%$). The lowest classification accuracies were obtained from the Parietal region ($78 \pm 7\%$). The classification accuracies increased when adding the Frontal region to the Central region ($87 \pm 4\%$); however, when adding more regions, the classification accuracies decreased ($85 \pm 4\%$).

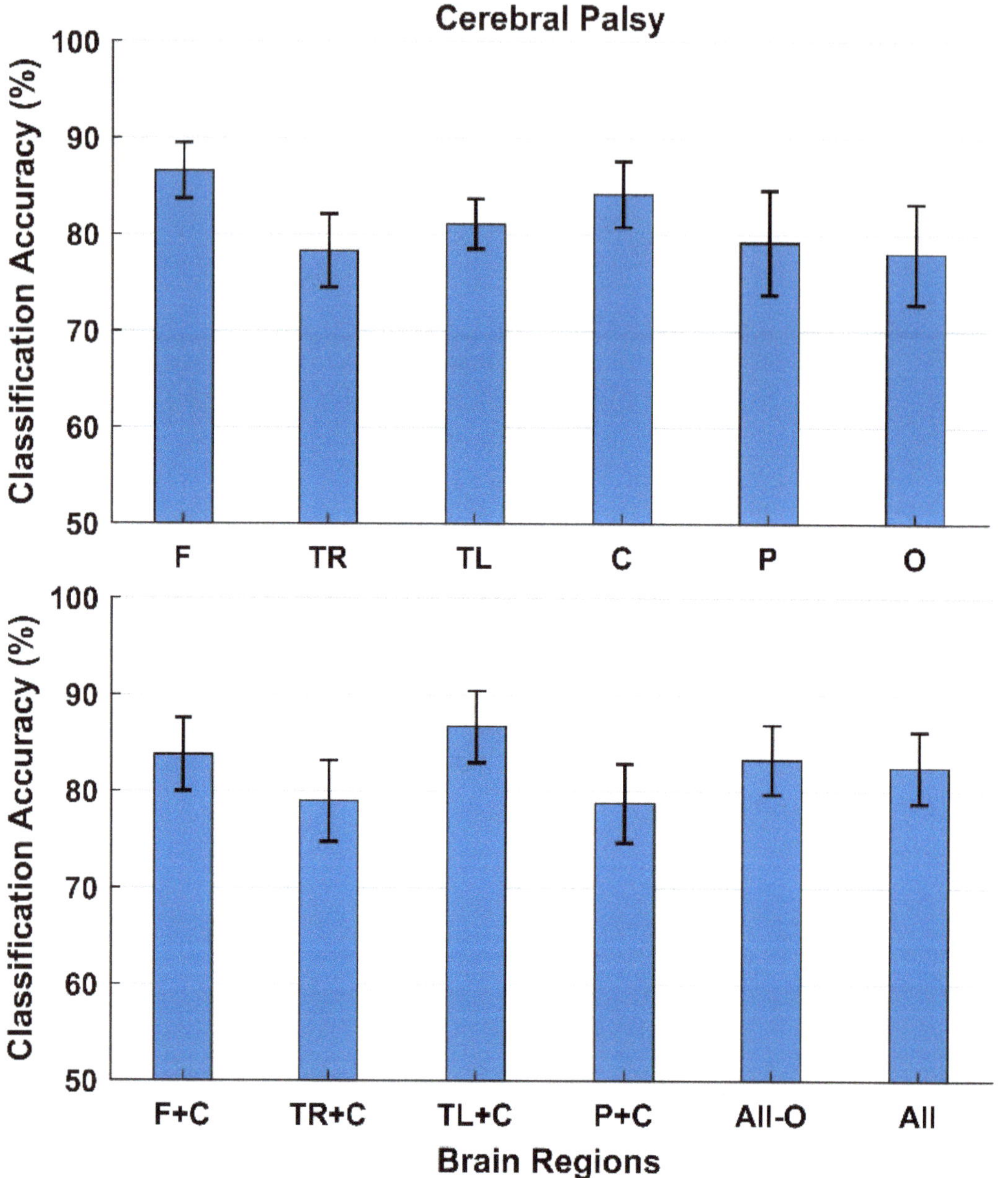

Figure 4. Classification between error and correct responses. The bars represent the mean across participants with cerebral palsy, and the standard error is shown as well. The top of the figure shows the classification accuracies obtained using each brain region individually, and the bottom of the figure shows the classification accuracies when combining different brain regions. F: Frontal, TR: Temporal Right, TL: Temporal Left, C: Central, P: Parietal, and O: Occipital.

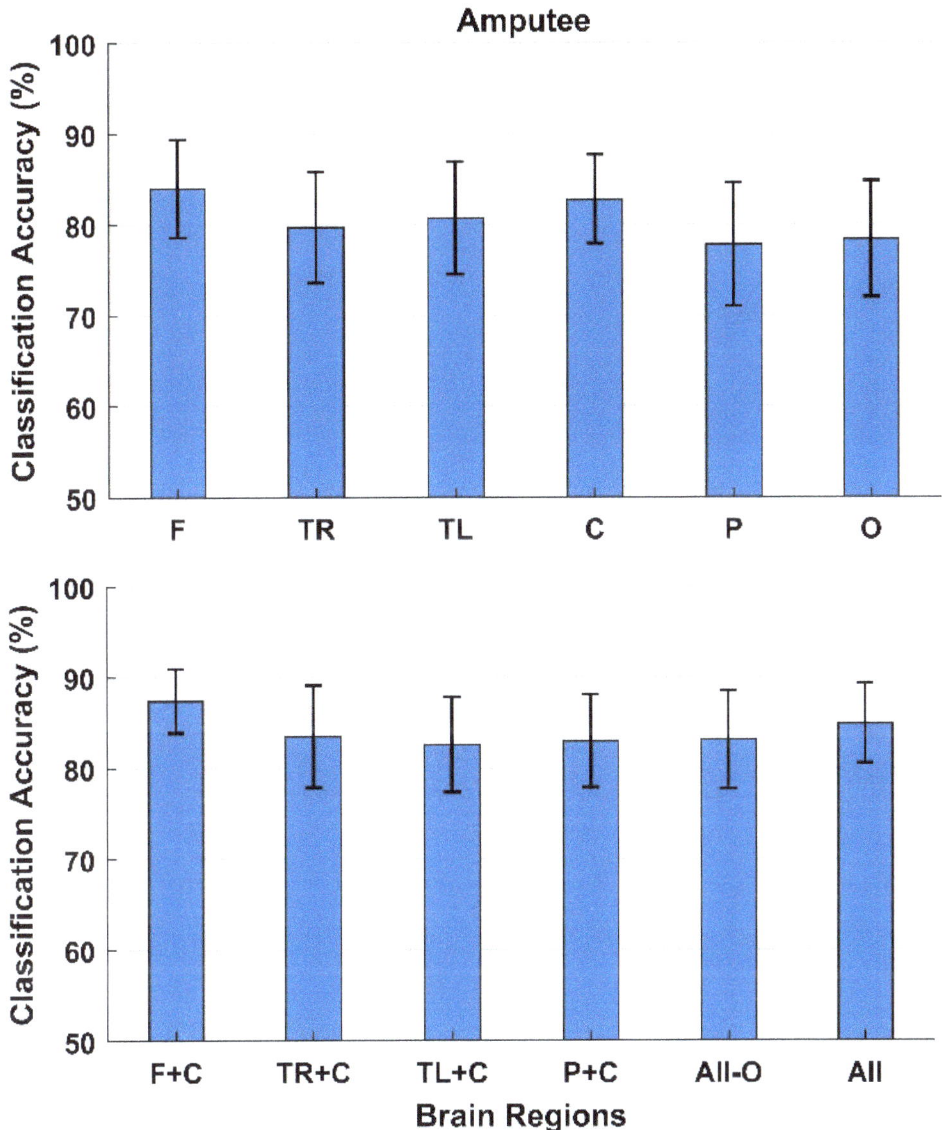

Figure 5. Classification between error and correct responses. The bars represent the mean across participants with an amputated limb, and the standard error is shown as well. The top of the figure shows the classification accuracies obtained using each brain region individually, and the bottom of the figure shows the classification accuracies when combining different brain regions. F: Frontal, TR: Temporal Right, TL: Temporal Left, C: Central, P: Parietal, and O: Occipital.

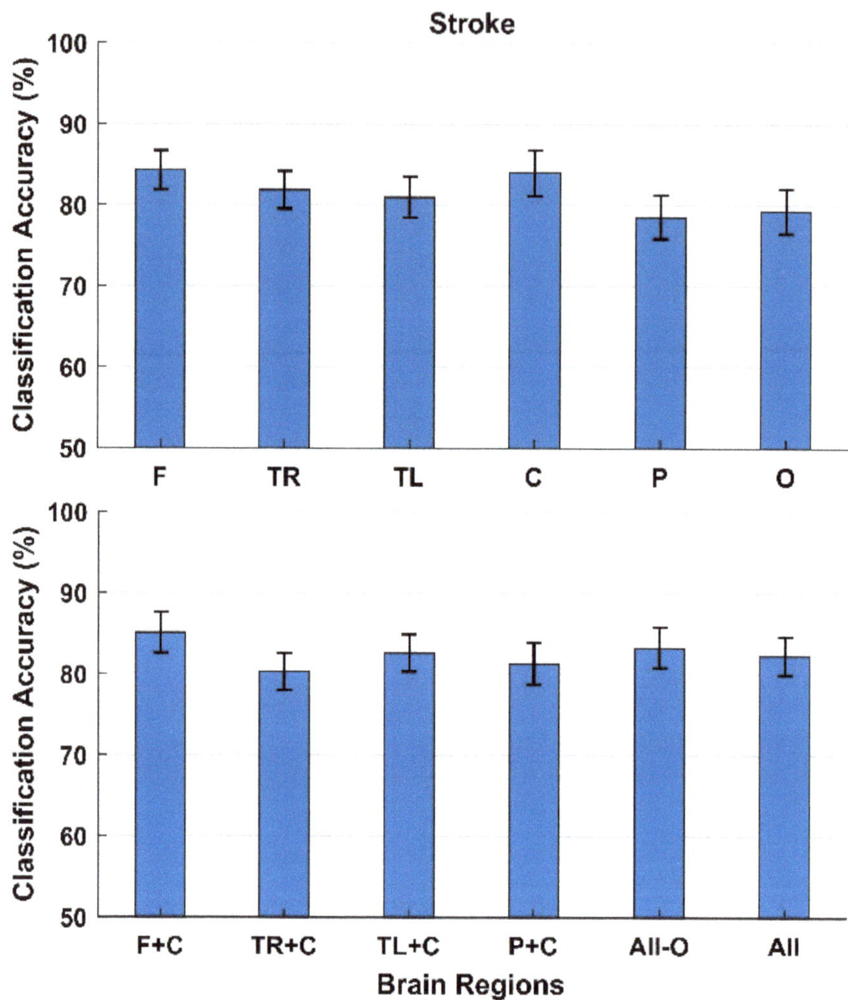

Figure 6. Classification between error and correct responses. The bars represent the mean across participants with a stroke, and the standard error is shown as well. The top of the figure shows the classification accuracies obtained using each brain region individually, and the bottom of the figure shows the classification accuracies when combining different brain regions. F: Frontal, TR: Temporal Right, TL: Temporal Left, C: Central, P: Parietal, and O: Occipital.

4. Discussion

In this study it was shown that error responses can be discriminated from correct responses from single-trial EEG recordings in participants with cerebral palsy, an amputation, and a stroke. The highest classification accuracies were obtained from the Frontal and Central brain regions from all participant groups with average classification accuracies of $84 \pm 4\%$, $87 \pm 4\%$, and $85 \pm 3\%$ for cerebral palsy, amputees, and stroke participants, respectively. This was also expected, since the Frontal and Central electrodes from the scalp are closest to the main neural generators associated with error processing such as the anterior cingulate cortex [37]. The classification accuracies only increased slightly when information from other brain areas were added to the Frontal and Central region. However, the classification accuracies from the different regions individually were all significantly higher than chance level, calculated with a significance level of 5% (between 62 and 66% for amputees/stroke and cerebral palsy, respectively) [38]. This could indicate that there is a high correlation between electrodes, which could be due to the fact that the EEG correlates of the error/correct responses are widely distributed over the scalp, potentially due to volume conduction.

The findings agree with several other studies that have reported the midline electrodes (especially FCz and Cz) to contain the highest error-related activity [24,28–31,34,35,39,40], but also Parietal areas [24,26–28,31–33,35] and the Occipital cortex have been associated with error-related activity [27,28,33,35]. It has been reported that the ErrP has the highest amplitudes around the midline channels but is still visible in the channels in the periphery furthest away from the midline, with a smaller amplitude though [27,33–35]. This may explain why all brain regions individually provide enough discriminative information to provide classification accuracies that are significantly higher than chance level; however, it has also been reported that the channels in the periphery lead to classification accuracies around chance level [29]. The fact that the ErrP can be observed in all channels, but with smaller amplitudes in the periphery, may also explain why adding additional regions to the Frontal and Central regions does not provide much additional discriminative information to the classification. This finding is also supported by previous studies that have found similar marginal increments/decrements in classification accuracies when multiple brain regions are used as input for classification [28,30]. It should also be noted that there is a different number of electrodes in the different brain regions in this study, which in itself could affect the classification accuracies although their spatial distribution probably matters most. It has been reported that the classification accuracies increase with a higher number of electrodes used for classification of ErrPs and motor imagery [29,33,41].

Despite this focus on populations with motor impairments, cerebral palsy, amputation, and stroke, it was still possible to decode ErrPs with accuracies similar to what has been reported previously in studies with able-bodied participants and participants with other types of motor impairments. This could be expected since the ErrP is elicited by the perception of an error, although the motor impairment could potentially cause lower expectations of one's own performance and hence affect the elicitation of the ErrP if the participant does not expect to be able to succeed in the task. However, this was not probed in the current study. The classification accuracies in similar studies have been reported to be in the range of 70–90% see e.g., [16,24,30,36,39,42–48]. However, it should be noted that bad epochs were rejected as part of the pre-processing in the current study, which is likely to have improved the decoding. These results may be optimistic as to what can be obtained in online decoding of errors. The approach of using MLP ANN has previously been shown to be useful for decoding ErrPs in stroke patients [16], and with this approach it is not necessary to extract features, since the entire epoch is used as input for the classifier. However, this classifier showed poor between-day and across-participant transfer [16], which suggests that calibration data for the classifier need to be collected every time it is going to be used. This can be a time-consuming process and it should be considered whether other, more generic approaches should be used [43,49–51], that can be individualized/adapted [17,52] to the user while an error monitoring/correction system is in use. Alternatively, it should be considered or whether a type of ErrP should be used in which multiple trials can be obtained more rapidly, such as observational ErrPs [53], or with higher error/correct ratios [25,54]. These considerations should be tested in future studies where online control of a BCI with error correction should be evaluated e.g., for error correction in myoelectric prosthetic control for people with an amputation, control of assistive devices such as wheelchairs, speller devices, or games for people with cerebral palsy or for adaptation of BCIs for stroke rehabilitation to account for fatigue or shifts in attention.

In future studies it could be relevant to perform source localization to better understand the origin of the error-related activity and investigate how it differs from the processing of feedback in general: i.e., the correct responses. It would also be relevant to perform channel selection to identify the optimal channels for identifying error-related activity and to investigate how much the relevant channels differ between users and across conditions. Another aspect that could be investigated is how to further optimize the decoding of ErrPs, which could be done using various techniques such as signal decomposition techniques and blind source separation for pre-processing the signals [55,56]. In this study, we only employed one specific type of neural network, but it is likely that other neural networks or tuning of them could yield better performance [57,58]. This could be tested systematically with different neural networks with inherent feature extraction, so no feature extraction with a priori knowledge is needed. Ideally, the classifier should have good generalization properties across days and users to avoid extensive calibration of it with user-specific data.

The results in this study also indicated that ErrPs could be decoded from the periphery, e.g., from non-hairy electrode locations around the ear. It could potentially allow the use of a more aesthetically appealing headset/electrode setup that also would not require hair wash after each use, which would be an important consideration for permanent BCI users [59,60]. This could be investigated in future studies.

5. Conclusions

In this study it was shown that ErrPs can be detected from single-trial EEG in participants with cerebral palsy, participants with an amputation, and participants with a stroke. The Frontal and Central brain regions were the most important ones, but it was also shown that other brain regions contributed with some discriminative information that increased the classification accuracy slightly. It was also shown that other brain regions beside the Frontal and Central regions could be used to classify ErrPs, this could be important in BCI applications where headsets are used that do not cover the Frontal or Central brain areas e.g., for more aesthetically pleasing headsets. Offline analyses were performed, but the findings should be validated with online error detection in future studies.

Author Contributions: Conceptualization, N.U., I.K.N., K.D. and M.J.; methodology, N.U., I.K.N., K.D. and M.J.; software, N.U.; validation, N.U., I.K.N. and M.J.; formal analysis, N.U.; investigation, N.U.; data curation, N.U.; writing—original draft preparation, N.U. and M.J.; writing—review and editing, N.U., I.K.N., K.D. and M.J.; visualization, N.U. and M.J.; supervision, I.K.N., K.D. and M.J.; project administration, N.U.; funding acquisition, M.J. All authors have read and agreed to the published version of the manuscript.

Funding: This research was partially funded by VELUX FONDEN, grant number 22357.

Institutional Review Board Statement: The study was conducted in accordance with the Declaration of Helsinki and approved by the Institutional Review Board of Allied Hospital, Faisalabad, Pakistan.

Informed Consent Statement: Informed consent was obtained from all subjects involved in the study.

Data Availability Statement: The data are not publicly available because of ethics and institutional policy.

Conflicts of Interest: The authors declare no conflict of interest.

References

1. Millán, J.R.; Rupp, R.; Müller-Putz, G.R.; Murray-Smith, R.; Giugliemma, C.; Tangermann, M.; Vidaurre, C.; Cincotti, F.; Kübler, A.; Leeb, R.; et al. Combining brain-computer interfaces and assistive technologies: State-of-the-art and challenges. *Front. Neurosci.* **2010**, *4*, 1–15. [CrossRef]
2. Wolpaw, J.R.; Birbaumer, N.; McFarland, D.J.; Pfurtscheller, G.; Vaughan, T.M. Brain-computer interfaces for communication and control. *Clin. Neurophysiol.* **2002**, *113*, 767–791. [CrossRef]
3. Daly, J.J.; Wolpaw, J.R. Brain-computer interfaces in neurological rehabilitation. *Lancet Neurol.* **2008**, *7*, 1032–1043. [CrossRef]
4. Grosse-Wentrup, M.; Mattia, D.; Oweiss, K. Using brain-computer interfaces to induce neural plasticity and restore function. *J. Neural Eng.* **2011**, *8*, 025004. [CrossRef]
5. Cervera, M.A.; Soekadar, S.R.; Ushiba, J.; Millán, J.D.R.; Liu, M.; Birbaumer, N.; Garipelli, G. Brain-computer interfaces for post-stroke motor rehabilitation: A meta-analysis. *Ann. Clin. Transl. Neurol.* **2018**, *5*, 651–663. [CrossRef]
6. Shibasaki, H.; Hallett, M. What is the Bereitschaftspotential? *Clin. Neurophysiol.* **2006**, *117*, 2341–2356. [CrossRef]

7. Mihajlovic, V.; Grundlehner, B.; Vullers, R.; Penders, J. Wearable, wireless EEG Solutions in daily life applications: What are we missing? *IEEE J. Biomed. Health Inform.* **2015**, *19*, 6–21. [CrossRef]
8. Leeb, R.; Perdikis, S.; Tonin, L.; Biasiucci, A.; Tavella, M.; Creatura, M.; Molina, A.; Al-Khodairy, A.; Carlson, T.; de Millán, J. Transferring brain-computer interfaces beyond the laboratory: Successful application control for motor-disabled users. *Artif. Intell. Med.* **2013**, *59*, 121–132. [CrossRef] [PubMed]
9. Chavarriaga, R.; Sobolewski, A.; Millán, J.D.R. Errare machinale est: The use of error-related potentials in brain-machine interfaces. *Front. Neurosci.* **2014**, *8*, 208. [CrossRef] [PubMed]
10. Kumar, A.; Gao, L.; Pirogova, E.; Fang, Q. A Review of Error-Related Potential-Based Brain-Computer Interfaces for Motor Impaired People. *IEEE Access* **2019**, *7*, 142451–142466. [CrossRef]
11. Parra, L.C.; Spence, C.D.; Gerson, A.D.; Sajda, P. Response error correction-a demonstration of improved human-machine performance using real-time EEG monitoring. *IEEE Trans. Neural Syst. Rehabil. Eng.* **2003**, *11*, 173–177. [CrossRef]
12. Kalaganis, F.P.; Chatzilari, E.; Nikolopoulos, S.; Kompatsiaris, I.; Laskaris, N.A. An error-aware gaze-based keyboard by means of a hybrid BCI system. *Sci. Rep.* **2018**, *8*, 13176. [CrossRef]
13. Cruz, A.; Pires, G.; Nunes, U.J. Double ErrP detection for automatic error correction in an ERP-based BCI speller. *IEEE Trans. Neural Syst. Rehabil. Eng.* **2017**, *26*, 26–36. [CrossRef]
14. Kumar, A.; Fang, Q.; Pirogova, E. The influence of psychological and cognitive states on error-related negativity evoked during post-stroke rehabilitation movements. *BioMedical Eng. OnLine* **2021**, *20*, 13. [CrossRef]
15. Kumar, A.; Fang, Q.; Fu, J.; Pirogova, E.; Gu, X. Error-related neural responses recorded by electroencephalography during post-stroke rehabilitation movements. *Front. Neurorobotics* **2019**, *13*, 107. [CrossRef]
16. Usama, N.; Niazi, I.K.; Dremstrup, K.; Jochumsen, M. Detection of Error-Related Potentials in Stroke Patients from EEG Using an Artificial Neural Network. *Sensors* **2021**, *21*, 6274. [CrossRef]
17. Dias, C.L.; Sburlea, A.I.; Breitegger, K.; Wyss, D.; Drescher, H.; Wildburger, R.; Müller-Putz, G.R. Online asynchronous detection of error-related potentials in participants with a spinal cord injury by adapting a pre-trained generic classifier. *J. Neural Eng.* **2020**, *18*, 046022. [CrossRef]
18. Roset, S.A.; Gant, K.; Prasad, A.; Sanchez, J.C. An adaptive brain actuated system for augmenting rehabilitation. *Front. Neurosci.* **2014**, *8*, 415. [CrossRef]
19. Spüler, M.; Bensch, M.; Kleih, S.; Rosenstiel, W.; Bogdan, M.; Kübler, A. Online use of error-related potentials in healthy users and people with severe motor impairment increases performance of a P300-BCI. *Clin. Neurophysiol.* **2012**, *123*, 1328–1337. [CrossRef]
20. Milekovic, T.; Ball, T.; Schulze-Bonhage, A.; Aertsen, A.; Mehring, C. Error-related electrocorticographic activity in humans during continuous movements. *J. Neural Eng.* **2012**, *9*, 026007. [CrossRef]
21. Milekovic, T.; Ball, T.; Schulze-Bonhage, A.; Aertsen, A.; Mehring, C. Detection of error related neuronal responses recorded by electrocorticography in humans during continuous movements. *PLoS ONE* **2013**, *8*, e55235.
22. Holroyd, C.B.; Praamstra, P.; Plat, E.; Coles, M.G. Spared error-related potentials in mild to moderate Parkinson's disease. *Neuropsychologia* **2002**, *40*, 2116–2124. [CrossRef]
23. Hakkarainen, E.; Pirilä, S.; Kaartinen, J.; van der Meere, J.J. Error detection and response adjustment in youth with mild spastic cerebral palsy: An event-related brain potential study. *J. Child. Neurol.* **2013**, *28*, 752–757. [CrossRef] [PubMed]
24. Ferrez, P.W.; Millán, J.D.R. Error-related EEG potentials generated during simulated brain-computer interaction. *IEEE Trans. Biomed. Eng.* **2008**, *55*, 923–929. [CrossRef]
25. Chavarriaga, R.; Millán, J.D.R. Learning from EEG error-related potentials in noninvasive brain-computer interfaces. *IEEE Trans. Neural Syst. Rehabil. Eng.* **2010**, *18*, 381–388. [CrossRef]
26. Dyson, M.; Thomas, E.; Casini, L.; Burle, B. Online extraction and single trial analysis of regions contributing to erroneous feedback detection. *Neuroimage* **2015**, *121*, 146–158. [CrossRef]
27. Shou, G.; Ding, L. Detection of EEG Spatial-Spectral-Temporal Signatures of Errors: A Comparative Study of ICA-Based and Channel-Based Methods. *Brain Topogr.* **2015**, *28*, 47–61. [CrossRef]
28. Shou, G.; Ding, L. EEG-based single-trial detection of errors from multiple error-related brain activity. In Proceedings of the 38th Annual International Conference of the IEEE Engineering in Medicine and Biology Society (EMBC), Lake Buena Vista, FL, USA, 16–20 August 2016; pp. 2764–2767.
29. Tong, J.; Lin, Q.; Xiao, R.; Ding, L. Combining multiple features for error detection and its application in brain-computer interface. *Biomed. Eng. Online* **2016**, *15*, 1–15. [CrossRef]
30. Vasios, C.E.; Ventouras, E.M.; Matsopoulos, G.K.; Karanasiou, I.; Asvestas, P.; Uzunoglu, N.K.; van Schie, H.T.; de Bruijn, E.R. Classification of event-related potentials associated with response errors in actors and observers based on autoregressive modeling. *Open Med. Inform. J.* **2009**, *3*, 32. [CrossRef]
31. Ventouras, E.M.; Asvestas, P.; Karanasiou, I.; Matsopoulos, G.K. Classification of Error-Related Negativity (ERN) and Positivity (Pe) potentials using kNN and Support Vector Machines. *Comput. Biol. Med.* **2011**, *41*, 98–109. [CrossRef]
32. Wilson, N.R.; Sarma, D.; Wander, J.D.; Weaver, K.E.; Ojemann, J.G.; Rao, R.P. Cortical topography of error-related high-frequency potentials during erroneous control in a continuous control brain-computer interface. *Front. Neurosci.* **2019**, *13*, 502. [CrossRef]
33. Yazmir, B.; Reiner, M. Neural Signatures of Interface Errors in Remote Agent Manipulation. *Neuroscience* **2021**. in press. [CrossRef]
34. Yazmir, B.; Reiner, M. I act, therefore I err: EEG correlates of success and failure in a virtual throwing game. *Int. J. Psychophysiol.* **2017**, *122*, 32–41. [CrossRef]

35. Yazmir, B.; Reiner, M. Neural correlates of user-initiated motor success and failure—A brain-computer interface perspective. *Neuroscience* **2018**, *378*, 100–112. [CrossRef]
36. Usama, N.; Leerskov, K.K.; Niazi, I.K.; Dremstrup, K.; Jochumsen, M. Classification of error-related potentials from single-trial EEG in association with executed and imagined movements: A feature and classifier investigation. *Med. Biol. Eng. Comput.* **2020**, *58*, 2699–2710. [CrossRef]
37. Hoffmann, S.; Falkenstein, M. Predictive information processing in the brain: Errors and response monitoring. *Int. J. Psychophysiol.* **2012**, *83*, 208–212. [CrossRef]
38. Müller-Putz, G.R.; Scherer, R.; Brunner, C.; Leeb, R.; Pfurtscheller, G. Better than random? A closer look on BCI results. *Int. J. Bioelectromagn.* **2008**, *10*, 52–55.
39. Iturrate, I.; Montesano, L.; Minguez, J. Task-dependent signal variations in EEG error-related potentials for brain-computer interfaces. *J. Neural Eng.* **2013**, *10*, 026024. [CrossRef]
40. Omedes, J.; Iturrate, I.; Minguez, J.; Montesano, L. Analysis and asynchronous detection of gradually unfolding errors during monitoring tasks. *J. Neural Eng.* **2015**, *12*, 056001. [CrossRef]
41. Zhu, K.; Wang, S.; Zheng, D.; Dai, M. Study on the effect of different electrode channel combinations of motor imagery EEG signals on classification accuracy. *J. Eng.* **2019**, *2019*, 8641–8645. [CrossRef]
42. Buttfield, A.; Ferrez, P.W.; Millan, J.R. Towards a robust BCI: Error potentials and online learning. *Neural Syst. Rehabil. Eng.* **2006**, *14*, 164–168. [CrossRef]
43. Bhattacharyya, S.; Konar, A.; Tibarewala, D.N.; Hayashibe, M. A generic transferable EEG decoder for online detection of error potential in target selection. *Front. Neurosci.* **2017**, *11*, 226. [CrossRef]
44. Omedes, J.; Schwarz, A.; Müller-Putz, G.R.; Montesano, L. Factors that affect error potentials during a grasping task: Toward a hybrid natural movement decoding BCI. *J. Neural Eng.* **2018**, *15*, 046023. [CrossRef]
45. Schmidt, N.M.; Blankertz, B.; Treder, M.S. Online detection of error-related potentials boosts the performance of mental typewriters. *BMC Neurosci.* **2012**, *13*, 19. [CrossRef]
46. Spüler, M.; Niethammer, C. Error-related potentials during continuous feedback: Using EEG to detect errors of different type and severity. *Front. Hum. Neurosci.* **2015**, *9*, 155.
47. Gao, C.; Li, Z.; Ora, H.; Miyake, Y. Improving error related potential classification by using generative adversarial networks and deep convolutional neural networks. In Proceedings of the 2020 IEEE International Conference on Bioinformatics and Biomedicine (BIBM), Seoul, Korea, 16–19 December 2020; pp. 2468–2476.
48. Chavarriaga, R.; Khaliliardali, Z.; Gheorghe, L.; Iturrate, I.; Millán, J.D.R. EEG-based decoding of error-related brain activity in a real-world driving task. *J. Neural Eng.* **2015**, *12*, 066028.
49. Aydarkhanov, R.; Uščumlić, M.; Chavarriaga, R.; Gheorghe, L.; del Millán, J.R. Spatial covariance improves BCI performance for late ERPs components with high temporal variability. *J. Neural Eng.* **2020**, *17*, 036030. [CrossRef]
50. Abu-Alqumsan, M.; Kapeller, C.; Hintermüller, C.; Guger, C.; Peer, A. Invariance and variability in interaction error-related potentials and their consequences for classification. *J. Neural Eng.* **2017**, *14*, 066015. [CrossRef]
51. Iturrate, I.; Chavarriaga, R.; Montesano, L.; Minguez, J.; Millán, J. Latency correction of event-related potentials between different experimental protocols. *J. Neural Eng.* **2014**, *11*, 036005. [CrossRef]
52. Schönleitner, F.M.; Otter, L.; Ehrlich, S.K.; Cheng, G. Calibration-Free Error-Related Potential Decoding with Adaptive Subject-Independent Models: A Comparative Study. *IEEE Trans. Med. Robot. Bionics* **2020**, *2*, 399–409. [CrossRef]
53. Kim, S.K.; Kirchner, E.A. Handling Few Training Data: Classifier Transfer Between Different Types of Error-Related Potentials. *IEEE Trans. Neural Syst. Rehabil. Eng.* **2016**, *24*, 320–332. [CrossRef] [PubMed]
54. Pezzetta, R.; Nicolardi, V.; Tidoni, E.; Aglioti, S.M. Error, rather than its probability, elicits specific electrocortical signatures: A combined EEG-immersive virtual reality study of action observation. *J. Neurophysiol.* **2018**, *120*, 1107–1118. [CrossRef] [PubMed]
55. Sadiq, M.T.; Yu, X.; Yuan, Z.; Fan, Z.; Rehman, A.U.; Li, G.; Xiao, G. Motor imagery EEG signals classification based on mode amplitude and frequency components using empirical wavelet transform. *IEEE Access* **2019**, *7*, 127678–127692. [CrossRef]
56. Kevric, J.; Subasi, A. Comparison of signal decomposition methods in classification of EEG signals for motor-imagery BCI system. *Biomed. Signal Processing Control* **2017**, *31*, 398–406. [CrossRef]
57. Sakhavi, S.; Guan, C.; Yan, S. Parallel convolutional-linear neural network for motor imagery classification. In Proceedings of the 2015 23rd European Signal Processing Conference (EUSIPCO), IEEE, Nice, France, 31 August–4 September 2015; pp. 2736–2740.
58. Sadiq, M.T.; Yu, X.; Yuan, Z. Exploiting dimensionality reduction and neural network techniques for the development of expert brain-computer interfaces. *Expert Syst. Appl.* **2021**, *164*, 114031. [CrossRef]
59. Nijboer, F.; Plass-Oude Bos, D.; Blokland, Y.; van Wijk, R.; Farquhar, J. Design requirements and potential target users for brain-computer interfaces—Recommendations from rehabilitation professionals. *Brain-Comput. Interfaces* **2014**, *1*, 50–61. [CrossRef]
60. Jochumsen, M.; Knoche, H.; Kidmose, P.; Kjær, T.W.; Dinesen, B.I. Evaluation of EEG Headset Mounting for Brain-Computer Interface-Based Stroke Rehabilitation by Patients, Therapists, and Relatives. *Front. Hum. Neurosci.* **2020**, *14*, 13. [CrossRef]

sensors

MDPI

Article

fNIRS-Based Upper Limb Motion Intention Recognition Using an Artificial Neural Network for Transhumeral Amputees

Neelum Yousaf Sattar [1,*], Zareena Kausar [1], Syed Ali Usama [1,*], Umer Farooq [1], Muhammad Faizan Shah [2], Shaheer Muhammad [3], Razaullah Khan [4] and Mohamed Badran [5]

[1] Department of Mechatronics and Biomedical Engineering, Air University, Main Campus, PAF Complex, Islamabad 44000, Pakistan; zareena.kausar@mail.au.edu.pk (Z.K.); maill2umer@gmail.com (U.F.)
[2] Department of Mechanical Engineering, Khwaja Fareed University of Engineering & IT, Rahim Yar Khan 64200, Pakistan; faizan.shah@kfueit.edu.pk
[3] Department of Computing, The Hong Kong Polytechnic University, Hung Hom, Hong Kong; postmuhammadshaheer@yahoo.com
[4] Institute of Manufacturing, Engineering Management, University of Engineering and Applied Sciences, Swat, Mingora 19060, Pakistan; razaullah@ueas.edu.pk
[5] Department of Mechanical Engineering, Faculty of Engineering and Technology, Future University in Egypt, New Cairo 11835, Egypt; mohamed.Badran@fue.edu.eg
*   Correspondence: neelumyousafsattar@gmail.com (N.Y.S.); ali.usama@mail.au.edu.pk (S.A.U.)

Abstract: Prosthetic arms are designed to assist amputated individuals in the performance of the activities of daily life. Brain machine interfaces are currently employed to enhance the accuracy as well as number of control commands for upper limb prostheses. However, the motion prediction for prosthetic arms and the rehabilitation of amputees suffering from transhumeral amputations is limited. In this paper, functional near-infrared spectroscopy (fNIRS)-based approach for the recognition of human intention for six upper limb motions is proposed. The data were extracted from the study of fifteen healthy subjects and three transhumeral amputees for elbow extension, elbow flexion, wrist pronation, wrist supination, hand open, and hand close. The fNIRS signals were acquired from the motor cortex region of the brain by the commercial NIRSport device. The acquired data samples were filtered using finite impulse response (FIR) filter. Furthermore, signal mean, signal peak and minimum values were computed as feature set. An artificial neural network (ANN) was applied to these data samples. The results show the likelihood of classifying the six arm actions with an accuracy of 78%. The attained results have not yet been reported in any identical study. These achieved fNIRS results for intention detection are promising and suggest that they can be applied for the real-time control of the transhumeral prosthesis.

Keywords: artificial neural network (ANN); functional near-infrared spectroscopy (fNIRS); machine learning; upper-limb prosthesis; transhumeral amputee

Citation: Sattar, N.Y.; Kausar, Z.; Usama, S.A.; Farooq, U.; Shah, M.F.; Muhammad, S.; Khan, R.; Badran, M. fNIRS-Based Upper Limb Motion Intention Recognition Using an Artificial Neural Network for Transhumeral Amputees. *Sensors* **2022**, *22*, 726. https://doi.org/10.3390/s22030726

Academic Editor: Steve Ling

Received: 26 November 2021
Accepted: 12 January 2022
Published: 18 January 2022

1. Introduction

Amputation refers to the removal of a human limb due to an illness, accident, or trauma. To overcome human limb loss, an artificial device (prosthetics) is provided [1]. The upper limb amputation is divided into five major types, as indicated in Figure 1. Amputees wear transhumeral prosthetic arms to substitute for the loss of elbow and lower portion of arm [2]. A human upper limb can perform seven different motions associated with joints in the arm. Three arm motions are mandatory for transhumeral prosthesis, including elbow extension–flexion, wrist supination–pronation, and hand opening and closing. Advances in the field of biomechatronic have opened new doors to expand the use and applications of prosthetic devices for amputees. However, the control of such prosthetic arms is new area for researchers to explore. Bio signals are preferably used for intention detection that further triggers the implementation of control.

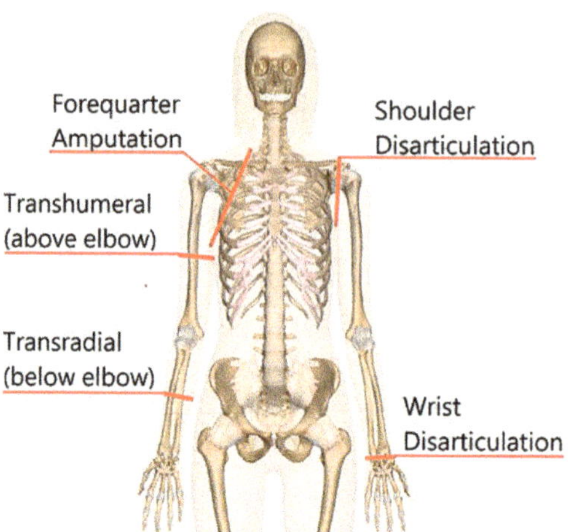

Figure 1. Levels of upper limb amputation [3].

For a long time, upper limb prostheses have been largely controlled using electromyographic (EMG) signals from remnant muscles. Various research studies [4] have considered sEMG for motion intention assessment and used in upper limb prosthetic control. In [5], a DEKA arm with three modular configurations was proposed for people suffering from transradial, transhumeral, and shoulder disarticulations. It utilizes sEMG along with a feet controller and pneumatic bladder for arms control. Lenzi et al. [6] proposed a 5-DoF transhumeral prosthesis for elbow, forearm, wrist, and grasping motions that used an EMG-based low-level controller. Researchers have additionally utilized many other biosensors to control prosthetic arms, such as mechanomyography (MMG) [7], inertial measurement unit (IMU) [8] and near-infrared spectroscopy (NIRS) [9]. Regardless of the above-mentioned developments, a gap exists in the simultaneous control of motions of multi-dimensional transhumeral prostheses.

Signal acquisition and processing are a great challenge in the control of above elbow amputation due to few or no amount of residual muscle and weak muscle activity [4,10,11]. Furthermore, remaining muscle sites for the prosthetic control are not physiologically identified to the distal arm functions [12]. In the past few years, the brain-machine interface (BMI) has appeared as a potential alternative that can offer an incredible opportunity to amputated individuals by empowering them to play out their daily routine [13,14]. It evades the muscles intentions. BMI systems are also implemented to restore motor functions called neuro-prosthesis. The latter assists motor disabled individuals achieve simple everyday tasks [15–17]. Quite a few modalities, EEG, MEG, and fMRI, have been considered for BMI applications for their capability to measure brain activities noninvasively. Optical brain imaging has been recently practiced in the BMI field, recognized as functional near-infrared spectroscopy (fNIRS) [18].

fNIRS is useful over the other mentioned modalities for BMI as portability, safety, low noise, and no susceptibility to electrical noise adds to the easy utilization of the system [19]. fNIRS measures hemodynamic response in the cerebral cortical tissue of the brain. The principle of fNIRS uses oxygenated hemoglobin (HbO) and deoxygenated hemoglobin (HbR). The optodes are sensitive to two dissimilar wavelengths in the near-infrared range, 700–1000 nm. This is known as an "optical window". The biological tissue is somewhat clear to light in this window. The light absorption by water molecules and hemoglobin is relatively low in this region. Henceforth, sensing the light pass through the brain tissue employing noninvasive signal acquisition is performed using an optical source/detector pair, which is placed on the scalp. A relative change in the concentration of HbO and HbR indicates neuronal activation relevant to the executed motion [13]. The attained brain responses relative to distinct motion may comprise of noises that pollute these recorded signals. The noise can be classified as physiological, experimental and instrumental noise [20]. These noises are removed from samples before converting them to magnitude by implementing the modified Beer–Lambert law [21]. The noise recorded because of a computer or neighboring environment is recognized as instrument noise. This noise typically has a high frequency (HF). The HF is separated by applying a low-pass filter. Experimental noise contains motion artefacts, such as head motions or optode dislocation from allotted positions. This generates spikes caused by a variation in light intensities. A study [18] used regularly advanced filtering techniques arbitrarily for noise reduction. Noise is physiologically fashioned [22] as a result of Mayer wave (~0.1 Hz), respiration (0.2~0.5 Hz), and heartbeat (1~1.5 Hz) This is large because of oscillations in blood pressure [23]. One of the chief BMI uses is to extract useful information from raw brain responses for a control–command generation [14]. The captured signals are refined in the four phases: signal preprocessing, feature extraction, classification, and control–command generation. In pre-processing, physiological and instrument artifact and noise are removed. Afterwards the filtration phase, feature extraction step in to gather detailed traits of the signal. Next, the extracted features are classified. The trained classifier is deployed generating control commands using the previously trained data samples [24,25].

Several researchers have embarked on the design and development of robotic arms. The configurations of robotic arms depend on the tasks to be performed by the human arm. The distinct comprehension of actuation approaches is employed. Earlier design approaches have focused on the mechanical issues of structures and the operation of the prosthetic arms [2]. Most of these prosthetic devices are controlled using unnatural methods, such as using the contraction of muscles of the opposite arm [4].

This research attempts to lay a foundation for a framework that offers functionality similar to the human arm, with an intuitive scheme of control. Therefore, by analyzing fNIRS signals to generate control commands for upper limb prosthetic devices, this current study proposes an ANN-based signal classification framework to recognize the intention of six upper limb motions of both healthy and transhumeral amputees. The novelty of the presented research is to generate six control commands using fNIRS for transhumeral amputees. To the best of the authors' knowledge, there is no existing literature for motion intention detection of six control commands using fNIRS for upper-limb prosthesis applications [1,2,7,14,26].

This paper is organized as follows. Section 2 describes the materials and methods deployed for this study. This includes data acquisition and signal processing. Section 3 consists of feature extraction and classification methods, whereas in Section 4, results are presented and discussed. This includes, filtration, channel selection, and classification accuracies. Section 5 is the conclusion.

2. Materials and Methods

In this section, details regarding the experimental procedure followed by the methodology used in signal acquisition and processing, feature extraction, classification, and control command generation are included. A block diagram representation of the methods used is presented in Figure 2.

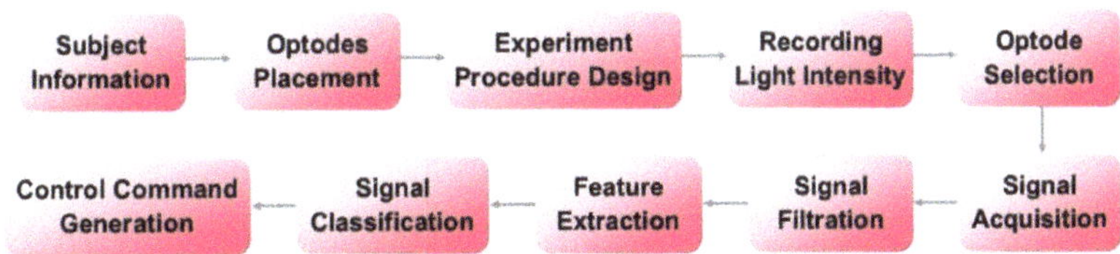

Figure 2. Methodology of the study.

2.1. Subject Information

The study included 15 healthy and three transhumeral amputated subjects among which the healthy subjects were dominant right-hand males. Female amputees are not included simply due to their unavailability. No subject had any psychological, neurological, or optical affliction in the past, as per the recommendation given in [27,28]. The subjects signed a written consent after being briefed about the experimental process. The demographics of transhumeral amputees are given in Table 1. The experiments were allowed by the Human Research Ethics Committee (HREC) of Air University Islamabad. The experiments were performed as per the standards issued by the recent declaration of Helsinki [29].

Table 1. Demographic characteristics of amputee subjects.

Amputee Title	A1	A2	A3
Gender	Male	Male	Male
Age	23	32	42
Amputated Side	Right	Left	Right
Residual Limb Length	14 cm	18 cm	10 cm
Time since Amputation	7 Months	24 Months	145 Months

2.2. Optode Placement

The fNIRS data were recorded from the NIRx Imager system, NIRsport (NIRx Medical Technologies, Germany), using an 8×8 sensor array positioned on the motor cortex region of the human head scalp. The fNIRS signals were acquired for six arm motions: elbow extension (E.E), elbow flexion (E.F), wrist pronation (W.P), and wrist supination (W.S), hand open (H.O), and hand close (H.C). The optodes were precisely placed on the motor cortex related areas on the 10–20 system that yielded 20 fNIRS channels (10 channels in each hemisphere). Figure 3a shows the position of fNIRS optodes on a healthy subject. Easy cap by NIRx technologies is specially made for optical brain imaging according to international standards [18]. The standard distance between source and detector is 3 cm, as illustrated in Figure 3b [8,22].

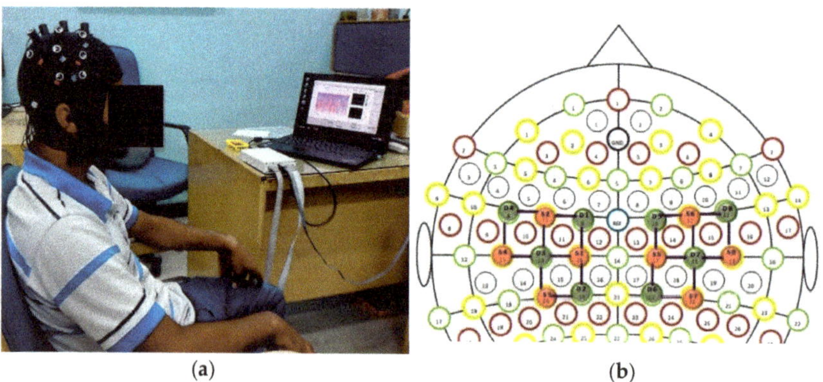

(a)   (b)

Figure 3. Signal acquisition environment and optode placement. (**a**) Experiment setup showing optode placement on the motor cortex of a healthy subject. (**b**) Eight sources (S) and eight detectors (D) were positioned on the subject's motor cortex region of the brain to record fNIRS signals with a separation of 3 cm resulting in twenty channels.

2.3. Experimental Procedure

The experimental procedure was designed for subjects to perform six motor imagery (MI) tasks. During MI tasks, subjects were instructed to think of performing one of the arm movements and refrain from any other action, such as muscle twitches. The individual subject was asked to perform MI, guided by the experimental team before starting the trials to make them aware of the experimental protocol [29]. During these tasks, the subjects were seated on a comfortable chair to remain relaxed. The chair was placed at an approximate distance of 90 cm from the screen so that the arm motion indications are noticeable and the computer screen backlight does not obstruct the optical sensors [18,30,31].

The experiment session began with an initial rest duration of 30 s to generate a baseline. After that, the routine for the motion was displayed on a computer monitor for subjects to follow. The experiment had two sessions. At first, all tasks were performed in sequence such that the arm motions were pre-defined. However, in the second step, all subjects performed similar arm motions but executed with random intentions. fNIRS logged all six tasks (E.E, E.F, W.P, W.S, H.O, and H.C). Each task/action comprised of ten-second trials with a rest session of 20 s. Each motion was repeated 10 times while in total 12 motions were performed by each subject. An experimental paradigm used in this study is described in Figure 4. A framework of the proposed study is illustrated in Figure 5.

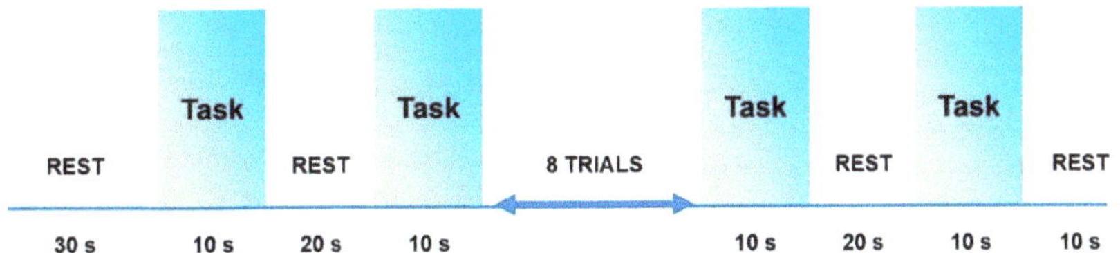

Figure 4. After an initial 30 s rest, each functional near-infrared spectroscopy block consists of 10 s activations and 20 s rests. The total experiment duration for acquiring fNIRS signals is 11 min and includes 12 trials in total.

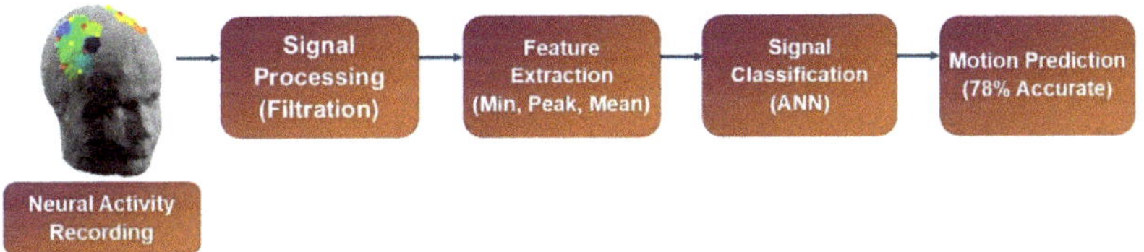

Figure 5. Flow diagram of fNIRS-based motion intention recognition for the transhumeral prosthesis.

After the signal acquisition, the signals were filtered using FIR filter. These filtered signals were then used to compute the hemodynamic responses using MBLL. Signal mean and peaks were extracted as a feature. The minimum values were also extracted to set the threshold for channel selection. These hemodynamic responses were then fed to the classifying network. Based on the training, the network predicted the motion class. All the details with mathematical equations and numeric values are briefly described in the next section.

2.4. Data Acquisition and Processing

This section includes signal acquisition, signal preprocessing, and statistical feature extraction. The signal classification algorithm is also contained within this section.

Before signal recording in fNIRS, an optical imaging technique, the light intensity values were recorded during the oxygenation and deoxygenation of human blood cells of the brain [32]. By using NIRx dual-tip optodes, the light concentration was measured at two values of wavelengths: 760 nm and 850 nm. The acquired light intensities are then processed in nirslab Using this application, one can truncate/remove unwanted data as well as infrequent gaps captured earlier during the acquisition process [2,33]. The dataset can be filtered and hemodynamic states can be computed in the same application as well [34].

2.4.1. Signal Acquisition

The fNIRS signals were acquired using a headset, a flexible cap made of a soft cloth with optodes referred to as EasyCap [35,36]. The experiments were performed with headset placement on the motor cortex in three ways: in the first setting, simply placing the cap; then, the optodes (on the cap) were fixed with spring grommet; and lastly, a complete black cap on top of second set was placed such that the optodes are not visible [35–37]. As soon as a subject was wearing the sensor cap, the optodes were calibrated. The results of the first setup are given in Figure 6b, while Figure 6a represents the outcome of the second set.

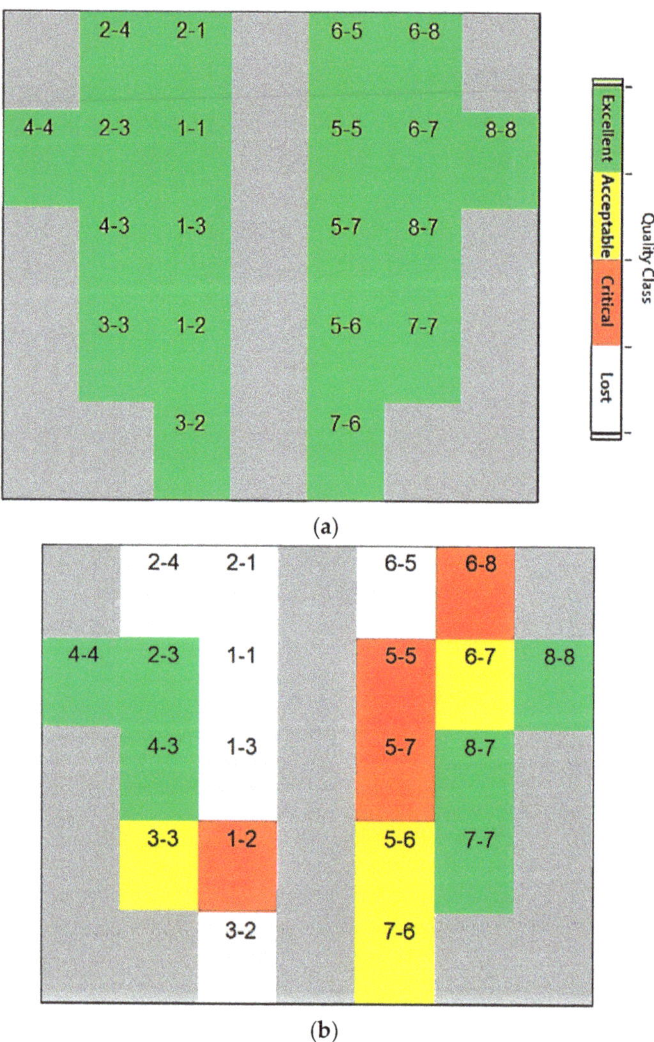

Figure 6. Optode status window. (**a**) Flawless optode connection with head scalp; (**b**) faulty optode connection. The signal quality class can be read from a color bar shown along with the optode status window.

The rectangles/boxes are the representation of the optodes whether source or detector. The intensity bar on the right is an indication of the optode data quality status. The box changes its color according to the physical status of the sensor to give an idea about which optode to settle on for better signals [38]. The "white color" portrays that there exists no connection between the optodes and the subject scalp. The "red color" indicates a "critical" connection, which directs that it requires the optodes to be adjusted. Commonly, an anomaly is detected because thick hair may be caught in the EasyCap cavity, and just putting the fNIRS optodes again helped to form a better connection. A "yellow color" represents an acceptable connection and the brain signals can be attained. The fNIRS data acquisition system settings were attuned by the system itself. It enhances the gain factor of each optode when the connection is acceptable. The system further saves these numeric values in a "conditions" file. This file can be of help later in the optode selection process. Finally, the "green color" demonstrates that the fNIRS optodes are flawlessly positioned on the subject's head. This also indicates that an outstanding tie is recognized between the sensing element and the scalp for data acquisition [38].

For the third case, when the data was continuously bad, although connections were fixed by giving a green signal during placement calibration, the dark noise tests were conducted [39]. This test examines the intensity of light that is incident on the optodes from the environment. Keeping in mind, the black covering cap cannot always be worn, dark noise was tested initially with a hypothesis that these special grommets make sure that the least amount of noise is induced to the sensors [40].

2.4.2. Signal Processing

As soon as the optodes were calibrated, the signal acquisition was started [34]. After that, the nirslab software comes with a fNIRS headset to differentiate between bad and good channels based upon the gain values as illustrated in Figure 7. The gain setting allows the exclusion of all channels that have a higher value than a specified value. The nirslab label 'bad' to any channel shows that it has a value, at either wavelength, equal to or greater than the threshold value that was specified [34,40]. This value is related to the light intensity of the environment in which the experiment is conducted. Then, the black covering cap was used to address the issue [41].

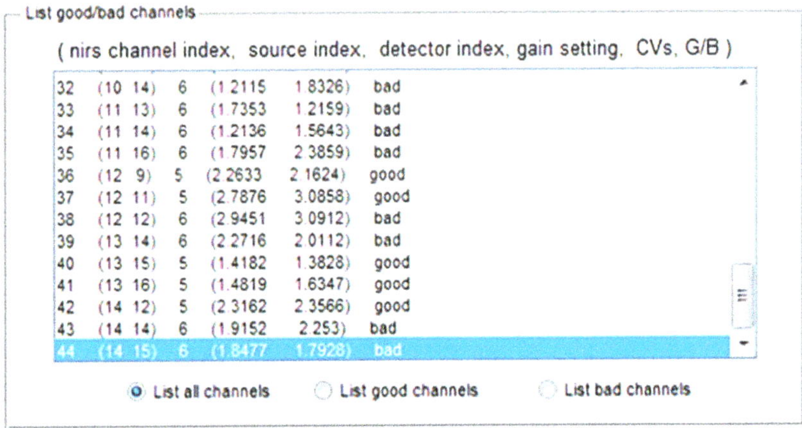

Figure 7. List of good/bad channels to remove bad channels from the analysis and signal classification.

The discontinuities/spikes caused by the cap placement are removed as illustrated in Figure 8. The number of lines indicates the signal acquired from each optode. Hence, a clean signal can be fed for further processing. The disturbed/noisy and clean signals can be seen in Figures 9 and 10, respectively.

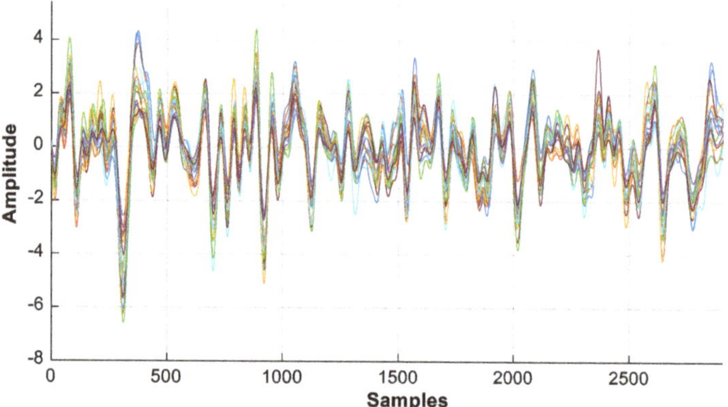

Figure 8. Visualization of recorded raw light intensity of each optode.

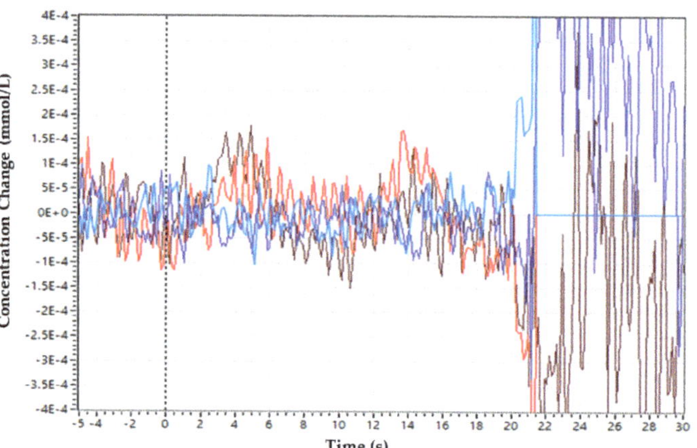

Figure 9. Disturbed/noisy signal before grommets and covering head cap is incorporated. Blue lines represent the data acquired from the detector while the red line represents the source signal.

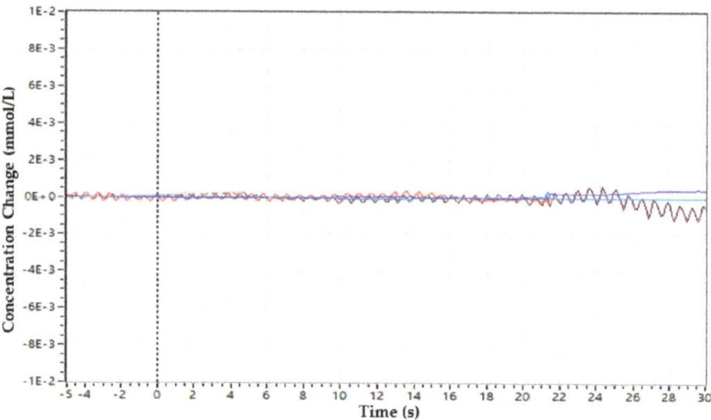

Figure 10. Clean signal after incorporation of grommets incorporated with covering head cap.

nirslab further provide an artifact removal method.

An acquired hemodynamic response after noise/spike artefact removal from a healthy subject is illustrated in Figure 11.

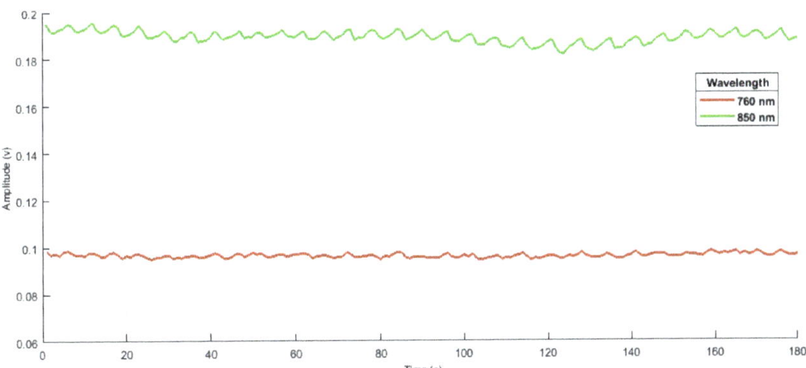

Figure 11. The fNIRS data obtained from a healthy subject according to the experimental protocol. The upper signal is from the source of 760 nm, and the lower one represents the 850 nm wavelength.

Additionally, the unwanted or disturbing segments in the signal can be removed according to the values of the threshold or depending on the gain factor incorporated by the machine earlier [8]. A band-pass filter was applied to further smooth the samples to compute hemodynamic states [42]. Filtered data at wavelength 750 nm are illustrated below in Figure 12.

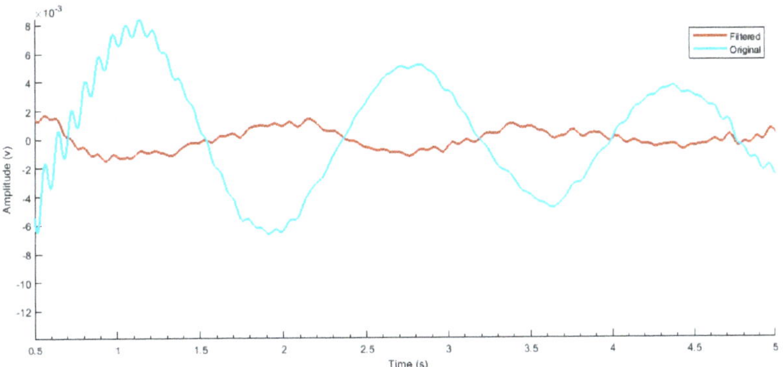

Figure 12. The filtered sample after the implementation of band-pass filter in a range of 0.01 Hz–0.2 Hz.

nirslab used "firls" and "filtfilt" MATLAB® instructions for filtering purposes. The "firls" proceeds parameters of linear-segment filtration [8]. The "filtfilt" employs filter parameters to samples. Then, a FIR is introduced. For FIR, a roll-off value states the size of the transition frequency band [34]. The mathematical formation of the filter is given in a set of equations, which are the Fourier transform of the truncated filter and are given in Equations (1) and (2):

$$H(\omega) = \frac{1}{2\pi} \int_{-\pi}^{\pi} H_d(\lambda) W(\omega - \lambda) d\lambda \tag{1}$$

$$h(n) = h_d(n) w(n) \tag{2}$$

The width of transition region between the band pass limits H increases with the width of main lobe W. It decides the steepness of transition amongst frequencies [43]. This value was by default set to 15 by the nirslab according to signal condition [44]. After filtration, hemodynamic states are computed and are then set to extract features from them.

3. Feature Extraction and Classification of Motion Intention Signals

The method of signal feature extraction performs a crucial part in the identification of the discriminatory information carried by the bio signals [32,44–46]. This section details the features extracted from the dataset and the details of the applied machine learning algorithm for motion classifications.

3.1. Feature Extraction

To execute control commands for six arm motions, features for the signal classification were extracted. For fNIRS brain signals, signal means (SM), signal peak (SP), and signal minimum (min) [43,47] were extracted for thresholding purposes. The signal mean was calculated as (3):

$$SM = \frac{1}{N}\Sigma_{i=1}^{N}X_i \tag{3}$$

where N represents the total data points and X_i represents the signal amplitude value. The signal peak was calculated using a signal amplitude variation between two head-to-head sections that exceeds a pre-defined threshold value to cut noise. It is given by (4):

$$SP = \Sigma_{i=1}^{N}f(|X_i - X_{i+1}|) \tag{4}$$

As per the existing literature [38], SM and SP offer improved control performance for fNIRS-based systems. However, regarding the stated possibility of an initial fNIRS signal dip, a (min) signal value was added as a feature [43]. The features were calculated from only selected optodes based on criteria described earlier in this section using a 2 s moving window. MATLAB® was employed to execute all the features computation.

3.2. Artificial Neural Network (ANN)

For the evaluation of the performance of acquired fNIRS signals, a widely used [38] classifier in pattern recognition was implemented, namely, artificial neural network (ANN). It uses various neuron layers to plan information starting with one circulation then onto the next one for better and enhanced results, i.e., returning less error [46,48]. A system called backpropagation assists ANN to form a bridge between input and output layers in which the corresponding labels/indicators are present [49]. The machine learning toolbox designed by MATLAB® for neural networks came into play for training the samples [50]. When using the toolbox, all you need to set is the number of hidden neurons in the layer of this artificial neural network [51]. The designed model then estimates the error of the probable output in contrast with the actual output. The network further explores the error to variate to adjust the weightage, to minimize the generated error for the next cycle, and this process continues unless the error approaches zero [46]. For the network, the rule activation function was applied, and the weights were randomly assigned by the toolbox. The results and ANN training particulars are presented in the next section. The comprehensive flow diagram of the ANN classifier is shown in Figure 13.

The ANN network had 2 hidden layers with 12 and 6 neurons. The output layer will give one definitive class, i.e., motion class, as defined in Figure 14. First, preprocessed information is passed through the first layers, which contain 128 filters with a kernel size of 12. The output from the first layer is 24 × 128 [52]. The second layer contains the same number of filters and a kernel size of six. The output from the second layer is 12 × 128. Subsequently, the global average pooling is applied between the output layer for which the Adam optimization method was deployed [53,54].

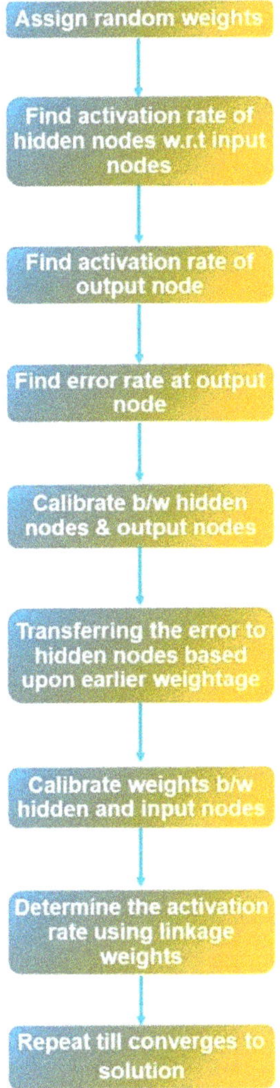

Figure 13. Flow diagram of the ANN classifier.

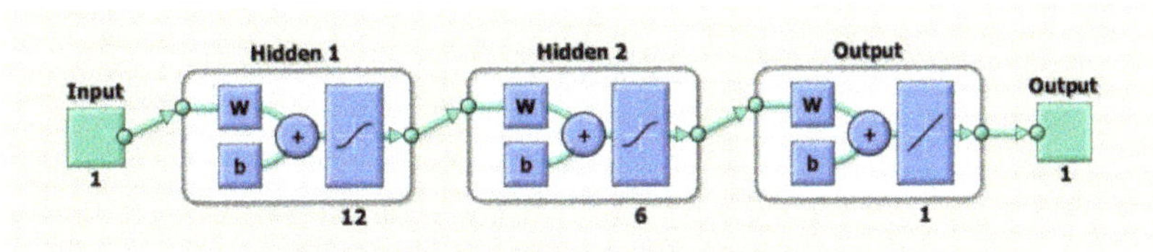

Figure 14. ANN network architecture.

After classifying the fNIRS signals, the trained model was tested. The obtained results are discussed in the next section.

4. Results and Discussion

In this study, fNIRS signals were acquired to generate control commands of human arm motions for the transhumeral amputee. The fNIRS hemodynamic responses are acquired employing optical sensors, i.e., optodes. These responses are used to drive a prosthetic arm for transhumeral amputees. For optical sensors, dark noise plays an important role. The NIRx Technologies designed a high-functioning device that can be used for real-time purpose. However, the difficulties have never been addressed before. For this study, the results were not only analyzed based upon the accuracy of the motion classification, but also the sustainability for real-time applications. The optode placement problem was addressed and researched upon and is detailed in previous sections. This section specifically presents the experimental results achieved using the proposed framework. Window sizing of various lengths and durations has been observed in the literature to detect fNIRS signals [31]. The period of 0–0.5 and 0–1 s was selected [55,56]. This split time window was used to inspect hemodynamic response. This then determines the optimal window, which then generates a command with a minimum amount of time.

4.1. Channel Selection

The electrical gain component that was adjusted to the absorption spectra is shown by number 6 in Figure 7. The photocurrent generated by the optical wave is amplified to a greater extent since this factor rises in value [41]. As the gain component increases, the signal-to-noise ratio of the input drops [42]. As a result, nirslab may identify channels with gain factors greater than a preset value and reject them from further processing and analysis. The ordered pair (1.8477 1.7928) reflects the values of a metric we employed to quantify the raw data's signal-to-noise ratio [45,46]. The coefficient of variation (CV) is the measure. Since there are different measuring wavelengths, two CV values are reported. The sampling frequency is 78.1 Hz. After the analysis presented earlier, a framework is presented here to overcome the issues contributing to the bad and unsteady signals from the fNIRS sensor [8,34].

4.2. Motion Classification Accuracy

The classification results were then analysed subject-wise and then the average accuracy of all subjects was calculated. The difference in the classification accuracy is evidence of the importance of signal acquisition and processing procedure. The reliability of the results is very well dependant on how the signals were acquired and further processed [8,47]. A noisy and disturbing signal resulted in error, while the refined signals were easy to handle by the classifier algorithms and hence a minimum error was generated [47,52,57]. Furthermore, a *t*-test was applied to student participants. This test checks the statistical significance of the attained results [54]. The confidence interval is specified at 95% ($p < 0.05$). A quantifiable comparison between healthy subjects and amputees was not possible due to a restricted number of amputees. However, the computed p-value is 0.0248, with a 95% confidence interval within healthy subjects.

The network used a sigmoid function for gradient descent. A total of 60% of the samples were fed for training and 20% samples were utilized by the network for testing and validation each. Each step was calculated in 320 µs with a 3 s epoch completion time. Then, a confusion matrix was extracted as soon as the training ends, which not only indicates the number of samples, those that were accurately classified, but also the false samples that generated an error. The number of hidden neurons was set to 20. Twelve neurons were present in each of the transitional hidden layers. The neuron number in the output layer is specified as six.

The healthy subject-wise accuracy is illustrated in Table 2, whereas Table 3 represents the accuracy of amputed subjects.

Table 2. Offline classification accuracies of fifteen healthy subjects using single features signal mean (SM), signal peak (SP), signal minimum (SMin) and waveform length features for EMG and fNIRS using LDA and ANN Classifiers.

Features	S1	S2	S3	S4	S5
SM	72.88	61.1	74.89	68.63	69.63
SP	67.85	68.89	76.9	69.03	70.03
SMin	74.99	63.84	73.49	68.13	69.13
	S6	S7	S8	S9	S10
SM	75.04	75.38	77.15	67.86	74.56
SP	72.22	72.28	65.38	75.77	69.76
SMin	77	74.44	71.15	71.75	66.62
	S11	S12	S13	S14	S15
SM	66.78	64.76	60.37	71.54	79.6
SP	69.74	69.56	65.47	71.94	59.81
SMin	66.87	70.61	64.74	69.22	67.6

Table 3. Offline classification accuracies of three amputee subjects using single features signal mean (SM), signal peak (SP), signal minimum (SMin), and waveform length features for EMG and fNIRS using LDA and ANN Classifiers.

Features	A1	A2	A3
SM	69.26	61.65	57.05
SP	68.91	60.18	57.72
SMin	55.1	50.05	51.93

As stated in Section 1, no work has been conducted for transhumeral amputees in non-invasive manner to generate six number of control commands. However, the studies that used fNIRS for some other applications are presented below in tabular form for accuracy and number of control commands comparison. The study comparison is illustrated in Table 4.

Table 4. Performance evaluation and comparison with existing classification models.

Technique	Learning Method	Time Response	Number of Control Commands	Classification Accuracy
TD features [58]	LDA	5.5 s	2	72.82%
FD features [59]	LDA/SVM	15 s	2	83%
Raw fNIRS [22]	ANN	4 s	4	58%
TD features [60]	SVM	0.5 s	6	68.1%
Proposed framework	ANN	320 µs	6	78.65%

It can be seen from the studies above that, as the number of control commands increase, the accuracy decrease. However, the time response of classifiers has no trend. It is due to the fact that these studies have been conducted on the signal set acquired by third parties via online forums. In the proposed framework, the signals were acquired and then analyzed. Based on the conditions during the signal acquisition process, further steps were taken, such as filtration and channel selection. This makes a difference, as documented by [22,58]. The results show the potential usability of the presented framework in real-time applications and is a step towards enhanced motion prediction in BMI applications.

5. Conclusions

In this research study, an fNIRS-based approach was investigated to recognize the motion intention of the human upper limb. The fNIRS signals are acquired from the motor cortex region of the brain using NIRSport from NIRx Technology. The fNIRS signals were acquired for six arm motions. These motions included elbow extension (E.E), elbow flexion (E.F), wrist supination (W.S), wrist pronation (W.P), hand open (H.O), and hand close (H.C). Channel selection was conducted based on the gain values computed during signal acquisition. An FIR filter was applied to filter the samples. Signal mean, signal peak and minimum value were computed as a feature set. ANN classifier was trained for motion intention prediction. On average, the motion intention prediction was 78% ($p < 0.05$) and 64% accurate for healthy and amputated subjects, respectively. The highest accuracy for an individual subject was recorded as 79.6%. A possible extension of the presented work includes the framework design for accuracy enhancement and eliminating the channel selection complications. The application of the presented approach with the increased number of arm motions, incorporating individuals of different age groups, and the implementation of generated control commands to control a prosthetic arm device in a real-time setting are some other directions for future work.

Author Contributions: Conceptualization, N.Y.S., Z.K. and S.A.U.; methodology, N.Y.S. and S.A.U.; software, U.F. and S.A.U.; validation, Z.K., N.Y.S., S.A.U., U.F., M.F.S. and S.M.; formal analysis, S.A.U., M.B. and R.K.; investigation, S.A.U. and N.Y.S.; resources, Z.K., M.B. and R.K.; data curation, Z.K. and S.A.U.; writing—original draft preparation, Z.K., N.Y.S. and S.A.U.; writing—review and editing, M.F.S., M.B., R.K., U.F., S.A.U. and S.M.; visualization, S.M., S.A.U. and N.Y.S.; supervision, Z.K.; project administration, Z.K. and N.Y.S.; funding acquisition, M.B., R.K. and S.M. All authors have read and agreed to the published version of the manuscript.

Funding: This research was funded by the Higher Education Commission of Pakistan, grant number 10702. This research was partially funded by Future University in Egypt, New Cairo, Egypt.

Institutional Review Board Statement: The Air University Human Research Ethics Committee (AU HREC) operates in compliance with The World Medical Association Declaration of Helsinki for Ethics Principles for Medical Research Involving Human Subjects.

Informed Consent Statement: I have read the Participant Information Sheet; have understood the nature of the research and why I have been selected. I have had the opportunity to ask questions and have them answered to my satisfaction. I have chosen to participate in this research voluntarily.

Data Availability Statement: It is a funded project and hence data is not publicly available. However, the data can be made available upon request.

Acknowledgments: We would like to show gratefulness to our friends who linked us with the transhumeral amputees. We would like to thank our funders i.e., HEC, Pakistan for their support.

Conflicts of Interest: The authors declare no conflict of interest. The funders had no role in the design of the study; in the collection, analyses, or interpretation of data; in the writing of the manuscript, or in the decision to publish the results.

References

1. Cordella, F.; Ciancio, A.L.; Sacchetti, R.; Davalli, A.; Cutti, A.G.; Guglielmelli, E.; Zollo, L. Literature Review on Needs of Upper Limb Prosthesis Users. *Front. Neurosci.* **2016**, *10*, 209. [CrossRef]
2. Ribeiro, J.; Mota, F.; Cavalcante, T.; Nogueira, I.; Gondim, V.; Albuquerque, V.; Alexandria, A. Analysis of Man-Machine Interfaces in Upper-Limb Prosthesis: A Review. *Robotics* **2019**, *8*, 16. [CrossRef]
3. Hussain, S.; Shams, S.; Khan, S.J. Impact of Medical Advancement: Prostheses. In *Computer Architecture in Industrial, Biomechanical and Biomedical Engineering*; IntechOpen: London, UK, 2019. [CrossRef]
4. Neelum, Y.S.; Kausar, Z.; Usama, S.A. Reference position estimation for prosthetic elbow and wrist using EMG signals. In *IOP Conference Series: Materials Science and Engineering*; IOP Publishing: Bristol, UK, 2019; Volume 635, p. 012031. [CrossRef]
5. Resnik, L.; Klinger, S.L.; Etter, K. The DEKA Arm: Its features, functionality, and evolution during the Veterans Affairs Study to optimize the DEKA Arm. *Prosthet. Orthot. Int.* **2014**, *38*, 492–504. [CrossRef]
6. Lenzi, T.; Lipsey, J.; Sensinger, J.W. The RIC Arm—A Small Anthropomorphic Transhumeral Prosthesis. *IEEE/ASME Trans. Mechatron.* **2016**, *21*, 2660–2671. [CrossRef]
7. Islam, M.A.; Sundaraj, K.; Ahmad, R.B.; Ahamed, N.U.; Ali, M.A. Mechanomyography sensor development, related signal processing, and applications: A systematic review. *IEEE Sens. J.* **2013**, *13*, 2499–2516. [CrossRef]
8. Bennett, D.A.; Goldfarb, M. IMU-Based Wrist Rotation Control of a Transradial Myoelectric Prosthesis. *IEEE Trans. Neural Syst. Rehabil. Eng.* **2017**, *26*, 419–427. [CrossRef] [PubMed]

9. Syed, U.A.; Kausar, Z.; Sattar, N.Y. Control of a Prosthetic Arm Using fNIRS, a Neural-Machine Interface. In *Data Acquisition-Recent Advances and Applications in Biomedical Engineering*; IntechOpen: London, UK, 2020.
10. Alshammary, N.A.; Bennett, D.A.; Goldfarb, M. Synergistic Elbow Control for a Myoelectric Transhumeral Prosthesis. *IEEE Trans. Neural Syst. Rehabil. Eng.* **2017**, *26*, 468–476. [CrossRef]
11. Sattar, N.Y.; Syed, U.A.; Muhammad, S.; Kausar, Z. Real-Time EMG Signal Processing with Implementation of PID Control for Upper-Limb Prosthesis. In Proceedings of the 2019 IEEE/ASME International Conference on Advanced Intelligent Mechatronics (AIM), Hong Kong, China, 8–12 July 2019; pp. 120–125. [CrossRef]
12. Jarrassé, N.; Nicol, C.; Touillet, A.; Richer, F.; Martinet, N.; Paysant, J.; de Graaf, J.B. Classification of phantom finger, hand, wrist, and elbow voluntary gestures in transhumeral amputees with sEMG. *IEEE Trans. Neural Syst. Rehabil. Eng.* **2016**, *25*, 71–80. [CrossRef] [PubMed]
13. Oda, Y.; Sato, T.; Nambu, I.; Wada, Y. Real-Time Reduction of Task-Related Scalp-Hemodynamics Artifact in Functional Near-Infrared Spectroscopy with Sliding-Window Analysis. *Appl. Sci.* **2018**, *8*, 149. [CrossRef]
14. Yamada, Y.; Suzuki, H.; Yamashita, Y. Time-Domain Near-Infrared Spectroscopy and Imaging: A Review. *Appl. Sci.* **2019**, *9*, 1127. [CrossRef]
15. Li, X.; Samuel, O.W.; Zhang, X.; Wang, H.; Fang, P.; Li, G. A motion-classification strategy based on sEMG-EEG signal combination for upper-limb amputees. *J. Neuroeng. Rehabil.* **2017**, *14*, 3. [CrossRef]
16. Bonilauri, A.; Intra, F.S.; Pugnetti, L.; Baselli, G.; Baglio, F. A Systematic Review of Cerebral Functional Near-Infrared Spectroscopy in Chronic Neurological Diseases—Actual Applications and Future Perspectives. *Diagnostics* **2020**, *10*, 581. [CrossRef]
17. Banville, H.; Falk, T. Recent advances and open challenges in hybrid brain-computer interfacing: A technological review of non-invasive human research. *Brain-Comput. Interfaces* **2016**, *3*, 9–46. [CrossRef]
18. Yao, L.; Meng, J.; Zhang, D.; Sheng, X.; Zhu, X. Combining Motor Imagery with Selective Sensation toward a Hybrid-Modality BCI. *IEEE Trans. Biomed. Eng.* **2013**, *61*, 2304–2312. [CrossRef]
19. Herold, F.; Wiegel, P.; Scholkmann, F.; Müller, N.G. Applications of functional near-infrared spectroscopy (fNIRS) neuroimaging in Exercise–Cognition science: A systematic, Methodology-Focused review. *J. Clin. Med.* **2018**, *7*, 466. [CrossRef] [PubMed]
20. Jian, C.; Deng, L.; Liang, L.; Luo, J.; Wang, X.; Song, R. Neuromuscular Control of the Agonist–Antagonist Muscle Coordination Affected by Visual Dimension: An EMG-fNIRS Study. *IEEE Access* **2020**, *8*, 100768–100777. [CrossRef]
21. Abitan, H.; Bohr, H.; Buchhave, P. Correction to the Beer-Lambert-Bouguer law for optical absorption. *Appl. Opt.* **2008**, *47*, 5354–5357. [CrossRef]
22. Herold, F.; Wiegel, P.; Scholkmann, F.; Thiers, A.; Hamacher, D.; Schega, L. Functional near-infrared spectroscopy in movement science: A systematic review on cortical activity in postural and walking tasks. *Neurophotonics* **2017**, *4*, 041403. [CrossRef]
23. Pfeifer, M.D.; Scholkmann, F.; Labruyère, R. Signal Processing in Functional Near-Infrared Spectroscopy (fNIRS): Methodological Differences Lead to Different Statistical Results. *Front. Hum. Neurosci.* **2018**, *11*, 641. [CrossRef]
24. Phinyomark, A.; Scheme, E. A feature extraction issue for myoelectric control based on wearable EMG sensors. In Proceedings of the 2018 IEEE Sensors Applications Symposium (SAS), Seoul, Korea, 12–14 March 2018; pp. 1–6. [CrossRef]
25. Farina, D.; Merletti, R.; Enoka, R.M. The extraction of neural strategies from the surface EMG. *J. Appl. Physiol.* **2004**, *96*, 1486–1495. [CrossRef]
26. Scholkmann, F.; Wolf, M. Measuring brain activity using functional near infrared spectroscopy: A short review. *Spectrosc. Eur.* **2012**, *24*, 6.
27. Rocon, E.; Gallego, J.A.; Barrios, L.; Victoria, A.R.; Ibánez, J.; Farina, D.; Negro, F.; Dideriksen, J.L.; Conforto, S.; D'Alessio, T.; et al. Multimodal BCI-mediated FES suppression of pathological tremor. In Proceedings of the 2010 Annual International Conference of the IEEE Engineering in Medicine and Biology, Buenos Aires, Argentina, 31 August–4 September 2010; pp. 3337–3340.
28. Pinti, P.; Aichelburg, C.; Gilbert, S.; Hamilton, A.; Hirsch, J.; Burgess, P.; Tachtsidis, I. A Review on the Use of Wearable Functional Near-Infrared Spectroscopy in Naturalistic Environments. *Jpn. Psychol. Res.* **2018**, *60*, 347–373. [CrossRef]
29. Lloyd-Fox, S.; Blasi, A.; Elwell, C. Illuminating the developing brain: The past, present and future of functional near infrared spectroscopy. *Neurosci. Biobehav. Rev.* **2010**, *34*, 269–284. [CrossRef] [PubMed]
30. World Medical Association. WMA Declaration of Helsinski–Ethical Principles for Medical Research Involving Human Subjects. *JAMA* **2013**, *310*, 2191–2194. [CrossRef]
31. Leeb, R.; Sagha, H.; Chavarriaga, R. Multimodal fusion of muscle and brain signals for a hybrid-BCI. In Proceedings of the 2010 Annual International Conference of the IEEE Engineering in Medicine and Biology, Buenos Aires, Argentina, 31 August–4 September 2010; pp. 4343–4346.
32. Buccino, A.P.; Keles, H.O.; Omurtag, A. Hybrid EEG-fNIRS asynchronous brain-computer interface for multiple motor tasks. *PLoS ONE* **2016**, *11*, e0146610.
33. Ortega, P.; Zhao, T.; Faisal, A.A. HYGRIP: Full-Stack Characterization of Neurobehavioral Signals (fNIRS, EEG, EMG, Force, and Breathing) During a Bimanual Grip Force Control Task. *Front. Neurosci.* **2020**, *14*, 919. [CrossRef]
34. Aryadoust, V.; Foo, S.; Ng, L.Y. What can gaze behaviors, neuroimaging data, and test scores tell us about test method effects and cognitive load in listening assessments? *Lang. Test.* **2021**, *39*, 56–89. [CrossRef]
35. Maira, G.; Chiarelli, A.M.; Brafa, S.; Libertino, S.; Fallica, G.; Merla, A.; Lombardo, S. Imaging System Based on Silicon Photomultipliers and Light Emitting Diodes for Functional Near-Infrared Spectroscopy. *Appl. Sci.* **2020**, *10*, 1068. [CrossRef]
36. Ramadan, R.A.; Vasilakos, A.V. Brain computer interface: Control signals review. *Neurocomputing* **2017**, *223*, 26–44. [CrossRef]

37. Kim, M. Shedding Light on the Human Brain. *Opt. Photon-News* **2021**, *32*, 26–33. [CrossRef]
38. Geissler, C.F.; Schneider, J.; Frings, C. Shedding light on the prefrontal correlates of mental workload in simulated driving: A functional near-infrared spectroscopy study. *Sci. Rep.* **2021**, *11*, 705. [CrossRef] [PubMed]
39. Lamberti, N.; Manfredini, F.; Baroni, A.; Crepaldi, A.; Lavezzi, S.; Basaglia, N.; Straudi, S. Motor Cortical Activation Assessment in Progressive Multiple Sclerosis Patients Enrolled in Gait Rehabilitation: A Secondary Analysis of the RAGTIME Trial Assisted by Functional Near-Infrared Spectroscopy. *Diagnostics* **2021**, *11*, 1068. [CrossRef]
40. Guo, W.; Sheng, X.; Liu, H.; Zhu, X. Toward an Enhanced Human–Machine Interface for Upper-Limb Prosthesis Control With Combined EMG and NIRS Signals. *IEEE Trans. Hum.-Mach. Syst.* **2017**, *47*, 564–575. [CrossRef]
41. Feng, N.; Hu, F.; Wang, H.; Gouda, M.A. Decoding of voluntary and involuntary upper-limb motor imagery based on graph fourier transform and cross-frequency coupling coefficients. *J. Neural Eng.* **2020**, *17*, 056043. [CrossRef] [PubMed]
42. Leff, D.; Orihuela-Espina, F.; Elwell, C.; Athanasiou, T.; Delpy, D.T.; Darzi, A.W.; Yang, G.-Z. Assessment of the cerebral cortex during motor task behaviours in adults: A systematic review of functional near infrared spectroscopy (fNIRS) studies. *NeuroImage* **2011**, *54*, 2922–2936. [CrossRef]
43. Borrell, J.A.; Copeland, C.; Lukaszek, J.L.; Fraser, K.; Zuniga, J.M. Use-Dependent Prosthesis Training Strengthens Contralateral Hemodynamic Brain Responses in a Young Adult with Upper Limb Reduction Deficiency: A Case Report. *Front. Neurosci.* **2021**, *15*, 693138. [CrossRef]
44. Matarasso, A.K.; Rieke, J.D.; White, K.; Yusufali, M.M.; Daly, J.J. Combined real-time fMRI and real time fNIRS brain computer interface (BCI): Training of volitional wrist extension after stroke, a case series pilot study. *PLoS ONE* **2021**, *16*, e0250431. [CrossRef]
45. Luo, J.; Shi, W.; Lu, N.; Wang, J.; Chen, H.; Wang, Y.; Lu, X.; Wang, X.; Hei, X. Improving the performance of multisubject motor imagery-based BCIs using twin cascaded softmax CNNs. *J. Neural Eng.* **2021**, *18*, 036024. [CrossRef]
46. Ang, K.K.; Guan, C.; Chua, K.S.G.; Ang, B.T.; Kuah, C.; Wang, C.; Phua, K.S.; Chin, Z.Y.; Zhang, H. A clinical study of motor imagery-based brain-computer interface for upper limb robotic rehabilitation. In Proceedings of the 2009 Annual International Conference of the IEEE Engineering in Medicine and Biology Society, Minneapolis, MN, USA, 3–6 September 2009; pp. 5981–5984. [CrossRef]
47. Wen, Y.; Avrillon, S.; Hernandez-Pavon, J.C.; Kim, S.J.; Hug, F.; Pons, J.L. A convolutional neural network to identify motor units from high-density surface electromyography signals in real time. *J. Neural Eng.* **2021**, *18*, 056003. [CrossRef]
48. Prôa, R.; Balardin, J.; de Faria, D.D.; Paulo, A.M.; Sato, J.R.; Baltazar, C.A.; Borges, V.; Silva, S.M.C.A.; Ferraz, H.B.; Aguiar, P.D.C. Motor Cortex Activation During Writing in Focal Upper-Limb Dystonia: An fNIRS Study. *Neurorehabilit. Neural Repair* **2021**, *35*, 729–737. [CrossRef]
49. Li, G.; Yuan, Y.; Ren, H.; Chen, W. fNIRS study of effects of foot bath on human brain and cognitive function. *J. Mech. Med. Biol.* **2021**, *21*, 2140022. [CrossRef]
50. Gusnard, D.A.; Raichle, M.E. Searching for a baseline: Functional imaging and the resting human brain. *Nat. Rev. Neurosci.* **2001**, *2*, 685–694. [CrossRef] [PubMed]
51. Gomez-Gil, J.; San-Jose-Gonzalez, I.; Nicolas-Alonso, L.F.; Alonso-Garcia, S. Steering a Tractor by Means of an EMG-Based Human-Machine Interface. *Sensors* **2011**, *11*, 7110–7126. [CrossRef]
52. Sitaram, R.; Zhang, H.; Guan, C.; Thulasidas, M.; Hoshi, Y.; Ishikawa, A.; Shimizu, K.; Birbaumer, N. Temporal classification of multichannel near-infrared spectroscopy signals of motor imagery for developing a brain–computer interface. *NeuroImage* **2007**, *34*, 1416–1427. [CrossRef]
53. Zimmermann, R.; Marchal-Crespo, L.; Edelmann, J.; Lambercy, O.; Fluet, M.C.; Riener, R.; Wolf, M.; Gassert, R. Detection of motor execution using a hybrid fNIRS-biosignal BCI: A feasibility study. *J. Neuroeng. Rehabil.* **2013**, *10*, 4. [CrossRef] [PubMed]
54. Yoo, S.-H.; Santosa, H.; Kim, C.-S.; Hong, K.-S. Decoding Multiple Sound-Categories in the Auditory Cortex by Neural Networks: An fNIRS Study. *Front. Hum. Neurosci.* **2021**, *15*, 211. [CrossRef]
55. Vélez-Guerrero, M.; Callejas-Cuervo, M.; Mazzoleni, S. Artificial Intelligence-Based Wearable Robotic Exoskeletons for Upper Limb Rehabilitation: A Review. *Sensors* **2021**, *21*, 2146. [CrossRef]
56. Medina, F.; Perez, K.; Cruz-Ortiz, D.; Ballesteros, M.; Chairez, I. Control of a hybrid upper-limb orthosis device based on a data-driven artificial neural network classifier of electromyography signals. *Biomed. Signal Process. Control.* **2021**, *68*, 102624. [CrossRef]
57. Holtzer, R.; Verghese, J.; Allali, G.; Izzetoglu, M.; Wang, C.; Mahoney, J.R. Neurological Gait Abnormalities Moderate the Functional Brain Signature of the Posture First Hypothesis. *Brain Topogr.* **2015**, *29*, 334–343. [CrossRef] [PubMed]
58. Su, Y.; Li, W.; Bi, N.; Lv, Z. Adolescents Environmental Emotion Perception by Integrating EEG and Eye Movements. *Front. Neurorobotics* **2019**, *13*, 46. [CrossRef] [PubMed]
59. Fazli, S.; Mehnert, J.; Steinbrink, J.; Curio, G.; Villringer, A.; Mueller, K.-R.; Blankertz, B. Enhanced performance by a hybrid NIRS–EEG brain computer interface. *NeuroImage* **2011**, *59*, 519–529. [CrossRef] [PubMed]
60. Witkowski, M.; Cortese, M.; Cempini, M.; Mellinger, J.; Vitiello, N.; Soekadar, S.R. Enhancing brain-machine interface (BMI) control of a hand exoskeleton using electrooculography (EOG). *J. Neuron. Rehabil.* **2014**, *11*, 165. [CrossRef] [PubMed]

Article

Classification of Individual Finger Movements from Right Hand Using fNIRS Signals

Haroon Khan [1], Farzan M. Noori [2], Anis Yazidi [3,4,5], Md Zia Uddin [6], M. N. Afzal Khan [7] and Peyman Mirtaheri [1,8,*]

1 Department of Mechanical, Electronics and Chemical Engineering, OsloMet-Oslo Metropolitan University, 0167 Oslo, Norway; haroonkh@oslomet.no
2 Department of Informatics, University of Oslo, 0315 Oslo, Norway; farzanmn@ifi.uio.no
3 Department of Computer Science, OsloMet-Oslo Metropolitan University, 0167 Oslo, Norway; anisy@oslomet.no
4 Department of Neurosurgery, Oslo University Hospital, 0450 Oslo, Norway
5 Department of Computer Science, Norwegian University of Science and Technology, 7491 Trondheim, Norway
6 Software and Service Innovation, SINTEF Digital, 0373 Oslo, Norway; zia.uddin@sintef.no
7 School of Mechanical Engineering, Pusan National University, Busan 46241, Korea; nasirafzal@pusan.ac.kr
8 Department of Biomedical Engineering, Michigan Technological University, Houghton, MI 49931, USA
* Correspondence: peymanm@oslomet.no

Citation: Khan, H.; Noori, F.M.; Yazidi, A.; Uddin, M.Z.; Khan, M.N.A.; Mirtaheri, P. Classification of Individual Finger Movements from Right Hand Using fNIRS Signals. *Sensors* **2021**, *21*, 7943. https://doi.org/10.3390/s21237943

Academic Editor: Tara Julia Hamilton

Received: 8 November 2021
Accepted: 26 November 2021
Published: 28 November 2021

Publisher's Note: MDPI stays neutral with regard to jurisdictional claims in published maps and institutional affiliations.

Abstract: Functional near-infrared spectroscopy (fNIRS) is a comparatively new noninvasive, portable, and easy-to-use brain imaging modality. However, complicated dexterous tasks such as individual finger-tapping, particularly using one hand, have been not investigated using fNIRS technology. Twenty-four healthy volunteers participated in the individual finger-tapping experiment. Data were acquired from the motor cortex using sixteen sources and sixteen detectors. In this preliminary study, we applied standard fNIRS data processing pipeline, i.e. optical densities conversation, signal processing, feature extraction, and classification algorithm implementation. Physiological and non-physiological noise is removed using 4th order band-pass Butter-worth and 3rd order Savitzky–Golay filters. Eight spatial statistical features were selected: signal-mean, peak, minimum, Skewness, Kurtosis, variance, median, and peak-to-peak form data of oxygenated haemoglobin changes. Sophisticated machine learning algorithms were applied, such as support vector machine (SVM), random forests (RF), decision trees (DT), AdaBoost, quadratic discriminant analysis (QDA), Artificial neural networks (ANN), k-nearest neighbors (kNN), and extreme gradient boosting (XG-Boost). The average classification accuracies achieved were 0.75 ± 0.04, 0.75 ± 0.05, and 0.77 ± 0.06 using k-nearest neighbors (kNN), Random forest (RF) and XGBoost, respectively. KNN, RF and XGBoost classifiers performed exceptionally well on such a high-class problem. The results need to be further investigated. In the future, a more in-depth analysis of the signal in both temporal and spatial domains will be conducted to investigate the underlying facts. The accuracies achieved are promising results and could open up a new research direction leading to enrichment of control commands generation for fNIRS-based brain-computer interface applications.

Keywords: functional near-infrared spectroscopy (fNIRS); finger-tapping; classification; motor cortex; machine learning

1. Introduction

Functional near-infrared spectroscopy (fNIRS) is a portable and non-invasive brain imaging modality for continuous measurement of haemodynamics in the cerebral cortex of the human brain [1]. Over the last decade, the method has gained popularity due to its acceptable temporal and spatial resolutions, and its easy-to-use, safe, portable, and affordable monitoring compared to other neuroimaging modalities [2]. fNIRS has been used to monitor a variety of cognitive activities, such as attention, problem-solving, working memory, and gait rehabilitation [3]. The underlying theory behind fNIRS functionality is based on optical spectroscopy and neurovascular coupling [1,4]. Optical spectroscopy uses the interaction of light with matter to measure certain characteristics of molecular structures, while neurovascular coupling defines the relationship between local neuronal activity and subsequent changes in cerebral blood flow due to cerebral activity [5–7]. It is known that most of the biological tissue is transparent to the near-infrared range (700–900 nm). The near-infrared window commonly used in fNIRS is 690–860 nm [8]. Haemoglobin is a protein that is responsible for delivering oxygen throughout the body via red blood cells. This protein is the major absorbent within the near-infrared range of light (def. 700–1100 nm). In summary, the continuous-wave fNIRS machine uses two near-infrared wavelengths to measure the relative change in oxygenated haemoglobin (ΔHbO) and deoxygenated haemoglobin (ΔHbR) in cerebral activation.

The most common brain areas studied in neuroimaging are the cerebral prefrontal and motor cortex, particularly for cognitive and motor tasks [9,10]. Since the beginning of the 19th century, the finger-tapping test has been used in various brain studies to assess the motor abilities and accessory muscular control [11]. Various brain and non-brain signals were obtained during the finger-tapping task to access the motor abilities and differentiated movements. Investigating finger movements is particularly important in the field of the brain-computer interface to decode the neurophysiological signal and generate control commands for external devices [9,12]. Individual finger movements were classified with an average accuracy of 85% using electromyogram (EMG) bio-signals while performing finger-tapping tasks [13]. Similarly, in another study using surface EMG, individual and combined finger movements were classified with an average accuracy of 98% on healthy and 90% in below-elbow amputee persons [14]. These higher classification accuracies of finger movements may be best for prosthetic hand development. Other modalities predicting dexterous individual finger movements include ultrasound imaging from the forehand and differentiating finger movements with a higher precision of 98% accuracy [15]. Most brain imaging modalities are limited to the movement of larger body parts, such as the upper and lower limbs. However, it is essential to decode dexterous functions from brain signals in case where other types of brain imaging are difficult to implement. Among invasive brain signals, electrocorticography (ECoG) was shown to differentiate between individual finger movements with acceptable classification accuracies [12,16,17]. However, to the best of the author's knowledge, only one study was found during a literature review that utilized noninvasive brain signals, i.e., electroencephalography (EEG) signals, to decode individual finger movements. The study found a broadband power increase and low-frequency-band power decrease in finger flexion and extension data when EEG power spectra were decomposed in principal components using principal component analysis (PCA). The average decoding accuracy over all subjects was 77.11% obtained with the binary classification of each pair of fingers from one hand using movement-related spectral changes and a support vector machine (SVM) classifier.

The prevalent motor execution task in fNIRS-based studies includes tapping of one or more fingers, single hand-tapping, both hand-tapping, right and left finger-tapping and hand-tapping. In the study, left and right index finger-tapping was distinguished with a classification accuracy of 85.4% using features from the vector-based phase and linear discriminant analysis [18]. In [19], three different tasks, i.e. right and left-hand unilateral complex finger-tapping, and foot-tapping, were performed. The classification accuracy achieved using SVM was 70.4% for the three-class problem. In single-trail classification for a motor imaginary with thumb and complex finger-tapping task achieves an average accuracy of 81% by simply changing the combination of a set of channels, time intervals, and features [20]. In [21] thumb and little finger were classified with an accuracy of 87.5% for ΔHbO data. Deep learning approaches are also becoming popular for the classification of these complex finger movements. In a study [22], using conditional generative adversarial networks (CGAN) in combination with convolutional neural networks (CNN), the left finger, right finger, and foot-tapping tasks were differentiated with higher classification accuracy of 96.67%. In one of the recent studies, left and right index finger-tapping were distinguished with a different tapping frequency using multilabeling and deep learning [23]. Different labels were assigned to right and left finger-tapping with different tapping frequencies labels such as rest, 80 bpm, and 120 bpm. With this complex combination using deep learning approach the average classification accuracy achieved was 81%. The aforementioned studies are difficult to compare since different models and finger-tapping exercises were conducted. However, according to the literature, the differentiation of finger movement patterns is very challenging using fNIRS. This fact is supported by legacy studies that show that there is no significant statistical difference between fNIRS signals recorded from primary- and pre-motor cortices during sequential finger-tapping and whole-hand grasping [24]. Furthermore, the dynamic relationship between the simultaneously activated brain regions during the motor task is becoming better understood. An interesting study conducted by Anwar et al. [25,26] describes the effective connectivity of the information flow in the sensorimotor cortex, premotor cortex, and contralateral dorsolateral prefrontal cortex during different finger movement tasks using multiple modalities such as fNIRS, fMRI, and EEG. It was found that there is an adequate bi-directional information flow between the cortices mentioned above. The study also concluded that, compared to fMRI, fNIRS is an attractive and easy to use alternative with an excellent spatial resolution for studying connectivity. In this perspective, multi-modal fNIRS-EEG is also an appealing alternative to fMRI. Hence, it is essential to study the flow and connectivity of individual finger movement from the motor cortex using fNIRS or multi-model integration of EEG-fNIRS. The multi-model EEG-fNIRS integration was shown to enhance classification accuracy [27], increase the number of control commands, and reduce the signal-processing time [4,28].

It has been unclear whether fNIRS signals have enough information to differentiate between individual finger movements. Some underlying limitations of fNIRS may be the reason for this drawback, such as comparatively low temporal resolution (1–10 Hz for commercially available portable devices), depth sensitivity of about 1.5 cm (depending upon source-detector distance, which is typically 3 cm), and spatial resolution up to 1 cm [29]. To shed light on this research area, the study is conducted to investigate the detection of individual finger-tapping tasks using fNIRS. Also, the study is a step forward towards understanding the dynamic relationship between the brain regions that are simultaneously activated during motor tasks. We believe that the advances made in sophisticated machine learning algorithms could help to identify individual finger movements from potential fNIRS signals. This study is structured and reported in accordance with the guidelines published in [30]. The following sections will address materials and methods (Section 2), results and discussion (Section 3) and conclusion (Section 4).

2. Materials and Methods

The section on materials and methods describes procedure followed during experimental design, data collection, and processing.

2.1. Participants

Twenty-four healthy right-handed participants, 18 males (M) and 6 females (F), selected from random university population participated in the experiment. The ages of the participants were for male (mean age ± standard deviation; range) (M = 30.44 ± 3.03; range: 24–34 years), and female (F = 29.17 ± 3.06; range: 24–34 years). The healthy young participants were selected in the age range of 25–35 years because the frequency of finger-tapping can vary between different age groups. The inclusion criterion for right-handedness was that the participants had to write with the right hand. The participants had normal vision or corrected to normal vision. Exclusion criteria include neurological disorders or limitation of motor abilities in any hands or finger. For ethical statements, please see Section 4.

2.2. Instrumentation

A continuous-wave optical tomography machine NIRScout (NIRx Medizintechnik GmbH, Germany) was used to spontaneously acquire brain data at one of the laboratories under the ADEPT (Advanced intelligent health and brain-inspired technologies) research group at Oslo Metropolitan University, Norway. The data acquisition used two wavelengths, i.e., 760 nm (λ_1) and 850 nm (λ_1) with a sampling rate of 3.9063 Hz.

2.3. Experimental Setup and Instructions

The experiment was performed in a relatively controlled environment. The environmental light, including monitor screen brightness, was shielded to minimise any influences during stimuli changes in the presentation. A black over-cap was used to further reduce the effect of surrounding light further, as shown in Figure 1C. The experiment was conducted in a noise-free room. A visual presentation of resting and task (finger-tapping corresponding to each finger) was displayed on the computer monitor to the participants. Before starting the actual experiment, the participants were given implicit instructions about the experimental protocol and procedure. Practice sessions were conducted before the experiment. The finger-tapping task was performed at a medium-to-fast pace but not with any specific frequency. The number of repetitions of experiments for each participant was dependent upon the comfort and convenience of the participants. No investigation was conducted during any inconvenience and discomfort experienced by the participant, resulting in unwanted signals such as frustration interference in brain signals. Data were acquired using commercial NIRx software NIRStar 15.1. The complete experimental setup is shown in Figure 1.

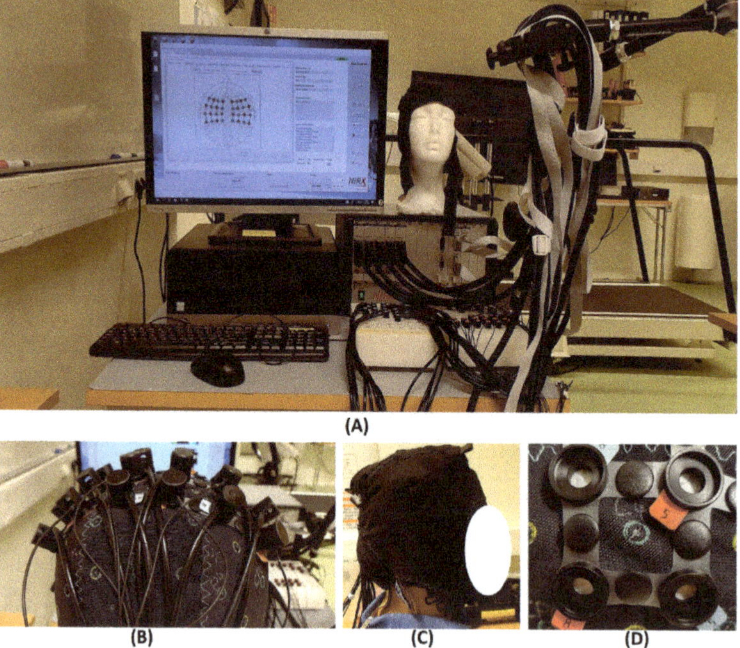

Figure 1. (**A**) Experimental setup; (**B**) optodes arrangement; (**C**) overcap to reduce external light; (**D**) optodes holder.

2.4. Experimental Design

The experiments were designed using blocks of rest (initial rest, final rest, and inter-stimulus rest) and task (thumb, index middle, ring, and little finger-tapping) of the right hand as shown in Figure 2. An optimal baseline period of 30 s was set up before and after the first and last task, respectively. The stimuli duration necessary to acquire an adequate and robust haemodynamic response corresponding to finger-tapping activity was set to 10 s [31]. The single experimental paradigm consists of three sessions of each finger tapping trial. The total length of an experiment was 350 s. The single trial includes 10 s rest followed by 10 s of the task. Experiments were repeated for each participant from one to three times in a single day depending upon his/her comfortability. The rest and task blocks were presented using NIRX stimulation software NIRStim 4.0.

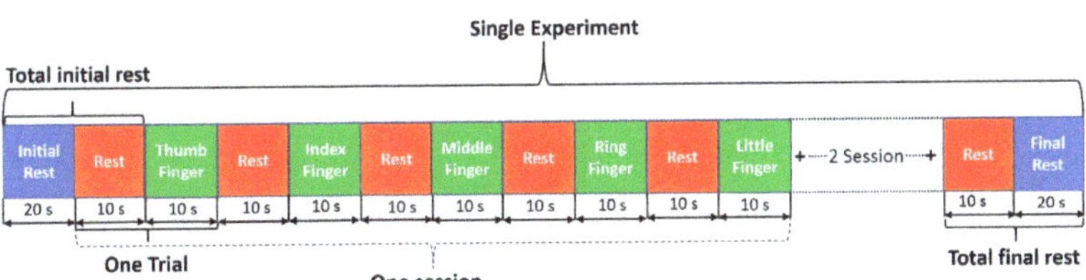

Figure 2. Experimental paradigm visualization. Single experiment consists of three sessions of each finger tapping trail. Single trial consists of 10 s task and 10 s finger tapping.

2.5. Brain Area and Montage Selection

Before placing the NIRScap on the participant's head, cranial landmarks (inion and nasion) were marked to locate C_z. The emitter and detector were placed in accordance with 10-5 international positioning layout. The distance between source and detector was kept at 3 cm using optode holders. Sixteen emitters and sixteen detectors were placed over the motor cortex in accordance with standard $motor_{16x16}$ montage available in NIRStar v15.2, as shown in Figure 3A,B. The source-detectors were placed over the frontal lobe (F1, F2, F3, F4, F5, F6, F7, and F8), frontal-central sulcus lobe (FC1, FC2, FC3, FC4, FC5, and FC6), central sulcus lobe (C1, C2, C3, C4, C5, C6), central-parietal lobe (CP1, CP2, CP3, CP4, CP5, and CP6), and the temporal-parietal lobe (T7, T8, TP7 and TP8). The data were collected from both the left and right hemispheres for further research work. However, in this particular work, only the channels of the left hemisphere were only further analysed.

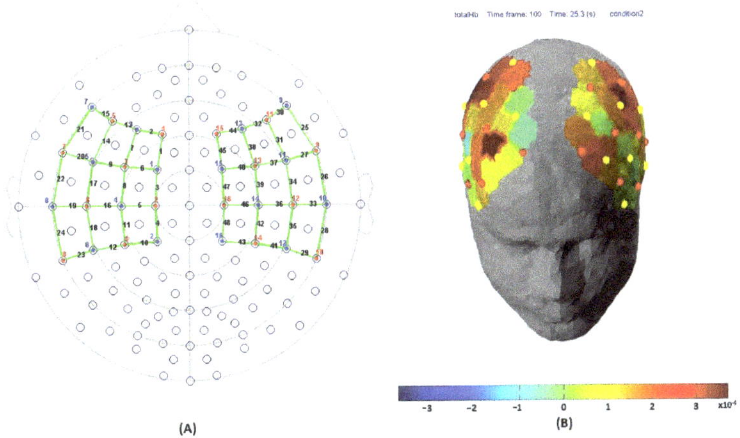

Figure 3. (**A**) Source-detector placement over motor cortex. Figure 3A Colour code: Red (source), Blue (detector), Green (channels), and black colour represent channel numbers. (**B**) Demonstration of total haemoglobin changes over motor cortex during index finger tapping.

2.6. Signal Prepossessing

Signal processing was performed using commercial fNIRS data processing software nirsLAB (v2019014) [32] and Matlab®. Signal were pre-processed in nirsLAB, for diverse tasks such as removing discontinuities, spikes, and truncation of the data points before and after the first and last stimuli appeared, respectively. Bad channels were identified using the criterion of the gain setting of three and coefficient of variation (CV) of 7.5% in nirsLAB. The coefficient of variation is equal to a hundred times the standard deviation divided by the mean value of the raw data measurements. A large value for CV is an indication of high noise. The gain setting was set to eight for all the data processed. Optical densities were converted into haemoglobin concentration change using Modified Beer–Lambert Law in nirsLAB (see details in Section 2.7).

2.7. Modified Beer–Lambert Law (MBLL)

The changes in optical densities were converted using MBLL into ΔHbO (Equation (1a)) and ΔHbR (Equation (1b)). The parameter for MBLL, such as the differential path length factor (DPF) and molar extinction coefficients (using standard W.B Gratzer spectrum) for ΔHbO and ΔHbR, are shown in Table 1. The molar concentration and MVO2Sat value are set as 75 μM and 70%, respectively.

$$\Delta HbO^i(k) = \frac{\left(\varepsilon_{\Delta HbR}^{\lambda_1}\frac{\Delta OD^{\lambda_2}(k)}{DPF^{\lambda_2}}\right) - \left(\varepsilon_{\Delta HbR}^{\lambda_2}\frac{\Delta OD^{\lambda_1}(k)}{DPF^{\lambda_1}}\right)}{l^i\left(\varepsilon_{\Delta HbR}^{\lambda_1}\varepsilon_{\Delta HbO}^{\lambda_2} - \varepsilon_{\Delta HbR}^{\lambda_2}\varepsilon_{\Delta HbO}^{\lambda_1}\right)} \tag{1a}$$

$$\Delta HbR^i(k) = \frac{\left(\varepsilon_{\Delta HbO}^{\lambda_2}\frac{\Delta OD^{\lambda_1}(k)}{DPF^{\lambda_1}}\right) - \left(\varepsilon_{\Delta HbO}^{\lambda_1}\frac{\Delta OD^{\lambda_2}(k)}{DPF^{\lambda_2}}\right)}{l^i\left(\varepsilon_{\Delta HbR}^{\lambda_1}\varepsilon_{\Delta HbO}^{\lambda_2} - \varepsilon_{\Delta HbR}^{\lambda_2}\varepsilon_{\Delta HbO}^{\lambda_1}\right)} \tag{1b}$$

where, ΔHbO^i and ΔHbR^i: concentration changes of ΔHbO and ΔHbR, $\varepsilon(\lambda)$: extinction coefficient corresponding to wavelengths and haemoglobin concentrations, ΔOD: variation in optical density at k^{th} sample, $DPF(\lambda)$: differential path length factor, i: i^{th} channel pair representation of emitter-detector, λ_1 and λ_2: two working wavelengths of fNIRS system, $\varepsilon_{HbR}^{\lambda_1}, \varepsilon_{\Delta HbO}^{\lambda_2}, \varepsilon_{\Delta HbR}^{\lambda_2}$ and $\varepsilon_{\Delta HbO}^{\lambda_1}$: extinction coefficients of ΔHbO and ΔHbR at two different wavelengths.

Table 1. Parameters for Modified Beer–Lambert Law (MBLL).

Wavelength (nm)	DPF (cm)	ΔHbO (1/cm) (moles/L)	ΔHbR (1/cm) (moles/L)
760	7.25	1466.5865	3843.707
850	6.38	2526.391	1798.643

2.8. Signal Filtration

The spontaneous contamination from physiological and non-physiological noise in fNIRS data, such as heart rate ($\simeq$1 Hz), respiration ($\simeq$0.2 Hz), Mayer waves ($\simeq$0.1 Hz), and very low frequency ($\leq$0.04, VLF) was removed by applying subsequent filters. Non-physiological noise refers to motion artefacts, measurements noise and machine drift due to the temperature changes in the optical system. The stimulation frequency for the given experimental paradigm was (1/20 s = 0.05 Hz). The stable 4th order band-pass Butter-worth filter with a low and high cut-off frequency of 0.01 Hz and 0.15 Hz, respectively [33], was applied to remove the noises. To avoid phase delay in filtering, the built-in MATLAB® command 'filtfilt' was used. Furthermore, smoothing of the fNIRS signal was done by applying the Savitzky-Golay filter with the optimal order and frame size recommended in [34]. In [34], the recommended filter order and frame size is three and nineteen, respectively, for a frequency band of 0.03–0.1 Hz. We used the same order and frame size because our band of frequencies are quite similar.

2.9. Feature Extraction

The most common statistical features (descriptive and morphological) used in fNIRS are signal mean, peak, minimum, Skewness, Kurtosis, variance, median, and peak-to-peak [35–38]. The window length was set to 10 s, which is equal to the task period. The descriptions of the extracted features are shown in Table 2 from ΔHbO data.

Table 2. Spatial feature extracted from ΔHbO.

Sr. No.	Statistical Feature	Mathematical Formulation/Description
1.	Signal Mean	Signal mean is calculated as: $\mu_w = \frac{1}{N_w} \sum_{k=k_L}^{k_U} \Delta HbX_w$ where, μ_w: Mean of window w: sample window N_w : Number of sample in the window k_L: Lower limit of the window k_U : Upper limit of the window ΔHbX_w: Stands for ΔHbO or ΔHbR
2.	Signal Peak (Signal maximum)	The feature select the maximum value in the window.
3.	Signal Minimum	The feature minimum value in the window.
4.	Signal Skewness	Signal skewness is calculated as: $skew_w = \frac{E_x(\Delta HbX_w - \mu_w)^3}{\sigma^3}$ where, E_x is the expectation, μ is the mean, and σ is the standard deviation of the haemoglobin ΔHbX_w
5.	Signal Kurtosis	Signal Kurtosis is calculated as: $Kurt_w = \frac{E_x(\Delta HbX_w - \mu_w)^4}{\sigma^4}$ where, E_x is the expectation, μ is the mean, and σ is the standard deviation of the haemoglobin ΔHbX_w
6.	Signal Variance	Signal variance is the measure of signal spread.
7.	Signal Median	Median is the value separating the higher half from the lower half of values in the time window.
8.	Peak-to-peak	Peak-to-peak is computed as the difference between the maximum to the minimum value in the time window.

2.10. Classification

Eight commonly used classifiers were evaluated to check the robustness of modern machine learning algorithms for decoding dexterous finger movements. The classifiers included Support vector machine (SVM), Random Forest (RF), Decision tree (DT), Adaboost, Quadratic discriminant analysis (QDA), Artificial neural networks (ANN), k-nearest neighbours (kNN), and Extreme Gradient Boosting (XGBoost). The different classifiers' parameters are shown in Table 3.

Table 3. Classifier parameters.

Classifiers	Parameters Setting
QDA	priors = None, reg_param = 0.0
AdaBoost	n_estimator = 10, random_state = 0, learning_rate = 1.0
SVM	Kernal = rbf, degree = 3, random_state = None
ANN	hidden layers = (5, 2), solver = 'lbfgs', random_state = 1, max_liter = 300,
Decision Tree	criterion = entropy, random_state = 0
kNN	n_neighbors = 5
Random Forest	n_estimators = 10, criterion = entropy, random_state = 0
XGBoost	booster = gbtree, verbosity = 1, nthread = maximum number of threads

2.11. Performance Evaluation

Each classifier was mostly evaluated using different performance measures, like accuracy, precision, recall, F1 score, receiver operating characteristic curve/ROC curve, and confusion matrix [39]. All these measures can be derived from the so-called true positives (TP), false positives (FP), true negatives (TN), and false negatives (FN). Reporting single metrics does not give us a complete understanding of the classifier behavior. Hence, it is important to at-least report a few of these parameters to gain a complete understanding of the classifier behaviour. In this study, we have reported accuracy, precision, recall and F1 score. Accuracy is the ratio between correctly classified points to the number of total point. The accuracy gives the probability of correct predictions of the model. However, in the case of highly imbalanced data sets, the model that deterministically classifies all the data as the majority class will yield higher classification accuracy, which makes this measure unreliable. The confusion matrix summarizes the predicted results in table format with visualisation of all the above-mentioned four parameters (TP, FP, TN, FN) of the classifiers. Precision and recall give us an understanding of how useful and complete are the results, respectively. F1 score is the harmonic mean of precision and recall. All these parameters are discussed in the results section, where we discuss the performance of the classifier in decoding individual finger-tapping.

3. Results and Discussion

In this study, we classified individual finger tapping of right-handed people using fNIRS signals. For that purpose, eight different spatio-statistical features were extracted from ΔHbO, as shown in Table 2. Furthermore, we also compared and evaluated the performance of different classifiers, such as SVM, RF, DT, Adaboost, QDA, ANN, kNN and XGBoost, as shown in Figure 4. Table 4 shows the four important performance measures among all of the subjects for the respective classifiers. It was noted that the kNN, RF and XGBoost classifiers yielded maximum classification accuracies, 0.75 ± 0.04, 0.75 ± 0.05, and 0.77 ± 0.06, respectively. We applied the student's t-test to validate whether or not these classifier's accuracies were statistically discriminant or not with respect to the rest of the classifiers. The p-values obtained among kNN, RF, and XGBoost were not statistically significant, since all the classifiers yielded a similar accuracy. On the other hand, the p-values using either classifiers kNN, RF or XGBoost versus all of the other classifiers were less than 0.05 for all ΔHbO signals, which establish the statistical significance of these classifiers performance. Previous studies showed that thumb finger-tapping gives a higher level of cortical activation among other fingers [40], which is also supported by our current study as shown in Figure 5f–h. Moreover, the highest peaks in ΔHbO signal which corresponds to higher brain activity during thumb finger-tapping can be seen in Figure 6.

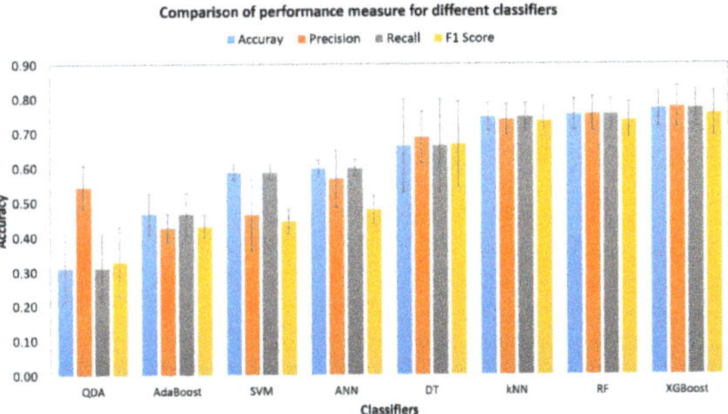

Figure 4. Comparison of different classifiers on basis of performance parameters (accuracy, precision, recall F1score).

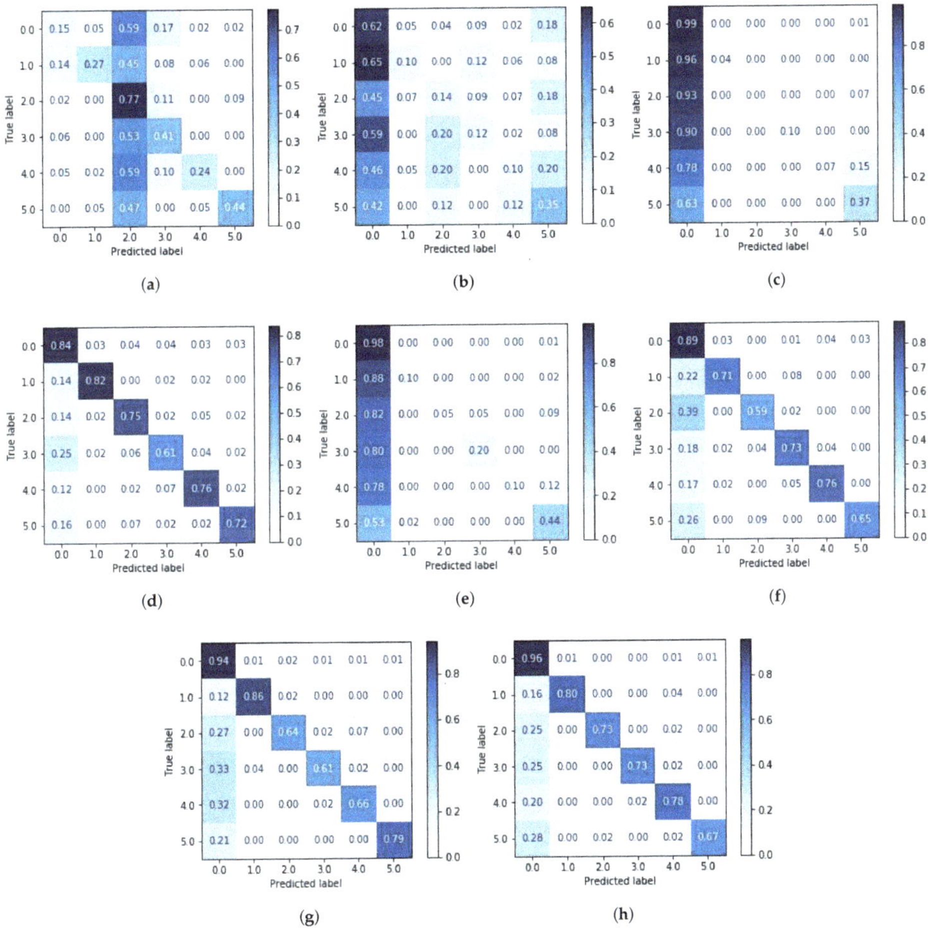

Figure 5. Confusion metrics for all classifiers for subject one (S01); Classes are labeled as '0', '1', '2', '3', '4' and '5', which stands for 'Rest', 'Thumb', 'Index', 'Middle', 'Ring', and 'Little' finger-tapping classes, respectively. (**a**) Quadratic discriminant analysis (QDA). (**b**) AdaBoost. (**c**) Support vector machine (SVM). (**d**) Decision tree (DT). (**e**) Artificial neural networks (ANN). (**f**) k-nearest neighbors (kNN). (**g**) Random forest (RF). (**h**) Extreme Gradient Boosting (XGBoost).

Table 4. Subject-wise comparison of classifiers performance parameters (accuracy, precision, recall, F1 score); 'S' stands for subject followed by number.

		S01	S02	S03	S04	S05	S06	S07	S08	S09	S10	S11	S12	S13	S14	S15	S16	S17	S18	S19	S20	S21	S22	S23	S24	Mean	STD
SVM	Accuracy	0.58	0.57	0.58	0.57	0.56	0.57	0.58	0.59	0.58	0.57	0.57	0.62	0.57	0.57	0.64	0.57	0.59	0.58	0.65	0.60	0.57	0.58	0.60	0.59	0.59	0.02
	Precision	0.65	0.32	0.34	0.48	0.32	0.49	0.34	0.62	0.53	0.39	0.41	0.46	0.48	0.41	0.67	0.40	0.35	0.43	0.49	0.44	0.45	0.50	0.52	0.65	0.46	0.10
	Recall	0.58	0.57	0.58	0.57	0.56	0.57	0.58	0.59	0.58	0.57	0.57	0.62	0.57	0.57	0.64	0.57	0.59	0.58	0.65	0.60	0.57	0.58	0.60	0.59	0.59	0.02
	F1 Score	0.47	0.41	0.43	0.42	0.41	0.42	0.43	0.45	0.45	0.43	0.41	0.49	0.43	0.42	0.55	0.42	0.44	0.43	0.52	0.45	0.42	0.44	0.50	0.46	0.45	0.04
RF	Accuracy	0.84	0.65	0.84	0.70	0.73	0.77	0.75	0.75	0.75	0.76	0.73	0.73	0.72	0.80	0.78	0.72	0.71	0.75	0.82	0.68	0.77	0.77	0.78	0.78	0.75	0.05
	Precision	0.84	0.63	0.85	0.70	0.73	0.77	0.75	0.75	0.75	0.77	0.73	0.73	0.72	0.80	0.80	0.73	0.70	0.75	0.82	0.67	0.77	0.78	0.78	0.78	0.75	0.05
	Recall	0.84	0.65	0.84	0.70	0.73	0.77	0.75	0.75	0.75	0.76	0.73	0.73	0.72	0.80	0.78	0.72	0.71	0.75	0.82	0.68	0.77	0.77	0.78	0.78	0.75	0.05
	F1 Score	0.83	0.61	0.83	0.67	0.72	0.75	0.73	0.74	0.74	0.75	0.70	0.72	0.70	0.78	0.77	0.70	0.69	0.73	0.81	0.65	0.75	0.76	0.77	0.77	0.74	0.05
DT	Accuracy	0.79	0.56	0.76	0.28	0.67	0.68	0.23	0.68	0.71	0.70	0.63	0.71	0.65	0.73	0.76	0.71	0.67	0.69	0.76	0.64	0.72	0.71	0.75	0.71	0.66	0.13
	Precision	0.79	0.56	0.76	0.49	0.67	0.69	0.53	0.68	0.71	0.70	0.63	0.72	0.65	0.74	0.76	0.71	0.68	0.69	0.78	0.64	0.72	0.72	0.75	0.71	0.69	0.07
	Recall	0.79	0.56	0.76	0.28	0.67	0.68	0.23	0.68	0.71	0.70	0.63	0.71	0.65	0.73	0.76	0.71	0.67	0.69	0.76	0.64	0.72	0.71	0.75	0.71	0.66	0.13
	F1 Score	0.79	0.56	0.75	0.32	0.67	0.69	0.27	0.68	0.71	0.70	0.63	0.71	0.65	0.74	0.76	0.71	0.67	0.69	0.77	0.64	0.72	0.71	0.75	0.71	0.67	0.13
AdaBoost	Accuracy	0.41	0.55	0.46	0.56	0.52	0.52	0.39	0.50	0.48	0.52	0.51	0.42	0.43	0.51	0.43	0.38	0.49	0.45	0.49	0.53	0.45	0.38	0.34	0.46	0.47	0.06
	Precision	0.40	0.41	0.46	0.33	0.46	0.43	0.39	0.44	0.39	0.39	0.44	0.46	0.38	0.44	0.46	0.38	0.44	0.44	0.53	0.41	0.41	0.41	0.47	0.45	0.43	0.04
	Recall	0.41	0.55	0.46	0.56	0.52	0.52	0.39	0.50	0.48	0.52	0.51	0.42	0.43	0.51	0.43	0.38	0.49	0.45	0.49	0.53	0.45	0.38	0.34	0.46	0.47	0.06
	F1 Score	0.40	0.44	0.46	0.42	0.46	0.45	0.38	0.46	0.43	0.43	0.45	0.43	0.40	0.46	0.43	0.37	0.45	0.44	0.50	0.45	0.42	0.39	0.38	0.43	0.43	0.03
QDA	Accuracy	0.28	0.22	0.31	0.28	0.24	0.42	0.23	0.41	0.20	0.21	0.24	0.28	0.32	0.25	0.58	0.32	0.31	0.30	0.36	0.26	0.34	0.24	0.56	0.28	0.31	0.10
	Precision	0.59	0.49	0.66	0.49	0.56	0.48	0.53	0.50	0.55	0.52	0.45	0.59	0.61	0.51	0.59	0.49	0.54	0.56	0.64	0.54	0.49	0.54	0.69	0.47	0.54	0.06
	Recall	0.28	0.22	0.31	0.28	0.24	0.42	0.23	0.41	0.20	0.21	0.24	0.28	0.32	0.25	0.58	0.32	0.31	0.30	0.36	0.26	0.34	0.24	0.56	0.28	0.31	0.10
	F1 Score	0.29	0.25	0.33	0.32	0.24	0.43	0.27	0.42	0.16	0.22	0.26	0.30	0.33	0.28	0.57	0.35	0.33	0.30	0.42	0.30	0.38	0.23	0.58	0.31	0.33	0.10
ANN	Accuracy	0.61	0.58	0.60	0.57	0.58	0.58	0.58	0.60	0.60	0.58	0.58	0.63	0.57	0.58	0.64	0.59	0.60	0.61	0.67	0.61	0.59	0.59	0.62	0.59	0.60	0.02
	Precision	0.69	0.42	0.56	0.54	0.48	0.54	0.34	0.69	0.67	0.62	0.61	0.54	0.60	0.52	0.60	0.52	0.52	0.60	0.64	0.57	0.56	0.62	0.58	0.58	0.57	0.08
	Recall	0.61	0.58	0.60	0.57	0.58	0.58	0.58	0.60	0.60	0.58	0.58	0.63	0.57	0.58	0.64	0.59	0.60	0.61	0.67	0.61	0.59	0.59	0.62	0.59	0.60	0.02
	F1 Score	0.52	0.43	0.48	0.45	0.44	0.44	0.43	0.48	0.49	0.44	0.45	0.53	0.46	0.44	0.54	0.46	0.46	0.50	0.59	0.48	0.48	0.48	0.55	0.48	0.48	0.04
kNN	Accuracy	0.80	0.65	0.78	0.71	0.69	0.77	0.74	0.74	0.74	0.73	0.72	0.74	0.72	0.78	0.78	0.70	0.73	0.76	0.82	0.68	0.77	0.76	0.79	0.77	0.75	0.04
	Precision	0.80	0.63	0.78	0.69	0.68	0.76	0.74	0.74	0.73	0.72	0.72	0.73	0.71	0.78	0.78	0.70	0.72	0.77	0.81	0.66	0.76	0.76	0.79	0.77	0.74	0.05
	Recall	0.80	0.65	0.78	0.71	0.69	0.77	0.74	0.74	0.74	0.73	0.72	0.74	0.72	0.78	0.78	0.70	0.73	0.76	0.82	0.68	0.77	0.76	0.79	0.77	0.75	0.04
	F1 Score	0.79	0.62	0.78	0.69	0.68	0.76	0.73	0.73	0.73	0.72	0.70	0.73	0.70	0.78	0.77	0.69	0.71	0.75	0.82	0.66	0.76	0.76	0.79	0.77	0.73	0.05
XGBoost	Accuracy	0.86	0.64	0.86	0.71	0.74	0.78	0.74	0.77	0.78	0.79	0.71	0.76	0.73	0.82	0.80	0.75	0.75	0.77	0.86	0.68	0.81	0.78	0.84	0.79	0.77	0.06
	Precision	0.87	0.62	0.86	0.72	0.74	0.79	0.74	0.78	0.79	0.79	0.72	0.76	0.73	0.83	0.80	0.76	0.75	0.77	0.85	0.66	0.82	0.79	0.84	0.79	0.77	0.06
	Recall	0.86	0.64	0.86	0.71	0.74	0.78	0.74	0.77	0.78	0.79	0.71	0.76	0.73	0.82	0.80	0.75	0.75	0.77	0.86	0.68	0.81	0.78	0.84	0.79	0.77	0.06
	F1 Score	0.86	0.58	0.85	0.69	0.72	0.77	0.72	0.76	0.76	0.77	0.69	0.75	0.72	0.81	0.78	0.73	0.73	0.75	0.85	0.64	0.80	0.77	0.83	0.78	0.75	0.07

Overall, it was noted that most of the classes were misclassified as a *rest* class, and KNNs were therefore unable to classify the index finger correctly. We tested kNNs on different neighbours (such as 5, 10, and 15), five of which performed better than others, whereas RFs performed poorly on classifying the middle finger. Similarly, like kNNs, we also tested RFs on different estimators and got the best results at 10 number of estimators. On the other hand, XGBoost only classified little fingers poorly. In general, KNNs, RFs, and XGBoost performed well.

One of the core objectives of the brain-computer interface is to achieve a maximum number of commands with good classification accuracy. If we look at the previous literature in the field of fNIRS demonstrates that most of the work utilized either two-class, three-class, or four-class classification . While classifying two commands using fNIRS-based brain signals Power et al. achieved an average classification accuracy of 0.56 for two tasks [41]. Hong et al., achieved an average classification accuracy of 0.75 for three commands [42]. Similarly, several studies have reported classification results for four-class classification as well [43]. To the best of the author's knowledge, this is the first work that has reported good accuracies for five class-classification in the field of fNIRS. In this work, the achieved classification accuracies are far above the chance level (i.e., 0.2), which shows that machine learning can result in a potential increase in the number of commands in the field of fNIRS-based brain imaging.

In future, the signals will be studied in depth to gain a better understanding and more precise understanding of the cortical hemodynamics response precisely. After all, the attributes of different brain regions and with repetition of trails could vary for the same experimental paradigm [44]. Selection of trails or active channels using the 3-gamma function, changing the window length, detection of initial dip, vector phase analysis, and optimal feature extraction are the future directions for data analysis that could help to increase the classification accuracy. Furthermore, deep learning approaches, including deep belief and convolutional neural networks models, could also help to increase classification accuracy [45]. Moreover, activation of the left and right finger-tapping is dominant in premotor and SMA areas comparative to motor execution finger-tapping [46]. In future work, we will focus on averaging over this region of interest to gain a better idea of which activation regions corresponding to different finger-tapping. Trail-to-trail variability in fNIRS signal for finger-tapping tasks could be reduced using seed correlation methods that can enhance the classification accuracy [47]. We also envisage to using estimation algorithms such as the q-step-ahead prediction scheme and the kernel-based recursive least squares (KRLS) algorithm to reduce the onset delay of the ΔHbO changes due to finger-tapping for real-time implementation in the BCI system [21,48–50]. In the study, we considered only ΔHbO data. The reason for selecting ΔHbO is that in the field of fNIRS-based brain imaging, although both ΔHbO and ΔHbR are indicators of cerebral blood flows. However, ΔHbO is more sensitive than ΔHbR [51,52]. As far as ΔHbT and cerebral oxygen exchange COE are concerned, the quantities are dependent on HbO and HbR [53]. In future, ΔHbR and total haemoglobin changes ΔHbT changes will also be considered in ordered to achieve understanding. Moreover, only left hemisphere channels were considered in the study. Investigating the dynamic relationship between the brain regions simultaneously activated during finger-tapping would be an interesting direction for the future study. In recent studies, different stimulation durations were investigated to find the appropriate duration that can shorten the command generation time [54]. Keeping in mind the findings of these studies, shorter stimulation durations will also be investigated in the future.

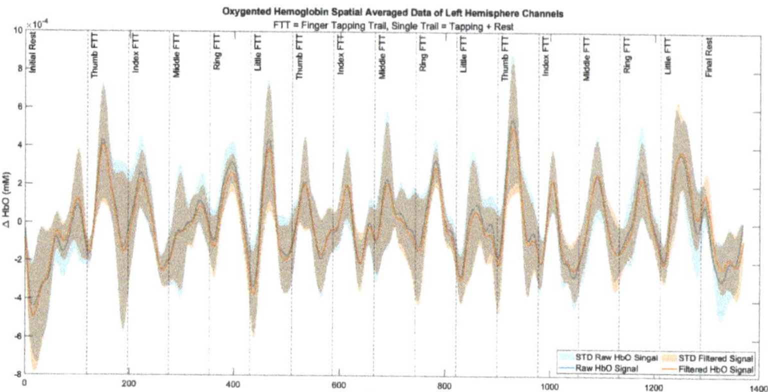

Figure 6. Oxygenated haemoglobin Signal for complete experimental trail.

4. Conclusions

Despite the outstanding performance of modern machine-learning algorithms, using functional near-infrared spectroscopy to classify movements from delicate anatomical structures, such as individual finger movements, is very challenging. This work presents a classification of individual finger movements (six classes) from the motor cortex. We have applied eight different classifiers, ranging from simple to sophisticated machine-learning algorithms. Quadratic discriminant analysis (QDA), AdaBoost, Support vector machine (SVM), Artificial neural networks (ANN), and Decision tree (DT) performed poorly, with an average classification accuracy of below 60%. On the other hand, other classifiers such as k-nearest neighbours (kNN), Random forest (RF) and Extreme Gradient Boosting (XGBoost) performed exceptionally well for such high-order data, with an average classification accuracy of 0.75 ± 0.04, 0.75 ± 0.05 and 0.77 ± 0.06, respectively. These are preliminary results from this novel research direction. In future, more in-depth analysis of the temporal and spatial domain will be conducted to understand the signals better. Achieving better classification accuracy could be a quantum leap for control command enrichment in brain-computer interface applications.

Author Contributions: Conceptualisation, H.K., P.M. and F.M.N.; methodology, H.K. and F.M.N.; analysis, H.K. and F.M.N.; suggestions and validation, P.M., A.Y., M.Z.U. and M.N.A.K.; writing—original draft preparation, H.K. and F.M.N.; writing—review and editing, P.M., A.Y., M.Z.U. and M.N.A.K.; supervision, P.M. and A.Y.; project administration, H.K. and P.M.; funding acquisition, P.M. All authors have read and agreed to the published version of the manuscript.

Funding: This research was funded and supported by the department of MEK, OsloMet-Oslo Metropolitan University and the Norwegian Research Council under a project titled 'Patient-Centric Engineering in Rehabilitation (PACER)' grant number 273599. Available online: https://prosjektbanken.forskningsradet.no/en/project/FORISS/273599?Kilde=FORISS&distribution=Ar&char (accessed on 1 November 2021).

Institutional Review Board Statement: The experiment was conducted according to the declaration of Helsinki. The study protocol and risk analysis were approved by the ethical committee of Oslo Metropolitan University. No objection certificate was obtained from Regional Committees for Medical and Health Research Ethics (REC) for experimental work (Ref. No. 322236).

Informed Consent Statement: Informed consent according to the Norwegian Centre for Research Data AS (NSD) of voluntary participation was given by all the participants before the experiment. The participant personal data is protected under NSD (Ref. No. 647457).

Data Availability Statement: Not applicable.

Conflicts of Interest: The authors declare no conflict of interest.

Abbreviations

The following abbreviations are used in this manuscript:

fNIRS	Functional Near-Infrared Spectroscopy
SVM	Support Vector Machine
RF	Random forest
DT	Decision Tree
QDA	Quadratic Discriminant Analysis
ANN	Artificial Neural Networks (ANN)
KNN	K-Nearest Neighbors (kNN)

References

1. Izzetoglu, M.; Izzetoglu, K.; Bunce, S.; Ayaz, H.; Devaraj, A.; Onaral, B.; Pourrezaei, K. Functional near-infrared neuroimaging. *IEEE Trans. Neural Syst. Rehabil. Eng.* **2005**, *13*, 153–159. [CrossRef]
2. Boas, D.A.; Elwell, C.E.; Ferrari, M.; Taga, G. Twenty years of functional near-infrared spectroscopy: Introduction for the special issue. *NeuroImage* **2014**, *85*, 1–5. [CrossRef] [PubMed]
3. Khan, R.A.; Naseer, N.; Qureshi, N.K.; Noori, F.M.; Nazeer, H.; Khan, M.U. fNIRS-based Neurorobotic Interface for gait rehabilitation. *J. Neuroeng. Rehabil.* **2018**, *15*, 1–17. [CrossRef] [PubMed]
4. Khan, H.; Naseer, N.; Yazidi, A.; Eide, P.K.; Hassan, H.W.; Mirtaheri, P. Analysis of Human Gait using Hybrid EEG-fNIRS-based BCI System: A review. *Front. Hum. Neurosci.* **2020**, *14*, 605. [CrossRef] [PubMed]
5. Villringer, A.; Chance, B. Non-invasive optical spectroscopy and imaging of human brain function. *Trends Neurosci.* **1997**, *20*, 435–442. [CrossRef]
6. Huneau, C.; Benali, H.; Chabriat, H. Investigating human neurovascular coupling using functional neuroimaging: A critical review of dynamic models. *Front. Neurosci.* **2015**, *9*, 467. [CrossRef]
7. Hendrikx, D.; Smits, A.; Lavanga, M.; De Wel, O.; Thewissen, L.; Jansen, K.; Caicedo, A.; Van Huffel, S.; Naulaers, G. Measurement of neurovascular coupling in neonates. *Front. Physiol.* **2019**, *10*, 65. [CrossRef]
8. Kumar, V.; Shivakumar, V.; Chhabra, H.; Bose, A.; Venkatasubramanian, G.; Gangadhar, B.N. Functional near infra-red spectroscopy (fNIRS) in schizophrenia: A review. *Asian J. Psychiatry* **2017**, *27*, 18–31. [CrossRef] [PubMed]
9. Naseer, N.; Hong, K.S. fNIRS-based brain-computer interfaces: A review. *Front. Hum. Neurosci.* **2015**, *9*, 3. [CrossRef]
10. Naseer, N.; Qureshi, N.K.; Noori, F.M.; Hong, K.S. Analysis of different classification techniques for two-class functional near-infrared spectroscopy-based brain-computer interface. *Comput. Intell. Neurosci.* **2016**, *2016*, 5480760. [CrossRef]
11. Ákos Jobbágy.; Harcos, P.; Karoly, R.; Fazekas, G. Analysis of finger-tapping movement. *J. Neurosci. Methods* **2005**, *141*, 29–39. [CrossRef]
12. Liao, K.; Xiao, R.; Gonzalez, J.; Ding, L. Decoding Individual Finger Movements from One Hand Using Human EEG Signals. *PLoS ONE* **2014**, *9*, e85192. [CrossRef]
13. Kondo, G.; Kato, R.; Yokoi, H.; Arai, T. Classification of individual finger motions hybridizing electromyogram in transient and converged states. In Proceedings of the 2010 IEEE International Conference on Robotics and Automation, Anchorage, AK, USA, 3–7 May 2010; pp. 2909–2915.
14. Al-Timemy, A.H.; Bugmann, G.; Escudero, J.; Outram, N. Classification of finger movements for the dexterous hand prosthesis control with surface electromyography. *IEEE J. Biomed. Health Inform.* **2013**, *17*, 608–618. [CrossRef]
15. Sikdar, S.; Rangwala, H.; Eastlake, E.B.; Hunt, I.A.; Nelson, A.J.; Devanathan, J.; Shin, A.; Pancrazio, J.J. Novel method for predicting dexterous individual finger movements by imaging muscle activity using a wearable ultrasonic system. *IEEE Trans. Neural Syst. Rehabil. Eng.* **2013**, *22*, 69–76. [CrossRef] [PubMed]
16. Samiee, S.; Hajipour, S.; Shamsollahi, M.B. Five-class finger flexion classification using ECoG signals. In Proceedings of the 2010 International Conference on Intelligent and Advanced Systems, Kuala Lumpur, Malaysia, 15–17 June 2010; pp. 1–4.
17. Flamary, R.; Rakotomamonjy, A. Decoding Finger Movements from ECoG Signals Using Switching Linear Models. *Front. Neurosci.* **2012**, *6*, 29. [CrossRef] [PubMed]
18. Nazeer, H.; Naseer, N.; Khan, R.A.A.; Noori, F.M.; Qureshi, N.K.; Khan, U.S.; Khan, M.J. Enhancing classification accuracy of fNIRS-BCI using features acquired from vector-based phase analysis. *J. Neural Eng.* **2020**, *17*, 056025. [CrossRef] [PubMed]
19. Bak, S.; Park, J.; Shin, J.; Jeong, J. Open-access fNIRS dataset for classification of unilateral finger-and foot-tapping. *Electronics* **2019**, *8*, 1486. [CrossRef]
20. Holper, L.; Wolf, M. Single-trial classification of motor imagery differing in task complexity: A functional near-infrared spectroscopy study. *J. Neuroeng. Rehabil.* **2011**, *8*, 1–13. [CrossRef] [PubMed]
21. Zafar, A.; Hong, K.S. Reduction of onset delay in functional near-infrared spectroscopy: Prediction of HbO/HbR signals. *Front. Neurorobotics* **2020**, *14*, 10. [CrossRef]
22. Wickramaratne, S.D.; Mahmud, M. Conditional-GAN Based Data Augmentation for Deep Learning Task Classifier Improvement Using fNIRS Data. *Front. Big Data* **2021**, *4*, 62. [CrossRef]
23. Sommer, N.M.; Kakillioglu, B.; Grant, T.; Velipasalar, S.; Hirshfield, L. Classification of fNIRS Finger Tapping Data With Multi-Labeling and Deep Learning. *IEEE Sens. J.* **2021**, *21*, 24558–24569. [CrossRef]

24. Kashou, N.H.; Giacherio, B.M.; Nahhas, R.W.; Jadcherla, S.R. Hand-grasping and finger tapping induced similar functional near-infrared spectroscopy cortical responses. *Neurophotonics* **2016**, *3*, 025006. [CrossRef] [PubMed]

25. Anwar, A.R.; Muthalib, M.; Perrey, S.; Galka, A.; Granert, O.; Wolff, S.; Heute, U.; Deuschl, G.; Raethjen, J.; Muthuraman, M. Effective connectivity of cortical sensorimotor networks during finger movement tasks: A simultaneous fNIRS, fMRI, EEG study. *Brain Topogr.* **2016**, *29*, 645–660. [CrossRef] [PubMed]

26. Vergotte, G.; Torre, K.; Chirumamilla, V.C.; Anwar, A.R.; Groppa, S.; Perrey, S.; Muthuraman, M. Dynamics of the human brain network revealed by time-frequency effective connectivity in fNIRS. *Biomed. Opt. Express* **2017**, *8*, 5326–5341. [CrossRef]

27. Cicalese, P.A.; Li, R.; Ahmadi, M.B.; Wang, C.; Francis, J.T.; Selvaraj, S.; Schulz, P.E.; Zhang, Y. An EEG-fNIRS hybridization technique in the four-class classification of alzheimer's disease. *J. Neurosci. Methods* **2020**, *336*, 108618. [CrossRef] [PubMed]

28. Hong, K.S.; Khan, M.J. Hybrid brain–computer interface techniques for improved classification accuracy and increased number of commands: A review. *Front. Neurorobotics* **2017**, *11*, 35. [CrossRef]

29. Quaresima, V.; Ferrari, M. Functional near-infrared spectroscopy (fNIRS) for assessing cerebral cortex function during human behavior in natural/social situations: A concise review. *Organ. Res. Methods* **2019**, *22*, 46–68. [CrossRef]

30. Yücel, M.A.; Lühmann, A.V.; Scholkmann, F.; Gervain, J.; Dan, I.; Ayaz, H.; Boas, D.; Cooper, R.J.; Culver, J.; Elwell, C.E.; et al. Best practices for fNIRS publications. *Neurophotonics* **2021**, *8*, 012101. [CrossRef]

31. Khan, M.A.; Bhutta, M.R.; Hong, K.S. Task-specific stimulation duration for fNIRS brain-computer interface. *IEEE Access* **2020**, *8*, 89093–89105. [CrossRef]

32. Santosa, H.; Zhai, X.; Fishburn, F.; Huppert, T. The NIRS brain AnalyzIR toolbox. *Algorithms* **2018**, *11*, 73. [CrossRef]

33. Pinti, P.; Scholkmann, F.; Hamilton, A.; Burgess, P.; Tachtsidis, I. Current Status and Issues Regarding Pre-processing of fNIRS Neuroimaging Data: An Investigation of Diverse Signal Filtering Methods Within a General Linear Model Framework. *Front. Hum. Neurosci.* **2019**, *12*, 505. [CrossRef] [PubMed]

34. Rahman, M.A.; Rashid, M.A.; Ahmad, M. Selecting the optimal conditions of Savitzky–Golay filter for fNIRS signal. *Biocybern. Biomed. Eng.* **2019**, *39*, 624–637. [CrossRef]

35. Hong, K.S.; Khan, M.J.; Hong, M.J. Feature extraction and classification methods for hybrid fNIRS-EEG brain-computer interfaces. *Front. Hum. Neurosci.* **2018**, *12*, 246. [CrossRef]

36. Naseer, N.; Noori, F.M.; Qureshi, N.K.; Hong, K.S. Determining optimal feature-combination for LDA classification of functional near-infrared spectroscopy signals in brain-computer interface application. *Front. Hum. Neurosci.* **2016**, *10*, 237. [CrossRef]

37. Noori, F.M.; Naseer, N.; Qureshi, N.K.; Nazeer, H.; Khan, R.A. Optimal feature selection from fNIRS signals using genetic algorithms for BCI. *Neurosci. Lett.* **2017**, *647*, 61–66. [CrossRef] [PubMed]

38. Qureshi, N.K.; Naseer, N.; Noori, F.M.; Nazeer, H.; Khan, R.A.; Saleem, S. Enhancing classification performance of functional near-infrared spectroscopy-brain–computer interface using adaptive estimation of general linear model coefficients. *Front. Neurorobotics* **2017**, *11*, 33. [CrossRef] [PubMed]

39. Elkan, C. *Evaluating Classifiers*; University of California: San Diego, CA, USA, 2012.

40. Jorge, A.; Royston, D.A.; Tyler-Kabara, E.C.; Boninger, M.L.; Collinger, J.L. Classification of individual finger movements using intracortical recordings in Human Motor Cortex. *Neurosurgery* **2020**, *87*, 630–638. [CrossRef] [PubMed]

41. Power, S.D.; Kushki, A.; Chau, T. Automatic single-trial discrimination of mental arithmetic, mental singing and the no-control state from prefrontal activity: Toward a three-state NIRS-BCI. *BMC Res. Notes* **2012**, *5*, 141. [CrossRef]

42. Hong, K.S.; Naseer, N.; Kim, Y.H. Classification of prefrontal and motor cortex signals for three-class fNIRS-BCI. *Neurosci. Lett.* **2015**, *587*, 87–92. [CrossRef]

43. Hong, K.S.; Santosa, H. Decoding four different sound-categories in the auditory cortex using functional near-infrared spectroscopy. *Hear. Res.* **2016**, *333*, 157–166. [CrossRef]

44. Kamran, M.A.; Jeong, M.Y.; Mannan, M. Optimal hemodynamic response model for functional near-infrared spectroscopy. *Front. Behav. Neurosci.* **2015**, *9*, 151. [CrossRef]

45. Ho, T.K.K.; Gwak, J.; Park, C.M.; Song, J.I. Discrimination of mental workload levels from multi-channel fNIRS using deep leaning-based approaches. *IEEE Access* **2019**, *7*, 24392–24403. [CrossRef]

46. Wu, S.; Li, J.; Gao, L.; Chen, C.; He, S. Suppressing systemic interference in fNIRS monitoring of the hemodynamic cortical response to motor execution and imagery. *Front. Hum. Neurosci.* **2018**, *12*, 85. [CrossRef] [PubMed]

47. Hu, X.S.; Hong, K.S.; Ge, S.S. Reduction of trial-to-trial variability in functional near-infrared spectroscopy signals by accounting for resting-state functional connectivity. *J. Biomed. Opt.* **2013**, *18*, 017003. [CrossRef] [PubMed]

48. Naseer, N.; Hong, K.S. Functional near-infrared spectroscopy based brain activity classification for development of a brain-computer interface. In Proceedings of the 2012 International Conference of Robotics and Artificial Intelligence, Rawalpindi, Pakistan, 22–23 October 2012; pp. 174–178. [CrossRef]

49. Khan, M.J.; Hong, K.S.; Bhutta, M.R.; Naseer, N. fNIRS based dual movement control command generation using prefrontal brain activity. In Proceedings of the 2014 International Conference on Robotics and Emerging Allied Technologies in Engineering (iCREATE), Islamabad, Pakistan, 22–24 April 2014; pp. 244–248. [CrossRef]

50. Xiao, J.; Xu, H.; Gao, H.; Bian, M.; Li, Y. A Weakly Supervised Semantic Segmentation Network by Aggregating Seed Cues: The Multi-Object Proposal Generation Perspective. *ACM J.* **2021**, *17*, 1–19. [CrossRef]

51. Hoshi, Y.; Kobayashi, N.; Tamura, M. Interpretation of near-infrared spectroscopy signals: A study with a newly developed perfused rat brain model. *J. Appl. Physiol.* **2001**, *90*, 1657–1662. [CrossRef]

52. Hu, X.S.; Hong, K.S.; Shuzhi, S.G.; Jeong, M.Y. Kalman estimator-and general linear model-based on-line brain activation mapping by near-infrared spectroscopy. *Biomed. Eng. Online* **2010**, *9*, 1–15. [CrossRef]
53. Zafar, A.; Hong, K.S. Neuronal activation detection using vector phase analysis with dual threshold circles: A functional near-infrared spectroscopy study. *Int. J. Neural Syst.* **2018**, *28*, 1850031. [CrossRef]
54. Khan, M.N.A.; Hong, K.S. Most favorable stimulation duration in the sensorimotor cortex for fNIRS-based BCI. *Biomed. Opt. Express* **2021**, *12*, 5939–5954. [CrossRef]

MDPI

St. Alban-Anlage 66

4052 Basel

Switzerland

www.mdpi.com

Sensors Editorial Office

E-mail: sensors@mdpi.com

www.mdpi.com/journal/sensors